CHEMISTRY AND CHEMICAL ENGINEERING FOR SUSTAINABLE DEVELOPMENT

Best Practices and Research Directions

Innovations in Physical Chemistry: Monograph Series

CHEMISTRY AND CHEMICAL ENGINEERING FOR SUSTAINABLE DEVELOPMENT

Best Practices and Research Directions

Edited by

Miguel A. Esteso, PhD

Ana Cristina Faria Ribeiro, PhD

A. K. Haghi, PhD

First edition published [2021]

Apple Academic Press Inc.
1265 Goldenrod Circle, NE,
Palm Bay, FL 32905 USA

4164 Lakeshore Road, Burlington,
ON, L7L 1A4 Canada

CRC Press
6000 Broken Sound Parkway NW,
Suite 300, Boca Raton, FL 33487-2742 USA

2 Park Square, Milton Park,
Abingdon, Oxon, OX14 4RN UK

First issued in paperback 2021

Library and Archives Canada Cataloguing in Publication

Title: Chemistry and chemical engineering for sustainable development : best practices and research directions / edited by Miguel A. Esteso, PhD, Ana Cristina Faria Ribeiro, PhD, A.K. Haghi, PhD.

Names: Esteso, Miguel A., editor. | Ribeiro, Ana Cristina Faria, editor. | Haghi, A. K., editor.

Description: Includes bibliographical references and index.

Identifiers: Canadiana (print) 20200280406 | Canadiana (ebook) 20200280511 | ISBN 9781771888707 (hardcover) | ISBN 9780367815967 (ebook)

Subjects: LCSH: Green chemistry. | LCSH: Chemical engineering. | LCSH: Materials—Environmental aspects.

Classification: LCC TP155.2.E58 C54 2020 | DDC 660.028/6—dc23

Library of Congress Cataloging-in-Publication Data

CIP data on file with US Library of Congress

ISBN: 978-1-77188-870-7 (hbk)
ISBN: 978-1-77463-908-5 (pbk)
ISBN: 978-0-36781-596-7 (ebk)

ABOUT THE EDITORS

Miguel A. Esteso, PhD
Emeritus Professor, Department of Analytical Chemistry, Physical Chemistry and Chemical Engineering, University of Alcalá, Spain

Miguel A. Esteso, PhD, is Emeritus Professor in Physical Chemistry at the University of Alcalá, Spain. He is the author of more than 300 published papers in various journals and conference proceedings, as well as chapters in specialized books. He has supervised several PhD theses and master degree theses. He is a member of the International Society of Electrochemistry (ISE), the Ibero-American Society of Electrochemistry (SIBAE), the Portuguese Society of Electrochemistry (SPE), and the Biophysics Spanish Society (SEB). He is on the editorial board of various international journals. His research activity is focused on electrochemical thermodynamics. He developed his postdoctoral work at the Imperial College of London (UK) and has made several research-stays at different universities and research centers, including the University of Regensburg, Germany; CITEFA (Institute of Scientific and Technical Research for the Armed Forces), Argentina; Theoretical and Applied Physicochemical Research Institute, La Plata, Argentina; Pontifical Catholic University of Chile, Chile; University of Zulia, Venezuela; University of Antioquia, Colombia; National University of Colombia, Bogota, Colombia; University of Coimbra, Portugal; and University of Ljubljana, Slovenia.

Ana Cristina Faria Ribeiro, PhD
Researcher, Department of Chemistry, University of Coimbra, Portugal

Ana Cristina Faria Ribeiro, PhD, is a researcher in the Department of Chemistry at the University of Coimbra, Portugal. Her area of scientific activity is physical chemistry and electrochemistry. Her main areas of research interest are transport properties of ionic and nonionic components in aqueous solutions. She has experience as a scientific adviser and teacher of different practical courses. Dr. Ribeiro has supervised master degree theses as well as some PhD theses and has been a theses jury member. She has been referee for various journals as well an expert evaluator of some of the research programs funded by the Romanian government through the National Council for Scientific Research. She has been a member of the organizing committee of scientific conferences and she is an editorial member of several journals. She is a member of the Research Chemistry Centre, Coimbra, Portugal.

A. K. Haghi, PhD

Professor Emeritus of Engineering Sciences, Former Editor-in-Chief, International Journal of Chemoinformatics and Chemical Engineering and Polymers Research Journal; Member, Canadian Research and Development Center of Sciences and Culture

A. K. Haghi, PhD, is the author and editor of 165 books, as well as 1000 published papers in various journals and conference proceedings. Dr. Haghi has received several grants, consulted for a number of major corporations, and is a frequent speaker to national and international audiences. Since 1983, he served as professor at several universities. He is former Editor-in-Chief of the *International Journal of Chemoinformatics and Chemical Engineering* and *Polymers Research Journal* and is on the editorial boards of many international journals. He is also a member of the Canadian Research and Development Center of Sciences and Cultures (CRDCSC), Montreal, Quebec, Canada. He holds a BSc in urban and environmental engineering from the University of North Carolina (USA), an MSc in mechanical engineering from North Carolina A&T State University (USA), a DEA in applied mechanics, acoustics, and materials from the Université de Technologie de Compiègne (France), and a PhD in engineering sciences from Université de Franche-Comté (France).

INNOVATIONS IN PHYSICAL CHEMISTRY: MONOGRAPH SERIES

This book series offers a comprehensive collection of books on physical principles and mathematical techniques for majors, non-majors, and chemical engineers. Because there are many exciting new areas of research involving computational chemistry, nanomaterials, smart materials, high-performance materials, and applications of the recently discovered graphene, there can be no doubt that physical chemistry is a vitally important field. Physical chemistry is considered a daunting branch of chemistry—it is grounded in physics and mathematics and draws on quantum mechanics, thermodynamics, and statistical thermodynamics.

Editors-in-Chief

A. K. Haghi, PhD
Former Editor-in-Chief, *International Journal of Chemoinformatics* and *Chemical Engineering and Polymers Research Journal*; Member, Canadian Research and Development Center of Sciences and Cultures (CRDCSC), Montreal, Quebec, Canada
E-mail: AKHaghi@Yahoo.com

Lionello Pogliani, PhD
University of Valencia-Burjassot, Spain
E-mail: lionello.pogliani@uv.es

Ana Cristina Faria Ribeiro, PhD
Researcher, Department of Chemistry, University of Coimbra, Portugal
E-mail: anacfrib@ci.uc.pt

BOOKS IN THE SERIES

- **Applied Physical Chemistry with Multidisciplinary Approaches**
  Editors: A. K. Haghi, PhD, Devrim Balköse, PhD, and
  Sabu Thomas, PhD

- **Biochemistry, Biophysics, and Molecular Chemistry: Applied Research and Interactions**
  Editors: Francisco Torrens, PhD, Debarshi Kar Mahapatra, PhD, and A. K. Haghi, PhD
- **Chemistry and Industrial Techniques for Chemical Engineers**
  Editors: Lionello Pogliani, PhD, Suresh C. Ameta, PhD, and A. K. Haghi, PhD
- **Chemistry and Chemical Engineering for Sustainable Development: Best Practices and Research Directions**
  Editors: Miguel A. Esteso, PhD, Ana Cristina Faria Ribeiro, and A. K. Haghi, PhD
- **Chemical Technology and Informatics in Chemistry with Applications**
  Editors: Alexander V. Vakhrushev, DSc, Omari V. Mukbaniani, DSc, and Heru Susanto, PhD
- **Engineering Technologies for Renewable and Recyclable Materials: Physical-Chemical Properties and Functional Aspects**
  Editors: Jithin Joy, Maciej Jaroszewski, PhD, Praveen K. M., and Sabu Thomas, PhD, and Reza Haghi, PhD
- **Engineering Technology and Industrial Chemistry with Applications**
  Editors: Reza Haghi, PhD, and Francisco Torrens, PhD
- **High-Performance Materials and Engineered Chemistry**
  Editors: Francisco Torrens, PhD, Devrim Balköse, PhD, and Sabu Thomas, PhD
- **Methodologies and Applications for Analytical and Physical Chemistry**
  Editors: A. K. Haghi, PhD, Sabu Thomas, PhD, Sukanchan Palit, and Priyanka Main
- **Modern Green Chemistry and Heterocyclic Compounds: Molecular Design, Synthesis, and Biological Evaluation**
  Editors: Ravindra S. Shinde, and A. K. Haghi, PhD
- **Modern Physical Chemistry: Engineering Models, Materials, and Methods with Applications**
  Editors: Reza Haghi, PhD, Emili Besalú, PhD, Maciej Jaroszewski, PhD, Sabu Thomas, PhD, and Praveen K. M.

- **Molecular Chemistry and Biomolecular Engineering: Integrating Theory and Research with Practice**
  Editors: Lionello Pogliani, PhD, Francisco Torrens, PhD, and A. K. Haghi, PhD
- **Physical Chemistry for Chemists and Chemical Engineers: Multidisciplinary Research Perspectives**
  Editors: Alexander V. Vakhrushev, DSc, Reza Haghi, PhD, and J. V. de Julián-Ortiz, PhD
- **Physical Chemistry for Engineering and Applied Sciences: Theoretical and Methodological Implication**
  Editors: A. K. Haghi, PhD, Cristóbal Noé Aguilar, PhD, Sabu Thomas, PhD, and Praveen K. M.
- **Practical Applications of Physical Chemistry in Food Science and Technology**
  Editors: Cristóbal Noé Aguilar, PhD, Jose Sandoval Cortes, PhD, Juan Alberto Ascacio Valdes, PhD, and A. K. Haghi, PhD
- **Research Methodologies and Practical Applications of Chemistry**
  Editors: Lionello Pogliani, PhD, A. K. Haghi, PhD, and Nazmul Islam, PhD
- **Theoretical Models and Experimental Approaches in Physical Chemistry: Research Methodology and Practical Methods**
  Editors: A. K. Haghi, PhD, Sabu Thomas, PhD, Praveen K. M., and Avinash R. Pai

CONTENTS

CONTRIBUTORS

E. P. Aparna
Research Scholar, School of Chemical Sciences, Mahatma Gandhi University, Kottayam, Kerala, India

Ana M. T. D. P. V. Cabral
Faculty of Pharmacy, University of Coimbra, 3000-295 Coimbra, Portugal

Gloria Castellano
Departamento de Ciencias Experimentales y Matemáticas, Facultad de Veterinaria y Ciencias Experimentales, Universidad Católica de Valencia San Vicente Mártir, Guillem de Castro-94, E-46001 València, Spain

K. S. Devaky
School of Chemical Sciences, Mahatma Gandhi University, Kottayam, Kerala, India

Miguel A. Esteso
U.D. QuímicaFísica, Universidad de Alcalá, 28871 Alcalá de Henares, Madrid, Spain

Diana M. Galindres
Universidad de los Andes, Department of Chemistry, Bogotá, Colombia
Centro de Química, Department of Chemistry, University of Coimbra, 3004-535 Coimbra, Portugal
U.D. QuímicaFísica, Universidad de Alcalá, 28871 Alcalá de Henares, Madrid, Spain

Peter Jurkovič
VIPO, Partizánske, Slovakia

V. I. Kodolov
Basic Research – High Educational Centre of Chemical Physics & Mesoscopics, Izhevsk, Russia
Kalashnikov Izhevsk State Technical University, Izhevsk, Russia

Ivan Michalec
Slovak University of Technology in Bratislava, Faculty of Materials Science and Technology, Trnava, Slovakia

T. M. Makhneva
Udmurt Federal Research Centre, Ural Division, Russian Academy of Sciences, Izhevsk, Russia

Milan Maronek
Slovak University of Technology in Bratislava, Faculty of Materials Science and Technology, Trnava, Slovakia

Divya Mathew
FDP Substitute, Department of Chemistry, St. Berchmans College, Changanassery, Kerala, India

Ján Matyašovský
VIPO, Partizánske, Slovakia

R. V. Mustakimov
Basic Research – High Educational Centre of Chemical Physics & Mesoscopics, Izhevsk, Russia
Kalashnikov Izhevsk State Technical University, Izhevsk, Russia

Igor Novák
Polymer Institute, Slovak Academy of Sciences, Bratislava, Slovakia

Sukanchan Palit
43, Judges Bagan, Post-Office - Haridevpur, Kolkata-700082, India

Yu. V. Pershin
Basic Research – High Educational Centre of Chemical Physics & Mesoscopics, Izhevsk, Russia

Lionello Pogliani
Facultad de Farmacia, Dept. de Química Física, Universitat de València,
Av. V.A. Estellés s/n, 46100 Burjassot (València), Spain

Ana C. F. Ribeiro
Centro de Química, Department of Chemistry, University of Coimbra, 3004-535 Coimbra, Portugal

M. Melia Rodrigo
U.D. Química Física, Universidad de Alcalá28871. Alcalá de Henares, Madrid, Spain

Carmen M. Romero
Universidad Nacional de Colombia, Bogotá, Colombia

Cecília I. A. V. Santos
Department of Chemistry, Coimbra University Centre, University of Coimbra,
3004-535 Coimbra, Portugal
U.D. Química Física, Universidad de Alcalá, 28871 Alcalá de Henares, Madrid, Spain

I. N. Shabanova
Basic Research – High Educational Centre of Chemical Physics & Mesoscopics, Izhevsk, Russia
Udmurt Federal Research Centre, Ural Division, Russian Academy of Sciences, Izhevsk, Russia

Ladislav Šoltés
Centre of Experimental Medicine, Institute of Experimental Pharmacology and Toxicology,
Bratislava, Slovakia

N. S. Terebova
Basic Research – High Educational Centre of Chemical Physics & Mesoscopics, Izhevsk, Russia
Udmurt Federal Research Centre, Ural Division, Russian Academy of Sciences, Izhevsk, Russia

Benny Thomas
Assistant Professor, Department of Chemistry, St. Berchmans College, Changanassery, Kerala, India

Elvin Thomas
School of Environmental Sciences, Mahatma Gandhi University, Priyadarsini Hills P.O.,
Kottayam 686560, Kerala, India

Francisco Torrens
Institut Universitari de Ciència Molecular, Universitat de València, Edifici d'Instituts de Paterna,
PO Box 22085, E-46071 València, Spain

V. V. Trineeva
Basic Research – High Educational Centre of Chemical Physics & Mesoscopics, Izhevsk, Russia
Udmurt Federal Research Centre, Ural Division, Russian Academy of Sciences, Izhevsk, Russia

Katarína Valachová
Centre of Experimental Medicine, Institute of Experimental Pharmacology and Toxicology,
Bratislava, Slovakia

Artur J. M. Valente
Centro de Química, Department of Chemistry, University of Coimbra, 3004-535 Coimbra, Portugal

Edgar F. Vargas
Universidad de los Andes, Department of Chemistry, Bogotá, Colombia

Luis M. P. Veríssimo
Centro de Química, Department of Chemistry, University of Coimbra, 3004-535 Coimbra, Portugal

D. K. Zhirov
Udmurt Federal Research Centre, Ural Division, Russian Academy of Sciences, Izhevsk, Russia

ABBREVIATIONS

ABH	analog black hole
AO	antioxidant
AOA	antioxidant activity
AP	atomic physics
BBs	building-blocks
BBR	blackbody radiation
BD	big data
BHs	black holes
BHF	Brueckner–Hartree–Fock
CA	cluster analysis
CAT	computerized axial tomography
CD	cyclodextrin
CERN	Conseil Européen pour la Recherche Nucléaire
CF	central field
CMB	cosmic microwave background
CN	coordination number
CPs	conservation principles
CT	critical temperature
DE	dark energy
DM	dark matter
DSC	differential scanning calorimetry
EF	electric fields
FRs	free oxygen radicals
FTIR	Fourier transform infrared spectroscopy
EFG	electric field gradient
EM	electromagnetic
EO	essential oil
EP	exclusion principle
FD	Fermi–Dirac
FM	ferromagnetism
GFTs	gauge fields theories
GHZ	Greenberger–Horne–Zeilinger
GS	ground state
GTR	general theory of relativity

GWs	gravitational waves
HEP	high-energy physics
HF	Hartree–Fock
HMs	hybrid materials
HR	Hawking radiation
HT	Hawking temperature
IHC	Israel–Hawking–Carter
ILs	ionic liquids
IPPs	ion pair potentials
JCs	Josephson currents
KN	kilonova
LDHs	layered double hydroxides
LHC	Large Hadron Collider
LIGO	Laser Interferometer Gravitational Wave Observatory
MA	meta-analysis
MF	magnetic fields
MOs	metal oxides
MOFs	metal–organic frameworks
MPs	medicinal plants
NSs	neutron stars
NCs	nanocomposites
NFs	nanofibres
NPs	nanoparticles
NRs	nanorods
NSs	nanostructures
NTs	nanotubes
NWs	nanowires
OMCs	ordered mesoporous Cs
PBHs	primordial black holes
PBI	polybenzimidazole
PCA	principal component analysis
PCT	Patent Co-operation Treaty
PEE	photoelectric effect
PMs	porous materials
PP	particle physics
PT	perturbation theory
PV	parity violation
PVA	polyvinyl alcohol
QFs	quantum fluctuations

QFI	quantum Fisher information
QM	quantum mechanics
RA	radioactivity
RPA	random-phase approximation
RR	Rydberg–Ritz
SCF	self-consistent field
SG	sol–gel
SM	standard model
SN	strong nuclear
SO	spin–orbit
SRs	selection rules
SSA	specific surface area
STEM	science–technology–engineering–mathematics
STLs	sesquiterpene lactones
SVs	symmetry violations
SWNTs	single-wall carbon NTs
TEM	transition electron microscopy
TFs	tide forces
TMs	traditional medicines
TRS	time-reversal symmetry
UC	unit cell
VDW	van der Waals
VMs	variational methods
WHs	wormholes
WL	wavelength
WMAP	Wilkinson Microwave Anisotropy Probe
WN	weak nuclear
WS	Weinberg–Salam
ZIFs	zeolitic imidazole frameworks

PREFACE

This volume brings together innovative research, new concepts, and novel developments in the application of new tools for chemical and materials engineers. It is an immensely research-oriented, comprehensive, and practical work. Postgraduate chemistry students would benefit from reading this book as it provides a valuable insight into chemical technology and innovations. It should appeal most to chemists and engineers in the chemical industry and research, who should benefit from the technological, scientific, and economic interrelationships and their potential developments. It contains significant research, reporting new methodologies, and important applications in the fields of chemical engineering as well as the latest coverage of chemical databases and the development of new methods and efficient approaches for chemists.

This volume should also be useful to every chemist or chemical engineer involved directly or indirectly with industrial chemistry. With clear explanations, real-world examples, this volume emphasizes the concepts essential to the practice of chemical science, engineering and technology while introducing the newest innovations in the field.

The book also serves a spectrum of individuals, from those who are directly involved in the chemical industry to others in related industries and activities. It provides not only the underlying science and technology for important industry sectors, but also broad coverage of critical supporting topics. Industrial processes and products can be much enhanced through observing the tenets and applying the methodologies covered in individual chapters.

This authoritative reference source provides the latest scholarly research on the use of applied concepts to enhance the current trends and productivity in chemical engineering. Highlighting theoretical foundations, real-world cases, and future directions, this book is ideally designed for researchers, practitioners, professionals, and students of materials chemistry and chemical engineering. The volume explains and discusses new theories and presents case studies concerning material and chemical engineering.

This book is an ideal reference source for academicians, researchers, advanced-level students, and technology developers seeking innovative research in chemistry and chemical engineering.

CHAPTER 1

LIMITING DIFFUSION COEFFICIENTS OF BOVINE SERUM ALBUMIN IN AQUEOUS SOLUTIONS OF SULFONATED RESORCINARENES

DIANA M. GALINDRES[1,2,3], ANA C. F. RIBEIRO[2*], EDGAR F. VARGAS[1], LUIS M. P. VERÍSSIMO[2], ARTUR J. M. VALENTE[2], and MIGUEL A. ESTESO[3]

[1]*Universidad de los Andes, Department of Chemistry, Bogotá, Colombia*

[2]*Centro de Química, Department of Chemistry, University of Coimbra, 3004-535 Coimbra, Portugal*

[3]*U.D. QuímicaFísica, Universidad de Alcalá, 28871 Alcalá de Henares, Madrid, Spain*

[*]*Corresponding author. E-mail: anacfrib@ci.uc.pt*

ABSTRACT

Limiting binary mutual diffusion coefficients for bovine serum albumin (BSA) were determined at 25°C in aqueous systems of 5,11,17,23-tetrakis-sulfonatomethylene-2,8,14,20-tetra (ethyl) resorcinarene (Na_4ETRA) and tetrasodium 5,11,17,23-tetrakissulfonatemethylen-2,8,14,20-tetra(propyl) resorcinarene (Na_4PRRA) at different concentrations, using the Taylor dispersion technique. The results are compared with the limiting binary mutual diffusion coefficients for BSA in water and discussed on the basis of salting-in effect. In addition, the hydrodynamic radii were also determined, contributing this way to a better understanding of the structure of such systems.

1.1 INTRODUCTION

Proteins are biologically important compounds and also have relevance in food, cosmetic, biomedical, and pharmaceutical industries. Among them, we have particular interest in albumin,[1] considered one of the most important proteins in the body because it has several important functions, such as substrate transport, buffering capacity, free radical scavenging, coagulation, and wound healing. Being a globular protein with useful functional properties, such as emulsifying and gelling, it possesses water-binding capacity and high nutritional value. Furthermore, it has also emerged as a versatile carrier for therapeutic and diagnostic agents, primarily for diagnosing and treating cancer, infectious diseases, and diabetes. In addition, over the past decades, it has been verified that an increasing number of albumin-based or albumin-binding drugs are used in clinical trials.[1,2]

Conversely, the rapid evolution of supramolecular chemistry during the last 30 years has offered to the scientific community a wide spread of host molecules, capable of including an equally vast number of guest molecules. These molecules that behave as carriers (e.g., resorcinarenes) are macrocycles that are characterized by their ability to preorganize and to accommodate guests in the cavities that can form, which allows them to behave as hosts and form host–guest complexes.[3,4]

The macrocycles studies in this work were tetrasodium 5,11,17,23-tetrakissulfonatemethylen-2,8,14,20-tetra(ethyl)resorcin[4]arene (Na_4ETRA) and tetrasodium5,11,17,23-tetrakissulfonatemethylen-2,8,14,20-tetra (propyl) resorcin[4]arene (Na_4PRRA)[4] were synthesized by a procedure similar to that described by Hogberg.[5–7]

However, despite considerable work,[8] the diffusion behavior of these systems is still scarce. Because this information is essential for the design of these systems in presence of biologically relevant compounds, such as albumin, we propose a comprehensive study of the limiting diffusion of albumin in aqueous solutions without and with sulfonated resorcinares. That is, binary mutual diffusion coefficients at infinitesimal concentration were determined at 25°C for albumin in aqueous solutions, and in solutions containing two resorcirarenes (Na_4ETRA and Na_4PRRA). The molecular structures of the corresponding monomers of tetrasodium5,11,17,23-tetrakissulfonatemethylen-2,8,14,20-tetra(ethyl)resorcin[4]arene ($ETRA^{4-}$) and tetrasodium 5,11,17,23-tetrakissulfonatemethylen-2,8,14,20-tetra(propyl) resorcin[4]arene ($PRRA^{4-}$) are reported in the literature.[4]

The obtained results are discussed in terms of the interactions "salting-in" effects. In addition, from these values, the hydrodynamic radii, R_h, have been also estimated.

In summary, with this work it has been possible to reach a better comprehensive understanding of the diffusion behavior of these systems involving albumin and resorcirarenes, as carriers.

1.2 TAYLOR DISPERSION MEASUREMENTS: CONCEPTS AND APPROXIMATIONS

1.2.1 *BINARY DIFFUSION*

Diffusion coefficient, D, in a binary system (i.e., with two independent components), may be defined in terms of the concentration gradient by phenomenological equations, representing the Fick's first and second laws (eqs 1.1 and 1.2),[9–10]

$$J_i = D\left(\frac{\partial c_i}{\partial x}\right) \tag{1.1}$$

$$\frac{\partial c}{\partial t} = \frac{\partial}{\partial x}\left(D\frac{\partial c}{\partial x}\right) \tag{1.2}$$

where J represents the flow of matter of component i across a suitable chosen reference plane per area unit and per time unit, in a one-dimensional system, and c is the concentration of solute in moles per volume unit at the considered point; D_F is the Fikian coefficient diffusion.

Measurements of diffusion coefficients of binary systems, D_F, can be obtained by using the Taylor dispersion technique. The theory of this technique has been described in detail in the literature[11–14] and, consequently, only a summary description of the apparatus and the procedure used in our study is presented here (Section 1.3.2).

1.2.2 *TERNARY DIFFUSION*

Diffusion in a ternary solution is described by the diffusion eqs (1.3) and (1.4)[11–14]

$$-(J_1)=(D_{11})_v \frac{\partial c_1}{\partial x}+(D_{12})_v \frac{\partial c_2}{\partial x} \tag{1.3}$$

$$-(J_2)=(D_{21})_v \frac{\partial c_1}{\partial x}+(D_{22})_v \frac{\partial c_2}{\partial x} \tag{1.4}$$

where J_1, J_2, $\frac{\partial c_1}{\partial x}$, and $\frac{\partial c_2}{\partial x}$ are the molar fluxes and the gradients in the concentrations of solute 1 and 2, respectively. The index v represents the volume-fixed frame of the reference used in these measurements. The main diffusion coefficients (D_{11} and D_{22}) give the flux of each solute produced by its own concentration gradient.

Extensions of the Taylor technique have been used to measure ternary mutual diffusion coefficients (D_{ik}) for multicomponent solutions. These D_{ik} coefficients, defined by eqs 1.3 and 1.4, were evaluated by fitting the ternary dispersion equation (eq 1.5) to two or more replicate pairs of peaks for each carrier stream.

$$V(t) = V_0 + V_1 t + V_{max} (t_R/t)^{1/2} \left[W_1 \exp\left(-\frac{12D_1(t-t_R)^2}{r^2 t} \right) + (1-W_1)\exp\left(-\frac{12D_2(t-t_R)^2}{r^2 t} \right) \right] \tag{1.5}$$

Two pairs of refractive index profiles, D_1 and D_2, are the eigen values of the matrix of the ternary D_{ik} coefficients. W_1 and $1 - W_1$ are the normalized pre-exponential factors.

In these experiments, small volumes, ΔV, of the solution, of composition $\bar{c}_1+\overline{\Delta c_1}$, $\bar{c}_2+\overline{\Delta c_2}$ are injected into carrier solutions of composition, $\bar{c}_1$ and $\bar{c}_2$ at time $t = 0$.

1.2.3 TRACER DIFFUSION OF SOLUTE 1 IN SOLUTIONS OF SOLUTE 2

A special case of ternary diffusion in solutions of solvent (0) + solute(1) + solute (2) arises if one of the solutes is present in trace amounts.

In the limit $C_1/C_2 \rightarrow 0$, for example, D_{11} is the tracer diffusion coefficient of solute 1 and D_{22} is the binary diffusion coefficient of solute 2 in the pure solvent. A concentration gradient in solute 2 cannot drive a coupled flux of solute 1 if the concentration of solute 1 is zero. Consequently, cross-coefficient D_{12} vanishes for the tracer diffusion of solute 2. D_{21}, the other cross-coefficient, however, is not necessarily zero, and can in fact be quite large, especially for mixed electrolyte solutes.

The tracer diffusion of solute 1 in solutions of solute 2 is measured by injecting small volumes of solution containing solutes 1 and 2 into carrier solutions of pure solute 1. Strong dilution of solute 1 with the carrier solution ensures its tracer diffusion. Under these conditions, once $D_{12} = 0$, the general expressions for ternary concentrations profiles simplify: $D_1 = D_{11}$ (the tracer diffusion coefficient of solute 1), $D_2 = D_{22}$ (the binary diffusion coefficient of solute 2). In those circumstances, due to the coupled diffusion of solute 2 caused by the concentration gradient in solute 1, tracer dispersion profiles for solute 1 generally resemble two overlapping Gaussian peaks of variance $r^2 t_R / (48D_{11})$ and $r^2 t_R / (48D_{22})$. The mathematical treatment involved for the computation of tracer (or limiting) diffusion coefficients has been described in detail elsewhere.[15–17]

The BSA/ETRA (or BSA/PRRA) systems should be considered a ternary system and we are actually measuring the tracer diffusion coefficients D_{11}^0 but not D_{12}^0, D_{21}^0 and D_{22}^0. However, in the present experimental conditions (i.e., flow and injected solutions of compositions $c_1 = 0$, and $c_2 = \bar{c}_2$, and $c_1 = \Delta c$, $c_2 = \bar{c}$, respectively) and also confirmed by the detector signal resembling a single normal distribution with variance $r^2 t_R/24D_{11}$, and not two overlapping normal distributions, we may consider the system as pseudo-binary and consequently take the measured parameter as the tracer diffusion coefficients of the BSA in aqueous solutions of ETRA (or PRRA). Support for this argument is given from other works.[15–17,18]

1.3 EXPERIMENTAL SECTION

1.3.1 MATERIALS

The macrocycles studied in this work were 5,11,17,23-tetrakissulfonato-methylene-2,8,14,20-tetra (ethyl) resorcinarene called in the short form C-tetra(ethyl)resorcin[4]arenesulphonated (Na_4ETRA) and tetrasodium 5,11,17,23-tetrakissulfonatemethylen-2,8,14,20-tetra(propyl)resorcinarene (Na_4PRRA). The syntheses of these compounds are well described in the literature.[4] Sodium chloride (Sigma-Aldrich, pro analysis > 0.99) was used without further purification. For these aqueous solutions, it was used Millipore-Q water (Table 1.1).

Bovine serum albumin lyophilized powder ≥96% was used without further purification. For these aqueous solutions, it was used Millipore-Q water (Table 1.1).

TABLE 1.1 Sample Descriptions.

Chemical name	Source	Mass fraction purity[a]
Na_4ETRA[b] Na_4PRRA		–
Bovine serum albumin lyophilized powder ≥96% (agarose gel electrophoresis)	Sigma-Aldrich	≥96%
H_2O	Millipore-Q water (1.82×10^5 Ω m at 25.0°C)	

[a]As stated by the supplier.

[b] See the description of the syntheses of Na_4ETRA and Na_4PRRA in Ref. [4].

1.3.2 TAYLOR TECHNIQUE MEASUREMENTS

This method is based on the dispersion of small amounts of solution injected into laminar carrier streams of solution of different compositions flowing through a long capillary tube.

At the start of each run, a 6-port Teflon injection valve (Rheodyne, model 5020) was used to introduce 63 mm^3 of solution into a laminar carrier stream of slightly different composition. A flow rate of 0.17 $cm^3\ min^{-1}$ was maintained by a metering pump (Gilson model Minipuls 3) to give retention times of about 1.1×10^4 s. The dispersion tube (length 32.799 (± 0.001) m) and the injection valve were kept at 25°C (± 0.01 K) in an air thermostat.

Dispersion of the injected samples was monitored using a differential refractometer (Waters model 2410) at the outlet of the dispersion tube. Detector voltages, $V(t)$, were measured at 5 s intervals with a digital voltmeter (Agilent 34401 A).

Binary diffusion coefficients were evaluated by fitting the dispersion equation

$$V(t) = V_0 + V_1 t + V_{max} (t_R/t)^{1/2} \exp[-12D(t - t_R)^2/r^2 t] \tag{1.6}$$

to the detector voltages. r is the internal radius of our Teflon dispersion tube and the additional fitting parameters are the mean sample retention time t_R, peak height V_{max}, baseline voltage V_0, and baseline slope V_1.

Solutions of BSA of different composition were injected into pure water and in solutions containing one resorcirarene (Na_4ETRA or Na_4PRRA). The more detailed description of the implicit experimental conditions can be found in the literature.[18]

1.4 RESULTS AND DISCUSSION

1.4.1 ANALYSIS OF DIFFUSION DATA

Table 1.2 shows the limiting values of diffusion coefficients for BSA at (D^0), at 25°C. These D^0 values were obtained from, at least, three independent runs, using different concentrations of the injected solutions in water. The uncertainty of these values is not larger than 3%.

TABLE 1.2 Limiting Binary Diffusion Coefficients of BSA, D^0,Using Water as Carrier Solution and Solutions of BSA at Different Concentrations, *c*, as Injection Solutions and at 25°C.

c (mol dm^{-3})	D^0_{BSA} (10^{-9}m^{2}s^{-1})
0.00003	0.212
0.00005	0.216
0.00006	0.218
0.00013	0.234
0.00025	0.259
0.00050	0.315

The diffusion experimental data D^0 values of the BSA in water (Table 1.2) were fitted using a least-squares method to a linear relationship and the value for the limiting diffusion coefficient at infinitesimal concentration, D^0 ,was obtained by extrapolation. That is, $D^0 = 0.205 \times 10^{-9}$ m^2 s^{-1}.

Mutual diffusion coefficients of BSA (D) in aqueous solutions at 25°C, but at finite concentrations, are shown in Table 1.3. D is the average value for each carrier solution determined from at least four profiles, generated by injecting samples both higher and lower concentrated than the carrier solution. The uncertainties were (± 0.01°C) in the temperature T, (± 0.001%) in the concentrations c and (1%–3%) in the value of D.

TABLE 1.3 Binary Diffusion Coefficients of BSA, D, in Aqueous Solutions at Different Concentrations, c, using Taylor Technique and at 25°C and F_T.

c (mol dm^{-3})	D_{BSA} (10^{-9}m^{2}s^{-1})	F_T
0	0.205[a]	1.000
5×10^{-5}	0.189	0.602
1×10^{-4}	0.198	0.630
2×10^{-4}	0.207	0.659

[a]This value was obtained by extrapolation of data shown in Table 1.2.

The observed decrease of the diffusion coefficient of BSA with concentration (that is, from infinitesimal to finite concentrations) may be interpreted on the basis of structural differences caused by this effect, affecting the motion of the species BSA.[19] The interactions between the monomers of BSA with water molecules are disrupted because the new aggregated species are formed, resulting of intramolecular hydrogen bonds between polar groups of monomers BSA. Due to their size, consequently these species will have lower mobility, and thus lower diffusion.

Two different effects, the ionic mobility and the gradient of the free energy, can control the diffusion process, considering that D is a product of both kinetic, F_M (or molar mobility coefficient of a diffusing substance) and thermodynamic factors, F_T[20]

$$D = F_T \times F_M \tag{1.7}$$

being

$$F_T == c\partial\mu / \partial c = \left(1+\frac{d\ln y_\pm}{d\ln c}\right) \tag{1.8}$$

where μ and $y_\pm$ represent the chemical potential and the thermodynamic activity coefficient of the solute, respectively. Using as an approximation the Nernst-Hartley equation (eq 1.9), eq (1.7) can be simplified and replaced by eq (1.9).

$$D = D_m^0\left(1+\frac{d\ln y_\pm}{d\ln c}\right) \tag{1.9}$$

On the other words, we can say that the variation in D is mainly due to the variation of F_T (attributed to the non-ideality in thermodynamic behavior), as shown in Table 1.3.

Thus, considering our experimental conditions (i.e., dilute solutions), and, consequently, assuming that some effects, such as variation of viscosity, dielectric constant, hydration, do not change with the concentration, we can conclude that the variation in D is mainly due to the variation of F_T (attributed to the non-ideality in thermodynamic behavior), and, secondarily, to the electrophoretic effect in the mobility factor, F_M.

Tables 1.4 and 1.5 present the limiting experimental diffusion coefficients, D, for aqueous system BSA/Na_4ETRAand BSA/Na_4PRRA at 25°C. The uncertainty of these values is also not greater than 3%. These values are compared with the previously published data of the limit diffusion coefficients for the BSA but in water, $D^0_{(inH2O)}$. The deviations are shown in Table 1.6.

TABLE 1.4 Limiting Diffusion Coefficients of BSA, D_i, in Aqueous Solutions of Na_4ETRA at Different Concentrations, c^a.

		D_i (10^{-9} m^2 s^{-1}) Flow solutions containing only Na_4ETRA at concentration c		
Concentration BSA in injection solutions, c_{BSA}	$c = 0$	$c = 0.0025$ (mol dm^{-3})	$c = 0.0050$ (mol dm^{-3})	$c = 0.0100$ (mol dm^{-3})
0.00005	0.216	0.104	0.105	0.094
0.0001	0.234	0.115	0.108	0.095
0.0002	0.259	0.132	0.112	0.096

[a]This table show the limiting values of D_i in which 70 μL of BSA at c_{BSA} + Na_4ETRA solution at concentration, c, was injected into Na_4ETRA solution at concentration, c, flowing solution.

TABLE 1.5 Tracer Diffusion Coefficients of Albumin in Aqueous Solutions of Na_4PRRA at Different Concentrations, c^a.

		D_i (10^{-9} m^2 s^{-1}) Flow solutions containing only Na_4ETRA at concentration	
Concentration BSA in injection solutions, c_{BSA}	$c = 0$	$c = 0.0050$ (mol dm^{-3})	$c = 0.0100$ (mol dm^{-3})
0.00005	0.216	0.089	0.101
0.0001	0.234	0.088	0.089
0.0002	0.314	0.086	0.080

[a]This table shows the limiting values of D_i in which 70 μL of BSA at c_{BSA} + Na_4PRRA solution at concentration, c, was injected into Na_4PRRA solution at concentration, c, flowing solution.

TABLE 1.6 Deviations Between the Limiting Diffusion Coefficients, $D^{\circ}_{(in\ Na4ETRA)}$, for BSA in Solutions Containing Na_4ETRA at Different Concentrations, *c*, $D^{\circ}_{(inNa4ETRA)}$, and the Limiting Values in Water $D^0_{(in\ H2O)}$ Obtained from the Taylor Technique at 25°C.

System	$(\Delta D^o/D)\%$ (c = 0.0025 mol dm^{-3})	$(\Delta D^o/D)\%$ (c_{flow} = 0.0050 mol dm^{-3})	$(\Delta D^o/D)\%$ (c_{flow} = 0.0100 mol dm^{-3})
BSA in solutions containing ETRA	−51.8	−51.4	−56.5
	−46.6	−53.8	−59.4
	−49.0	−56.8	−62.0

TABLE 1.7 Deviations Between the Limiting Diffusion Coefficients, $D^0_{(in\ Na4ETRA)}$, for BSA in Solutions Na_4PRRA at Different Concentrations, *c*, $D^0_{(inNa4PR\ RA)}$, and the Limiting Values in $D^0_{(in\ H2O)}$ Obtained from the Taylor Technique at 25°C.

System	**(ΔD°/D)%** **(c_{flow} = 0.0050 mol dm^{-3})**	**(ΔD°/D)%** **(c_{flow} = 0.0100 mol dm^{-3})**
BSA in solutions containing Na_4ETRA	−58.8	−53.2
	−62.4	−66.2
	−66.8	−69.1

From the negative deviations observed in Tables 5.6 and 5.7, we can say that added Na_4ETRA and Na_4PRRA at different concentrations, lead to the decrease of the diffusion coefficient of BSA from 51% to 69%, as well as the gradient of chemical potential (eq 1.9), indicating us the presence of strong interactions between aggregates BSA and resorcirarenes entities.

1.4.2 HYDRODYNAMIC RADII

Hydrodynamic radius values, R_h, of the BSA species in aqueous solutions of Na_4ETRA and Na_4PRRA at infinitesimal concentration can be estimated from Stokes–Einstein eq (1.10)[9,10] (Table 1.8), which considers the solvent as a continuum characterized by its bulk viscosity value. That is,

$$D^0 = \frac{k_B T}{6\pi \eta_0 R_h} \tag{1.10}$$

where k_B and η_0 are the Boltzmann's constant and the viscosity of water at absolute temperature, *T*, respectively. Despite this equation is only approximated (e.g., particles are considered perfectly spherical and are solely subject to solvent friction), it can be used to estimate the radius of the moving species, since BSA molecules are large enough when compared with the water molecules.

TABLE 1.8 Hydrodynamic Radii for BSA Species in Water ($R_{hin\ H2O}$) and in Aqueous Na_4ETRA ($R_{hin\ ETRA}$) and Na_4PRRA($R_{hin\ PRRA}$) Solutions at 25°C.

$R_{hinwater}$ (nm)[a]	$R_{hinETRA}$ (nm)[b]	$R_{hinPRRA}$ (nm)[c]
1.2	2.6	–
	2.4	2.7
	2.6	2.3

[a]Values estimated from eq (1.10), using for D^0, the D value estimated by extrapolation of data shown in Table 1.2.
[b]Values estimated from eq (1.10),using for D^0, the D values estimated by extrapolation of data for each flow solution, shown in Table 1.4.
[c]Values estimated from eq. (1.10),using for D^0, the D values estimated by extrapolation of data for each flow solution, shown in Table 1.5.

Table 1.8 presents the values of hydrodynamic radii for BSA species in water, and in aqueous Na_4ETRA and Na_4PRRA solutions at 25°C. From this table, we can see that the deviations between the dydrodynamic radii of the BSA species in aqueous solutions with and without Na_4ETRA (or Na_4PRRA) are positives. That is, $R_{h(in\ Na4ETRA\ or\ Na4PRRA\ )} > R_h\ (_{inwater})$. The higher hydrodynamic radii for BSA in aqueous Na_4ETRA and Na_4PRRA solutions can be related to the hydration of the BSA due to the fact that a hydrodynamic radius is a result not only of the particle size, but also of the solvent effects. Thus, this observed increase in the hydrodynamic radius of the BSA can be therefore attributed to capture of water molecules in the hydration sphere of the BSA solute caused by the strong interactions between Na_4ETRA (or Na_4PRRA) and the polar groups of the BSA.[19]

1.5 CONCLUSIONS

The limiting mutual diffusion coefficients for BSA in water and in aqueous solutions of Na_4ETRA and Na_4PRRA at 25°C were measured.

From negative deviations (D^0 (in Na_4ETRA or Na_4PRRA) $<D^0$ (in H_2O)), we can conclude that the presence of Na_4ETRA and Na_4PRRAenhances the interactions of BSA and water molecules, being more favored. This fact can also be interpreted on the basis of the salting-in effect. The observed behavior

could be a consequence of strong interactions between sodium and $ETRA^{4-}$ ($PRRA^{4-}$) ions and the water molecules, and, secondly, interactions between sodium and chloride ions with the charged groups of the BSA, which could reduce the electrostrictive effect of the water molecules, decreasing the value of the diffusion coefficient of the BSA.

These data may be useful once they provide transport data necessary to model the diffusion for various applications, such as pharmaceutical and biological applications.

ACKNOWLEDGMENTS

The authors in Coimbra are grateful for funding from "The Coimbra Chemistry Centre," which is supported by the Fundação para a Ciência e a Tecnologia (FCT), Portuguese Agency for Scientific Research, through the programs UID/QUI/UI0313/2019 and COMPETE. The author in Colombia are grateful for funding from Universidad de los Andes and the Instituto Colombiano para el Desarrollo de la Ciencia y la Tecnología COLCIENCIAS, Doctorado Nacional 6172.

KEYWORDS

- **diffusion coefficient**
- **sulfonated resorcinares**
- **bovine serum albumin**
- **Taylor dispersion**
- **transport properties**

REFERENCES

1. Elsadek, B.; Kratz, F. Impact of Albumin on Drug Delivery—New Applications on the Horizon. *J. Controll. Release* **2012,** *157*, 4–28. doi.org/10.1016/j.jconrel.2011.09.069.
2. Kratz, F. Albumin as a Drug Carrier: Design of Prodrugs, Drug Conjugates and Nanoparticles. *J. Control. Release* **2008,** *132*, 171–183. doi.org/10.1016/j.jconrel.2008.05.010.
3. Ribeiro, A. C. F.; Esteso, M. A. Transport Properties for Pharmaceutical Controlled-Release Systems: A Brief Review of the Importance of Their Study in Biological Systems. *Biomolecules* **2018,** *178*. doi.org/10.3390/biom8040178.

4. Galindres, D. M.; Ribeiro, A. C. F.; Valente, A. J. M.;. Esteso, M. A; Sanabria, E.; Vargas, E. F. Verissimo, L. M. P.; Leaist, D. G. Ionic Conductivities and Diffusion Coefficients of Alkyl Substituted Sulfonatedresorcinarenes in Aqueous Solutions. *J. Chem. Thermodynamics* **2019,** *133*, 222–228. doi.org/10.1016/j.jct.2019.02.018.
5. Hogberg, A. G. S. Two Stereoisomeric Macrocyclic Resorcinol-acetaldehyde Condensation Products. *J. Org. Chem.* **1980,** *45*, 4498–4500.
6. Kazakova, E. K.; Makarova, N. A.; Ziganshina, A. U.; Muslinkina, L. A. Novel Water-Soluble Tetrasulfonatomethylcalix [4] resorcinarenes. *Tetrahedron Lett.* **2000**, *41*, 10111–10115. doi: 10.1016/S0040-4039(00)01798-6.
7. Sanabria, E.; Esteso, M. A.; Pérez-redondo, A. Vargas, E.; Maldonado, M. Synthesis and Characterization of Two Sulfonatedresorcinarenes: A New Example of a Linear Array of Sodium Centers and Macrocycles *Molecules* **2015**, *20* (6), 9915–9928. doi:10.3390/molecules20069915.
8. Leaist, D. G.; Hao, L. Diffusion in Buffered Protein Solutions: Combined Nernst–Planck and Multicomponent Fick Equations. *J. Chem. Soc., Faraday Trans.* **1998,** *94*, 3527–3780.
9. Robinson, R. A.; Stokes, R. H. *Electrolyte Solutions*, 2nd Ed., Butterworths: London, 1959.
10. Tyrrell, H. J. V.; Harris, K. R. *Diffusion in Liquids: A Theoretical and Experimental Study*. Butterworths: London, 1984.
11. Aris, R.On the Dispersion of a Solute in a Fluid Flowing through a Tube. *Proc. R. Soc. L.* **1956,** *235*, 67–77. doi:10.1098/rspa.1956.0065.
12. Barthel, J.; Gores, H. J.; Lohr, C. M.; Seidl, J. J. Taylor Dispersion Measurements at Low Electrolyte Concentrations. I. Tetraalkylammonium Perchlorate Aqueous Solutions. *J. Solut. Chem.* **1996,** *25*, 921–935.
13. Loh, W. Taylor Dispersion Technique for Investigation of Diffusion in Liquids and Its Applications. *Quim. Nova* **1997,** 20, 541–545.
14. Callendar, R.; Leaist, D. G. Diffusion Coefficients for Binary, Ternary, and Polydisperse Solutions from Peak-Width Analysis of Taylor Dispersion Profiles. *J. Solut. Chem.* **2006,** *35*, 353–379. doi: 10.1007/s10953-005-9000-2.
15. Grossmann, T.; Winkelmann, J. Ternary Diffusion Coefficients of Cyclohexane plus Toluene plus Methanol by Taylor Dispersion Measurements at 298.15 K. Part 1. Toluene-Rich Area. *J. Chem. Eng. Data* **2009,** *54*, 405–410. doi:10.1021/je800444e.
16. Leaist, D. G.; Hao, L. Tracer Diffusion of Some Metal Ions and Metal–EDTA Complexes in Aqueous Sodium Chloride Solutions. *J. Chem. Soc. Faraday Trans.* **1994,** *90*, 133–136. doi:10.1039/FT9949000133.
17. Leaist, D. G. Coupled Tracer Diffusion Coefficients of Solubilizates in Ionic Micelle Solutions from Liquid Chromatography. *J. Solut. Chem.* **1991,** *20*, 175–186.
18. Rodriguez, M.; Verissimo, L. M. P.; Barros, M. C. F.; Rodrigues, D. F. S. L.; Rodrigo, M. M.; Miguel, A.; Esteso, C. M.; Romero, A. C. F. Ribeiro Limiting Values of Diffusion Coefficients of Glycine, Alanine,α-amino Butyric Acid, Norvaline and Norleucine in a Relevant Physiological Aqueous Medium. *Eur. Phys. J. E.* **2017,** *40* (21), 1–5. doi: 10.1140/epje/i2017-11511-y.
19. He, S.; Huang, M.; Ye, W.; Chen, D.; He, S.; Ding, L.; Yao, Y.; Wan, L.; Xu, J.; Miao, S. Conformational Change of Bovine Serum Albumin Molecules at Neutral pH in Ultra-Diluted Aqueous Solutions. *J. Phys. Chem. B* **2014,** *118*, 12207−12214.Doi:10.1021/jp5081115|.
20. Onsager, L.; Fuoss, R. M. *J. Phys. Chem.* **1932,** *36*, 2689.

CHAPTER 2

NOTES ON THE BAROMETRIC FORMULA

LIONELLO POGLIANI

Facultad de Farmacia, Dept. de Química Física,
Universitat de València, Av. V.A. Estellés s/n,
46100 Burjassot (València), Spain
E-mail: liopo@uv.es

ABSTRACT

The barometric formula, relating the pressure $p(z)$ of an isothermal, ideal gas of molecular mass m at some height z to its pressure at sea level, p_0, is discussed. Three mathematical derivations of the formula are given together with three generalizations, like nonisothermal atmosphere, nonconstant gravitational field, and Earth rotation. A generalization of the barometric formula for negative heights is discussed. Some related historical aspects are also given.

2.1 A BIT OF HISTORY

Galileo's disciple Evangelista Torricelli (1608–1647) in 1643–1644 devised a decisive experiment with the help of an ad hoc setup consisting of two long glass tubes (ca. 1.2 m), sealed at one end, and a bowl of mercury. In Figure 2.1 is instead a modern version of the barometer used by Torricelli (for the original drawing, and for more details see Ref. 1). In it, P_{Hg} is the pressure exerted by the (76 cm) column of mercury (Hg), while the pressure exerted by the atmosphere is P_{air}. Torricelli carried out his celebrated mercury column experiment in collaboration with another disciple of Galileo, Vincenzo Viviani (1622–1703). It should be borne in mind that up to then ruled Aristotelian physics that denied the existence of vacuum (nature's abhorrence of a vacuum). The purposes of the experiment were: (1) to confirm the existence of a vacuum, (2) to show the existence of air pressure, and (3) to display the variations of pressure with weather.

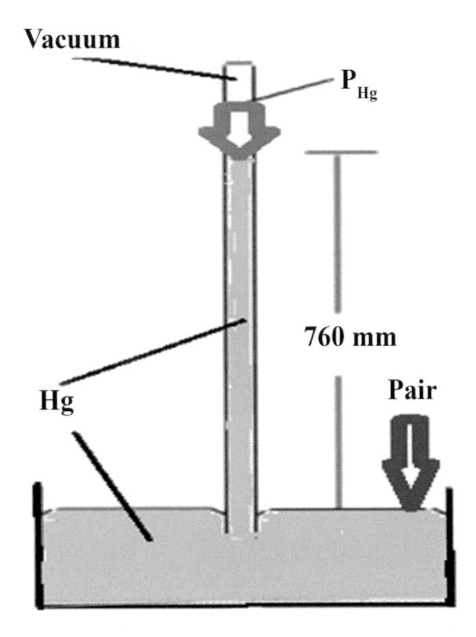

FIGURE 2.1 Torricelli's experiment at sea level.

Torricelli was well aware of the great importance of his experiment, though he chose, for fear of the inquisition, not to publicize it outside a small circle of friends and colleagues. Thanks to the exchange of scientific letters and to scientific travelers such as the French monk Marin Mersenne, it became well known throughout Europe as the *Italian Experiment*, but the name of its author was revealed only after Torricelli's death. Of all experiments done to confirm his findings, the one by the French polymath Blaise Pascal (1623–1662) was decisive. Pascal recorded the height of a column of mercury as a function of altitude, as following the French scientist, on the one hand, the height should decrease with an increase in altitude, as less air exerts weight on top of a mountain than at its base; on the other hand nature's abhorrence of a vacuum must be the same at both places. It should be remarked that variations of air density with altitude were mentioned by Torricelli in a letter to Ricci.[1] Pascal, after some experiments carried out, under his suggestion, by his brother-in-law, Florin Périer, in 1648, decided to repeat himself the experiment in Paris, at the St. Jacques tower (ca. 52 m height), that still exists, and has at its foot a statue of Pascal with a barometer!

A quantitative relation (barometric formula), that is, the well-known exponential dependence of pressure on height could only be obtained after the discovery of Boyle's law ($P \propto 1/V$, Oxford, 1662), and was first recognized in 1686 by the English physicist and astronomer Edmund Halley (1656–1742), also from Oxford University. Much later, the French mathematician Pierre-Simon de Laplace (1749–1827) explicitly obtained the barometric formula (and extensions of it) in his *Traité de Mécanique Céleste.*[1] This is the reason why the barometric formula is called (sometimes) Laplace's formula. The barometer (name coined by Boyle) was soon used for the measurement of altitude, although the results were subject to error, owing to local pressure changes and temperature and composition variations, as the real atmosphere is not in strict thermodynamic equilibrium.

2.2 THE BAROMETRIC FORMULA

The following barometric formula[1] relates the pressure $p(z)$ of an isothermal, ideal gas of molecular mass m at some height z to its pressure $p(0)$ at height $z = 0$, where g is the acceleration of gravity, k the Boltzmann constant, and T the temperature. In spite of its simplicity, namely, the assumption of isothermal temperature, it applies reasonably well to the lower troposphere (for altitudes up to 6 km, the error is less than 5%), and also to the stratosphere, up to 20 km (with $T = 217$ K, and with $H = kT/mg_0$ = scale height)

$$p(z) = p_0\exp(-mg_0z/kT) = p_0\exp(-z/H) \tag{2.1}$$

Actually the assumption of constant temperature is not the only one on which this formula is based. Some other assumptions worth citing that are not implicit in the formula are: the atmosphere is not uniform in molecular mass throughout, the gravitational field is not uniform, atmospheric gases do not behave as ideal gases, conditions of equilibrium do not hold. The formula does not consider Earth rotation and assumes that pressure is just the weight of an overlying column of air, rather than the more accurate conic section, that is, locally the Earth is considered flat.

Actually, some of these drawbacks, like the nonuniform gravitational field, temperature gradient, nonequilibrium system, and the Earth rotation, are examined in Ref. 1 and in a subsequent paper,[2] and some of them will be discussed here. It is interesting to notice how this formula is useful even today, suffice to cite the many studies that cite the two papers[1,2] that are the basis of the present mini review. The nearly hundreds citations cover all sorts of scientific domains: chemistry, climatology, ecology, education, electronics, engineering, history of science, informatics, medicine, physics, physical chemistry, statistical mechanics, and thermodynamics. Some of them[3–28] are cited for the curious reader who wishes to deepen the subject.

2.3 THE MATHEMATICS

We discuss now different mathematical formalisms behind the barometric formula. Each one highlights diverse aspects of the problem.

2.3.1 HYDROSTATICS FORMULATION

From the perfect gas equation $pV = NkT$, that is, $N/V = p/kT$, where N is the number of molecules contained in the volume V, one obtains the following mass density $\rho_m(z)$ at a given height z,

$$\rho_m(z) = Nm/V(z) = mp(z)/kT \tag{2.2}$$

If this gas is contained in a vessel of height h, and if g and T are constant, then at equilibrium the pressure is given by eq (2.3) where the first integral is the mass of the gas in a column of a vessel of unit area that extends from z to h,

$$p(z) = p(h) + g\int_z^h \rho_m(u)\,du = p(h) + \frac{gm}{kT}\int_z^h p(u)\,du \tag{2.3}$$

Solving eq (2.3), we retrieve eq (2.1). Actually, a differential balance of forces can be written from the start, as is common practice. In mechanical equilibrium, the opposite forces acting in a column of air of unit area between z and $z+dz$ must be equal (with successive insertion of eq 2.2):

$$p(z + dz) + \rho_m(z)\,gdz = p(z) \tag{2.4}$$

$$dp/dz = mgp/kT \tag{2.5}$$

Taking $p_0 = 1.0$ atm, and $g = 9.8$ ms^{-2}, we get $M_0 = p_0/g_0 = 1.0$ kg cm^{-2} for the air mass per unit area. This mass is exponentially distributed in height, according to eq (2.1). Were air an incompressible fluid (as mercury approximately is), its density would not vary with height. Assuming that in such a case the density was that for zero height, the total height of the air column, H (the scale height), would be

$$H = M_0/\rho_{m0} = [p_0/g_0] \times [mp_0/\mathrm{kT}]^{-1} = kT/mg_0 \tag{2.6}$$

For our atmosphere with $T = 290$ K, $H = 8.5$ km (the so-called scale height). Such a value overlaps with the value estimated by E. Halley[1] by using the ratio of mercury to air densities, which is 10,800/1, and in this case we also have: $H = 762 \times 10{,}800 = 8.3$ km.

2.3.2 STOCHASTIC FORMULATION

Let us treat now the case of a dilute suspension of tiny particles in a liquid that also obey the barometric formula. To this conclusion had already arrived the French physicist Jean Perrin in 1909 in his studies of the Brownian motion.[1] He noticed that a suspension in water of tiny tree resin spherical particles ($0.2 < r < 0.5$ μm) behaved as a miniature atmosphere and could be studied with an optical microscope. This kind of motion (Brownian motion) is roughly described by the diffusion (or Fokker–Planck) equation, for timescales larger than the decay time of the particle's velocity autocorrelation. Therefore, such an equation applies also to the dilute gas, but on a much coarser, but still microscopic, scale. With the above restrictions, the probability density function $w(z,t)$ of a diffusing particle under the action of a constant gravitational field directed along the negative z direction obeys the following Smoluchowski equation (a special kind of Fokker–Planck equation),[1,30,31]

$$\partial w/\partial t = D\partial^2 w/\partial z^2 + c\partial w/\partial z \tag{2.7}$$

Here, D is the diffusion coefficient, $D = kT/f$, f being the drag coefficient ($= 6\pi\eta r$ for a macroscopic sphere of radius r in a fluid of viscosity η), and $c = m'g/f$, m' being the apparent mass of the particle. Owing to the buoyancy of the liquid $m' = m(1 - \rho_L/\rho_P)$, where m is the true mass of the particle, and ρ_L = liquid's density, and ρ_P = particle's density (see note 16 in Ref. 1). The time evolution of $w(z,t)$ is subject to the initial condition, and to the boundary conditions, respectively,

$$w(z, 0) = \delta(z - z_0), \text{ and } D\partial w/\partial z + cw = 0 \text{ at } z = 0 \text{ for all } t \geq 0 \tag{2.8}$$

These conditions mean that the particle starts its motion at $z = z_0$ (1st condition), and cannot cross the plane $z = 0$ (bottom of the vessel, 2nd condition). With these conditions then $w(z,t)$ is given by

$$w(z,t) = Y(z,t) + \frac{c}{D\sqrt{\pi}} e^{-cz/D} \int_{l}^{\infty} \exp(-x^2)\, dx$$

$$Y(z,t) = \frac{1}{2\sqrt{\pi Dt}} \left\{ exp\left[-\frac{(z-z_0)^2}{4Dt} \right] + exp\left[-\frac{(z+z_0)^2}{4Dt} \right] \right\} exp\left[-\frac{c}{2D}(z-z_0) - \frac{c^2}{4D}t \right] \tag{2.9}$$

$$l = \frac{(z + z_0 - ct)}{2\sqrt{Dt}} \tag{2.10}$$

At short times, a Gaussian-like curve is obtained, as it is the case for free diffusion; however, an asymmetry soon develops, owing to gravity. Finally, at $t \to \infty$, only the last term survives, yielding the exponential density. This density is easily obtained by setting $\partial w/\partial t = 0$ in eq (2.7) and solving the resulting ordinary differential equation. For the case of particles suspended in a liquid, the exponential function can be used to explain sedimentation: while small particles may be approximately homogeneously distributed in a liquid, the aggregates formed when they coalesce, will have masses high enough to compress, so to speak, the exponential function, yielding a thin layer at the bottom. Conversely, for a fixed particle mass, sedimentation can still be made to occur by increasing the acceleration g, as is done in centrifuges.

2.3.3 STATISTICAL FORMULATION

The most straightforward derivation of the barometric formula is perhaps from Boltzmann's distribution,[32]

$$P(\mathbf{r},\mathbf{v})=\frac{exp\left(-\frac{E(\mathbf{r},\mathbf{v})}{kT}\right)}{\int_V exp\left(-\frac{E(\mathbf{r},\mathbf{v})}{kT}\right)d\mathbf{r}d\mathbf{v}} \tag{2.11}$$

Here V is the phase-space volume, and $P(\mathbf{r}, \mathbf{v})$ is the joint equilibrium distribution function for position and velocity. For three-dimensional motion in a constant gravitational field, and with $v = |\mathbf{v}|$, we have

$$E(\mathbf{r}, \mathbf{v}) = mv^2/2 + mgz \tag{2.12}$$

Thus, the final form of eq (2.11) is the following eq (2.13),

$$P(\mathbf{r},\mathbf{v})=\frac{1}{L^2}\left(\frac{m}{2\pi kT}\right)^{3/2}\left(\frac{mg}{kT}\right)exp\left(-\frac{mv^2}{2kT}\right)exp\left(-\frac{mgz}{kT}\right) \tag{2.13}$$

Here L is the (very large) linear dimension of the container, and $w(z) = \iiint P(\mathbf{r}, \mathbf{v})d\mathbf{v}dxdy$ is again the exponential density.

2.4 GENERALIZING THE FORMULA

In the present section three generalization of the barometric formula are presented that allow to amplify the range of applicability of the formula.

2.4.1 THE ATMOSPHERE IS NOT ISOTHERMAL

Temperatures throughout the troposphere, the lowest layer of Earth's atmosphere, cover a wide range of values. This layer is heated from below, and it is warmest at the bottom near Earth's surface (average temperature = 15°C), and coldest at its top (−57°C), and, typically, the temperature drops about 6.5°C per km (= β) increase in altitude. The troposphere extends from Earth's surface up to a height of 6 km (in polar regions in winter) to 17 km (middle latitudes) above sea level, the average altitude being 13 km. Most of the mass (about 75%–80%) of the atmosphere is in this region, and almost all weather occurs within it. The sunlight heats the Earth's surface that radiates the heat back into the adjacent atmosphere. Temperatures in the troposphere, both at the surface and at various altitudes, do vary based on latitude, season, time of day or night, regional weather conditions, and in some circumstances, the temperature at the top of the troposphere can be as low as −80°C. The phenomenon known as "temperature inversion" means that the temperature in some part of the troposphere gets warmer with increasing altitude, contrary to the normal situation.[32] Consider the case of uniform gravitational field, g_0, with a vertical temperature gradient, where the temperature changes linearly with height, as in the following eq (2.14), which is a good approximation for the troposphere,

$$T(z) = T_0 - \beta \cdot z \tag{2.14}$$

Here β [K/km] is a positive constant. Now, by the aid of eq (2.3) or (2.5) we obtain,

$$p(z) = p_0[1 - \beta \cdot z/T_0]^{mg_0/k\beta} \tag{2.15}$$

Equation (2.15) is a good model for the pressure dependence on altitude up to 11 km with: $g_0 = 9.8$ m·s^{-2}, $p_0 = 10^5$ Pa, $T_0 = 288$ K (15°C), and $\beta = 6.5$ K/km.

2.4.2 NONCONSTANT G

This constant varies with latitude, longitude, and elevation. These variations are rather small and lead to a minor error. The general case for an isothermal atmosphere where the acceleration g depends on altitude z, is given by the following equation,

$$p(z) = p_0 exp\left(-\int_0^z \frac{mg(u)}{kT} du\right) \tag{2.16}$$

According to the law of gravitation, and noting that the mass of the atmosphere is quite small compared to Earth's mass we have,

$$mg(z) = G\frac{Mm}{(R+z)^2} = G\frac{Mm}{R^2(1+z/R)^2} = g_0\frac{m}{(1+z/R)^2} \tag{2.17}$$

Here, M = mass of the Earth, and at the Earth's surface: $g_0 = GM/R^2$. Now, inserting eq (2.17) into eq (2.16), we have eq (2.18) that integrates into eq (2.19),

$$p(z) = p_0 exp\left(-\frac{mg_0}{kT_0} \int_0^z \frac{du}{(1+u/R)^2} \right) \tag{2.18}$$

$$p(z) = p_0 exp\left[-\frac{mg_0 R}{kT_0} \left(1 - \frac{1}{1+z/R} \right) \right] \tag{2.19}$$

Now, with $1/H = mg_0/kT_0$ we obtain,

$$p(z) = p_0 exp\left[-\frac{R}{H} \left(1 - \frac{R}{R+z} \right) \right] = p_0 exp\left[\left(-\frac{1}{1+z/R} \right) \frac{z}{H} \right] \tag{2.20}$$

Calculations show that a noticeable difference between data obtained from eqs (2.1) and (2.20) shows up only for $z > 0.01R \approx 64$ km ($\approx 8\%$, $T_0 = 288$K), that is, well above the stratosphere. Following eq (2.20) pressure approaches a nonzero value for $z \to \infty$, which is physically questionable. This aspect shows that a static and isothermal atmosphere is intrinsically unstable. A deeper analysis of this problem is discussed in Ref. 1.

2.4.3 EARTH ROTATION

Suppose that Earth atmosphere and solid Earth rotate with a unique angular velocity ω (for the troposphere this is quite appropriate), this means that due to this rotation, the weight of gas is not the same at the pole than at the equator. At the pole, acceleration of gravity is described by eq (2.17), that is, $g(z) = g_0(1+z/R)^{-2}$. At the equator, instead, the weight is decreased by the centrifugal force $m\omega^2(R+z)$, that is, the effective acceleration is smaller than the acceleration of gravity and obeys eq (2.21), where $\Phi(z)$ is given by eq (2.22). Assuming, $\omega = 7.27\cdot10^{-5}$ rad/s, $g_0 = 9.8$ m$\cdot$s^{-2}, and $R = 6.4\cdot10^6$ m, we have $\omega^2R/g_0 \approx 3.45\cdot10^{-3}$. It is clear from eq (2.21) that the centrifugal force becomes important (change in gravity acceleration $\geq 1\%$) for $(z) \geq 0.01$, that is, for $z/R \geq 0.43$, which leads to a large $z = 2700$ km.

$$g(z) = g_0(1+z/R)^{-2} - \omega^2 R(1+z/R) = g_0(1+z/R)^{-2}\left[1-\Phi(z)\right] \tag{2.21}$$

$$\Phi(z) = \frac{\omega^2 R}{g_0}(1+z/R)^3 \tag{2.22}$$

If it is assumed that atmosphere and Earth rotate as a whole, irrespective of height, the upper limit of the atmosphere at the equator can be obtained from eq (2.21) by setting $g(z) = 0$. Then, $\Phi(z) = 0 \rightarrow z/R = 5.6$, and the result is a meaningless z = 36,000 km. In fact, the density of outer space is attained for altitudes lower than 1000 km. Escape of molecules, atoms, and ions from the upper atmosphere occurs by thermal and photochemical mechanisms still in the presence of a significant inward force. Combining eqs (2.21) and (2.16), and reminding that $H^{-1} = mg_0/kT_0$ we obtain for the barometric equation in the case of an isothermal atmosphere,

$$p(z) = p_0 exp\left(-\frac{z}{H}\int_0^z (1+u/R)^{-2}\left[1-\Phi(u)\right]du\right) \quad (2.23)$$

Integrating this expression we finally obtain,

$$p(z) = p_0 exp\left(-\frac{z}{H}\left[\frac{1}{1+z/R}-\frac{\omega^2 R}{g_0}(1+z/2R)\right]\right) \quad (2.24)$$

Calculations show a noticeable difference (≈ 2%) between data obtained from eqs (2.24), and (2.20) for $z > 0.01R \approx 64$ km, that is, well above the stratosphere. The major problem in applying the barometric formula to the real atmosphere, however, derives from the fact that the atmosphere is not in equilibrium.

2.5 PRESSURE INSIDE THE EARTH

Suppose a shaft is drilled down to the center of the Earth, notwithstanding the technical impossibility of this feat, namely, owing to the immense pressures and temperatures that exist inside the Earth, and to the physical state of its inner layers, it is interesting to imagine what would be the depth dependence of air pressure within this imaginary shaft. Let us first have a short excursus throughout history about this topic (a more detailed excursus with appropriate references see Ref. 2).

2.5.1 A BIT OF HISTORY

The motion of an object (neglecting drag) dropped in a bottomless shaft was again considered by Hooke in 1679. The main point under discussion was the effect of Earth's rotation on the trajectory. Hooke obtained the correct result qualitatively: The object should oscillate like a pendulum, describing an ellipse. In fact, an object dropped in a shaft connecting the poles of a homogeneous and spherical Earth behaves as a one-dimensional harmonic oscillator and strictly obeys Hooke's "law," although this is not the present standard pedagogical example. In 1882 the respected French civil engineer and applied mathematician É. Collignon (1831–1897) speculated on the possibility of travel between cities by means of long linear tunnels inside the Earth, in a kind of partial free-fall planetary subway, for which the transit time *in the absence of drag* is 42 min, independently of the location of the two cities. An account of his ideas, published on a semi-humorous tone in the scientific periodical *La Nature*, is suggestively entitled "From Paris to Rio de Janeiro in 42 minutes and 11 seconds" In it, the effect of pressure is discussed, and it is considered an insurmountable problem. Numerical estimates of the enormous pressures at several depths (but provided with no computational details) are given, but differ from the calculations given below by several orders of magnitude.

2.5.2 THE MATHEMATICS

Assuming for simplicity that air temperature and Earth's density are both uniform, eq (2.18) applies, where the acceleration of gravity now is,

$$g(z) = \frac{4}{3}\pi G\rho(R+z) = g_0(1+z/R) \tag{2.25}$$

Equation (2.18), with $-R < z < 0$, and with eq (2.25) now, after integration, becomes (reminding that, $g_0 = GM/R^2$, and $1/H = mg_0/kT_0$),

$$p(z) = p_0 exp\left[-\left(1+\frac{z}{2R}\right)\frac{z}{H}\right] \tag{2.26}$$

As the reader can notice the dependence is similar to eq (2.1), apart from the multiplicative factor (<1) in the argument of the exponential that slightly reduces the variation, owing to the decrease of g with depth. The deepest gold mines in South Africa attain a depth of 3.9 km, ca., for which one obtains $p = 1.6$ atm, in good agreement with the observations. While eq (2.26) predicts a pressure of 1000 atm for $z = -58$ km, for $z = -R$, instead, the calculated pressure becomes: $p(-R) = p_0\exp(R/2H) = 10^{165}p_0$, clearly, a meaningless value, as the air for pressures of few tons of atm ceases to behave as an ideal gas. A more detailed calculation with the van der Waals equation is discussed in Ref. 2.

2.6 LAST REMARKS

Whenever we take a formula that is valid only for an ideal gas under conditions of equilibrium and apply it to real cases, it is not a surprise if it does not fit the data perfectly. Nevertheless, pressure seldom departs from the average value by more than a few percent, and within this restriction the barometric equation given by eq (2.1) does its job. The nonideality of a gas has instead a much more dramatic influence on the barometric equation at negative heights.

It should be remarked that the atmosphere as a whole is never in a state of equilibrium, as it continuously exchanges mass and energy with its surroundings and this is why there is weather. To better understand this last topic, we should know something about the Bernoulli's principle of fluid dynamics[33] that is though valid only for ideal fluids. It states that an increase in the speed of a fluid occurs simultaneously with a decrease in pressure or a decrease in the fluid's potential energy, it can be applied to various types of fluid flow, and there are different forms of Bernoulli's equation for different types of flow. In Ref. 2 and 5 a full discussion of this principle is given.

KEYWORDS

- **barometric formula**
- **pressure**
- **mathematics**
- **generalization**
- **ideal gas**

REFERENCES

1. Berberan-Santos, M. N.; Bodunov, E. N.; Pogliani, L. On the Barometric Formula. *Am. J. Phys.* **1997,** *65*, 404–412.

2. Berberan-Santos, M. N.; Bodunov, E. N.; Pogliani, L. On the Barometric Formula Inside the Earth. *J. Math. Chem.* **2010,** *47*, 991–1004.
3. Hall, D. S.; Standish, T. E.; Behazin, M.; Keech, P. G. Corrosion of Copper-Coated Used Nuclear Fuel Containers Due to Oxygen Trapped in a Canadian Deep Geological Repository. *Corros. Eng. Sci. Techn.* **2018,** *53*, 309–315.
4. Eymüller, C.; Wanninger, C.; Hoffmann, A.; Reif. W. Semantic Plug and Play - Self-Descriptive Modular Hardware for Robotic Applications. *Int. J. Semant. Comput.* **2018,** *12*, 559–577.
5. Alonso-González, E.; et al. Daily Gridded Datasets of Snow Depth and Snow Water Equivalent for the Iberian Peninsula from 1980 to 2014. *Earth Syst. Sci. Data* **2018,** *10*, 303–315.
6. Hall, D. S.; Standish, T. E.; Behazin, M.; Keech, P. G. Corrosion of Copper-Coated Used Nuclear Fuel Containers Due to Oxygen Trapped in a Canadian Deep Geological Repository. *Int. J. Corros. Process. Corros. Contr.* **2018,** *53*, 309–315.
7. Zhao, F.; Luo, H.; Zhao, X.; Pang, Z.; Park, H. HYFI: Hybrid Floor Identification Based on Wireless Fingerprinting and Barometric Pressure *IEEE T. Ind. Info.* **2017,** *13*, 330–341.
8. Akinnubi, R. T.; Adeniyi, M. O. Modeling of Diurnal Pattern of Air Temperature in a Tropical Environment: Ile-Ife and Ibadan, Nigeria. *Model. Earth Syst. Environ.* **2017,** *3*, 1421–1439.
9. Massé, F.; Gonzenbach, R.; Paraschiv-Ionescu, A.; Luft, A.R.; Aminian, K. Wearable Barometric Pressure Sensor to Improve Postural Transition Recognition of Mobility-Impaired Stroke Patients. *IEEE Trans. Neural. Syst. Rehabil. Eng.* **2016,** *24*, 1210–1217.
10. Wua, Z.; Zhou, X.; Liu, X.; Ni, Y.; Zhao, K.; Peng, F.; Yang, L. Investigation on the Dependence of Flash Point of Diesel on the Reduced Pressure at High Altitudes. *Fuel* **2016,** *181*, 836–842.
11. Garcia-Diez, R.; Gollwitzer, C.; Krumrey, M. Nanoparticle Characterization by Continuous Contrast Variation in Small-Angle X-Ray Scattering with a Solvent Density Gradient. *J. Appl. Cryst.* **2015,** *48*, 20–28.
12. Fajardo, S.; Frankel, G. S. Gravimetric Method for Hydrogen Evolution Measurements on Dissolving Magnesium. *J. Electrochem. Soc.* **2015,** *162*, C693–C701.
13. Cleasby, I. R.; Wakefield, E. D.; Bearhop, S.; Bodey, T. W.; Votier, S. C.; Hamer, K. C. Three-Dimensional Tracking of a Wide-Ranging Marine Predator: Flight Heights and Vulnerability to Offshore Wind Farms. *J. Appl. Ecol.* **2015,** *52*, 1474–1482.
14. Xia, H.; Wang, X. ; Qiao, Y.; Jian, J.; Chang, Y. Using Multiple Barometers to Detect the Floor Location of Smart Phones with Built-in Barometric Sensors for Indoor Positioning. *Sensors* **2015,** *15*, 7857–7877.
15. Spinoni, J.; et al. Climate of the Carpathian Region in the Period 1961–2010: Climatologies and Trends of 10 Variables. *Int. J. Climatol.* **2015,** *35*, 1322–1341.
16. Fabien Massé, F.; Gonzenbach, R. R.; Arami, A.; Paraschiv-Ionescu, A.; Luft, A. R.; Aminian, K. Improving Activity Recognition Using a Wearable Barometric Pressure Sensor in Mobility-Impaired Stroke Patients. *J. NeuroEng. Rehab.* **2015,** *12*, 1–15.
17. Ignaccolo, R.; Franco-Villoria, M.; Fassò, A. Modelling Collocation Uncertainty of 3D Atmospheric Profiles. *Stoch. Environ. Res. Risk. Assess.* **2015,** *29*, 417–429.
18. Massé, F.; Bourke, A. K.; Chardonnens, J. ; Paraschiv-Ionescu, A.; Aminian, K. Suitability of Commercial Barometric Pressure Sensors to Distinguish Sitting and Standing Activities for Wearable Monitoring. *Med. Eng. Phys.* **2014**, *36*, 739–744.

19. Igoe, D. P.; Parisi, A.; Carter, B. Smartphone-Based Android App for Determining UVA Aerosol Optical Depth and Direct Solar Irradiances. *Photochem. Photobiol.* **2014**, *90*, 233–237.
20. Marthews, T. R.; Malhi, Y.; Iwata, H. Calculating Downward Longwave Radiation Under Clear and Cloudy Conditions Over a Tropical Lowland Forest Site: An Evaluation of Model Schemes for Hourly Data. *Theor. Appl. Climatol.* **2012,** *107*, 461–477.
21. Dubinova, A. A: Exact Explicit Barometric Formula for a Warm Isothermal Fermi Gas. *Tech. Phys.,* **2009,** *54*, 210–213.
22. Monti, D.; Ariano, P.; Distasi, C.; Zamburlin, P.; Bernascone, S.; Ferraro, M. Entropy Measures of Cellular Aggregation. *Physica A* **2009,** *388*, 2762–2770.
23. Bottecchia, O. L. The Barometric Formula as Resource for Teaching Chemistry. *Quím. Nova* **2009,** *32*, 1965–1970.
24. Fink, J. K. Equilibrium. In *Physical Chemistry in Depth.* Springer: Berlin, Heidelberg, 2009.
25. López-Moreno, J. I.; Goyette, S.; Beniston, M. Impact of Climate Change on Snowpack in the Pyrenees: Horizontal Spatial Variability and Vertical Gradients. *J. Hydrol.* **2008,** *374*, 384–396.
26. Bohm, T. D.; Griffin, S. L.; DeLuca Jr., P. M.; DeWerd, L. A. The Effect of Ambient Pressure on Well Chamber Response: Monte Carlo Calculated Results for the HDR 1000 Plus. *Med. Phys.* **2005**, *32*, 1103–114.
27. Alin, S. R.; Johnson T. C. Carbon Cycling in Large Lakes of the World: A Synthesis of Production, Burial, and Lake-Atmosphere Exchange Estimates. *Global Biogeochem. Cy.*, **2007,** *21*, 1–12, doi: 10.1029/ 2006GB002881.
28. Pantellini, F. G. E. A Simple Numerical Model to Simulate a Gas in a Constant Gravitational Field. *Am. J. Phys*. **2000,** *68*, 61–68.
29. Chandrasekhar, S. Stochastic Problems in Physics and Astronomy. *Rev. Mod. Phys.* **1943,** *15*, 1–91.
30. Van Kampen, N. G. *Stochastic Processes in Physics and Chemistry*. 2nd ed; Springer: Berlin, 1985; pp. 195–210.
31. Reif, F. *Fundamentals of Statistical and Thermal Physics*. McGraw-Hill: New York, 1965.
32. https://www.windows2universe.org/earth/Atmosphere/troposphere_temperature.html&edu=high.
33. https://en.wikipedia.org/wiki/Bernoulli%27s_principle.

CHAPTER 3

GREEN TECHNOLOGY PRODUCTS FOR SUSTAINABLE DEVELOPMENT

ELVIN THOMAS

School of Environmental Sciences, Mahatma Gandhi University, Priyadarsini Hills P.O., Kottayam 686560, Kerala, India
E-mail: elvinthomas1992@gmail.com

ABSTRACT

In this chapter we have developed a brief review of the history of green technology for sustainable development along with an update of its current status. We have also presented new insights in green technology products with sustainable development.

3.1 INTRODUCTION

With increasing population and urbanization raising the living standards of an average human, the damage inflicted upon the environment through depletion of the world's stock of fossil fuels for energy and the subsequent greenhouse gas (GHG) emissions is monstrous. Society's approach to every product available as well as manufactured with a cradle-to-grave philosophy of extract–process–consume–dispose has led to the accumulation of large amounts of waste subsequently polluting the environment. This led to the search for clean and sustainable energy supplies as well as ways to reduce waste and maximize resource efficiency, all of which gave birth to green technology. Green technology encompasses all products, services, and practices that take into account the long-term as well as short-term impacts it causes to the environment. It thereby implements technology in a manner that minimizes environmental impact as well as resource consumption without compromising on the economic output. Both green chemistry and green engineering seek to maximize efficiency and minimize health and environmental hazards throughout the chemical production process providing a framework for the development of green technology products (Figure 3.1). Green products as defined by the Commission of European Communities (2001) are products that "use less resources, have lower impacts and risks to the environment and prevent waste generation already at the conception stage."[1] They are developed with the aim to reduce waste and maximize resource efficiency by making products that can be fully reclaimed and reused, which in effect prevents pollution of the environment.

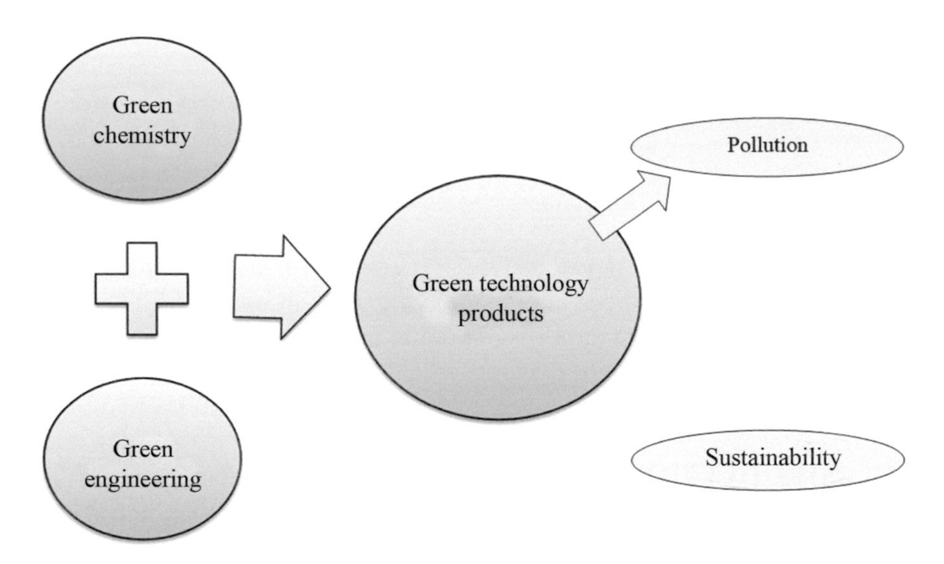

FIGURE 3.1 General overview of green technology products.

Facing the harsh realities of climate change and global warming, now more than ever before the world is striving to be green through the use of green technology. The green technology and sustainability market is now a multibillion business anticipating to grow from $8.7 billion in 2019 to $28.9 billion by 2024.[2] A majority of the current population assume green technology to be the future or, at the least, a fairly novel technology. Though the degree of widespread interest in the use of green technology products to abate the consequences of climate change is quite recent, the history of green technology is not so new. Green technology has been used by humans since time immemorial, but it is only since the 1990s that people started taking it seriously. This would be clear as the present chapter, developed with an intention to outline a brief review of the history of green technology, an update of its current status, and an attempt to prognosticate its future, unfolds.

3.2 GREEN TECHNOLOGY DURING PREHISTORIC TIMES

The power that comes with technology has fascinated human minds ever since the Paleolithic or Old Stone Age. The Oldowan stone tools formed almost 2.5 million years ago by *Homo habilis*, ancestors of *Homo sapiens*, can be considered as the first green technology product. During the Middle Paleolithic, several other tools were developed mainly the Acheulean stone tools, the Levallois technique-based flake tools, and Mousterian stone tools. These miniature stone tools, a major milestone in human evolutionary history, qualified all the modern-day criteria required of a green product. Even though it is not clear when humans began the controlled use of fire for light and warmth, archaeological findings provide evidence of prehistoric humans' controlled use of fire almost 1 million years ago. This marked the beginning of human dependence on bioenergy. Settlements developed only when the climatic conditions made it impossible for them to live in the open. The early settlements of the Paleolithic humans consisted of caves and open-air encampments that were made using wood, bones, and animal skin (probable leftovers from their hunting expeditions). All these settlements had a hearth inside and were near water sources generally. Many of the caves were also found to be inhabited by humans on a cyclic or seasonal basis for more than 10,000 years.[3] These settlements were thus in perfect harmony with the environment. The use of organic fibers in the manufacture of strings and ropes has also been indirectly evidenced in the Paleolithic sites thereby shedding light on the Stone Age fiber technology.[4] During the Upper Paleolithic, the ability of clay to retain its shape when dried and baked resulted in the discovery of ceramic products, which is being used to date as enablers of clean and green technologies. As the human population intensified, maximum resources began to be extracted from the hunted animal. Grease rendered from the bones along with a wick was used to light and bring warmth to the caves. Leather and animal skin were used as clothes, belts, straps, and shoes. The invention of boats or simple vessels (bamboo shafts) and seaworthy watercraft was another major development during the final phases of this epoch some 40,000 years ago. Thus, it can be concluded that human beings lived up to their utmost ability as an integral part of nature till the very end of Upper Paleolithic. During the Early and Middle Mesolithic, changes in the environment also brought changes in the fauna. With the food resources then available, hunting became much more difficult than in the previous period and so the social groups turned out to be smaller during this period. Subsequently, there occurred a transition from the use of stone tools to small blades and fishing tools made from bones or deer antlers. People began depending more on bird eggs, snails, and mollusks, and also started harvesting more edible plants, fruits, and roots. The settlements became more widely spread than in the Paleolithic. Between seven and eight millennia BP, during the Late Mesolithic, the fixed nature of settlements and population became more marked. Domestication of certain species (sheep and goats) appeared, and pottery-making techniques were soon acquired. Selective hunting and selective harvesting were practiced during this period for the sustainability of resources. As every tool of the prehistoric people who were hunter-gatherers was made using natural resources at their disposal, unintentionally or not they were leading the so-called "green life." With the life expectancy of the average human during this period being only 30 years and in exceptional cases 40 years, the fact that human groups were still too few to disturb the biological balance of the environment must be noted.[5] During this era, most of the energy needs were satisfied by the limited power of human metabolism along with the inefficient use of fire.

The Neolithic or the New Stone Age began about 12,000 years ago making a gradual transition from the hunter-gatherer lifestyle to a more settled way of life by domesticating plants and animals. This period is also described as the "age of building" marking the transition from Old Stone Age dwellings to New Stone Age permanent buildings. Increased population and intensive farming during this period required the need for land which led to large-scale forest clearance, often through burning thereby causing long-term impacts on the pristine forested environment. Large-scale digging of ditches and channels as well as the building of dikes was evidenced as early as toward the end of the ninth millennium BP, thereby solving issues related to irrigation of excessively dry land and the drainage of excessively wet or marshy lands. Archaeological evidence confirms that rainwater harvesting, now considered a green building practice, was practiced almost 9000 years ago in Jordan and later in many parts of the ancient world.[6] This period also witnessed remarkable development in the field of textiles with the ready availability of fibers from plant stems such as reeds, maguey or cabuya, rushes, bulrushes, and flax (linen) as well wool from sheep. Clothes made from these fibers may be considered as the first sustainable clothing.

In addition, many of these fabrics were colored using natural dyes mainly obtained from vegetables, minerals, and insects or animals making them all the more eco-friendly. The first use of geothermal energy by humans might date back to the Paleolite period (Quaternary times up to 14,000 B.C.E.) when humans discovered the advantages of warm springs and began to use them. However, the first systematic use of geothermal heat, according to archaeological evidence, began some 11,000 years B.C.E. on a Japanese islands and some 5000 years B.C.E. on the Asian continent. Humans were found to settle in the vicinity of geothermally active places, where they could bathe, rest, or use hydrothermal or volcanic products. The use of thermal energy and geothermal by-products (e.g., thermo-mineral mud) for therapeutic and cosmetic purposes abounded in Greek islands. Prehistoric people knew the healing properties of thermal waters and mud to alleviate stress and arthritis, to cure wounds and other skin ailments. The land chosen for the cultivation was often the rich soils close to volcanic zones, regions with thermal springs, fumaroles, gas exhalations, mud lakes, kaolin, sulfur, iron oxide, boron, and many other hydrothermal deposits. The cooking of food indirectly using the heat of the earth was also practiced by prehistoric people. Underground mining and metal working of flint, chert, copper, and gold also occurred during this era. Travois a wooden load-bearing frame structure consisting of two poles whose one end is fastened to a horse or dog and the other end splayed apart was used to haul loads by indigenous people of the Plains Aboriginals of North America. Another groundbreaking technology was the invention of the wheel during the 6th millennium BP. Though at first used for pottery, it later became an essential part of the wagon to which draught animals were harnessed for long-distance transportation, and this use of the wheel spread very fast in the ancient world. Though not as efficient as the present-day green vehicles, these wagons may be considered as the origin of green transportation systems. Even though boats were invented long ago, it was 5000 years ago that sails were invented in Egypt, and humans began using the driving force of the wind for transportation.[5] This was the very first time humans began harnessing the power of wind, a major milestone in the development of green technology. Archaeological pieces of evidence point to the "bow-drill fire-making" grasped by humans almost 6000 years ago making them skilled in making fire. Biomass began to be intentionally collected, dried, and used to create fire for cooking, warmth, to clear land, smelt ores, and treat clay artifacts for china, bricks, and tiles.[7] The swape or shadoof, which worked with a lever mechanism, was the first irrigation tool developed by the Mesopotamians around 3000 B.C.E. for lifting water, which was then later adopted by several ancient civilizations. The construction of sun-dried mud bricks around 3100 B.C.E. by ancient Egyptians turns out to be the origin of green building materials as we know of now. The sun was also used to dry and preserve food as well as to procure salt from seawater. Saltwater was collected in solar ponds and after the water was evaporated by the sun, the salt remained in the pond. This technology is still in use around the world. Thus, the Neolithic Revolution brought about radical changes in diet, adoption of a sedentary way of life, and, most importantly, increased social stratification between different categories developed during this period. While the hunter-gatherer's community shared equally among themselves the resources available, the transition to food production put an end to this and instead replaced it with the competition to possess as much in the way of resources and other "prestige objects" possible. What was till then a self-sufficient green economy soon declined.[12] The first great energy transition to extrasomatic energy sources occurred during this period with the domestication of draught animals, development of sail for transportation, harnessing of fire and solar energy for the production of metal tools, and other durable materials such as bricks.

3.3 HISTORY AND PRESENT STATUS OF GREEN TECHNOLOGY

The advent of Metal Age around 3500 B.C.E. also witnessed the beginnings of writing and the development of systems of weight and measures which resulted in the formation of urban, state-based societies with developed economies. Trade and industry played an important role in the development of ancient Metal Age civilizations. Metal smelting increased the demand for wood thereby leading to extensive deforestation which led to the complete disappearance of forests in many parts of the world. As the energy needs of the people multiplied further, they began to depend on traditional sources of renewable energy to supplement their energy needs.

3.3.1 ANIMAL POWER

The domestication of wild animals dramatically increased the amount of energy available to humans. Horses, donkeys, and oxen were used to draw the wheeled vehicles such as carts and wagons. In Asia, elephants have been tamed and used for travel, transport, and war since ancient times. Other animals that have been used for travel and transport through the ages are dogs, llamas, reindeer, camels, and water buffalo. The plow was used for the first time to make use of animal energy in the preparation of fields, thereby providing a functional link between animal keeping and crop production. At present, while the draught animal power (DAP) is expanding in Africa, in Asia and Latin America, it is persistent. Even in highly developed countries such as Spain, Portugal, and Greece, DAP remains an important energy source in small farms. In the United States, many farms profitably run by Amish farmers solely depend on animal power. Animal power, though an old technology, has many benefits. It is a renewable source of energy, generally affordable and easily accessible, to smallholder farmers who are responsible for much of the world's food production. Animal power can be used as a sustainable and environment-friendly technology for rural development.

3.3.2 HYDROPOWER

The ability to utilize the power that moving water provides (hydropower) was a major green technological revolution. The history of hydropower starts as early as classical antiquity. It was used as the major source of energy prior to being replaced by coal during the early industrial revolution.

3.3.2.1 WATERWHEELS

The ancient Greeks have often been credited with the development of waterwheels though it was a work of art rather than of science. Waterwheels invented by the Romans around 600–700 B.C.E. consisted of the wooden wheel, powered by water flow and fitted with buckets that lifted water for irrigating nearby lands. Ox-, camel-, and human-driven waterwheels were also used for simple irrigation purposes. Waterwheels basically consist of two designs: a horizontal wheel with a vertical axle and a vertical wheel with a horizontal axle. The latter is further divided into undershot, overshot, and breastshot waterwheels according to where the water hits the wheel. The first ancient Greek waterwheel was an undershot waterwheel as the water passed underneath the wheel. Toward the fifth century C.E., the sudden shortage of human energy (slaves) and animal labor forced the Roman Empire to seek new energy sources. This search led to the dependence on waterwheels to perform certain works that once the slaves completed. Watermills utilize the flow of water to spin the waterwheels which then powers a drive shaft that would then complete mechanical tasks such as grinding of grains to produce flour, crushing olives to produce oil, sawing, and moving bellows and water hammers for metallurgy. These watermills later became a major energy source for textile production processes such as spinning, weaving, cleaning, trimming, as well as cloth thickening. They were also used in the paper mills and mining industry before the industrial revolution. The early steel manufacturing industries of Europe during the 13th century even used watermills for hammering.

The overshot waterwheels developed during the Middle Ages. The first scientific evaluation of the waterwheels was made by a British civil engineer, John Smeaton, during the mid-1700s. His findings led to the development of the breastshot waterwheels which were capable of harnessing significantly larger flow rates than was possible with the overshot model. The conversion to water power of the spinning jenny, the mechanical machine used for cotton cloth manufacturing, in 1769, commenced the development of several water-powered mills in England, France, and later in the United States. Though by 1830 in England, the use of coal replaced the use of hydropower in cotton spinning mills; in France and the United States, hydropower remained the major source of energy.[8] In 1824, the use of curved buckets on undershot waterwheels to extract the maximum fluid momentum from the fluid stream was first demonstrated by Jean-Victor Poncelet, a French engineer. He also adjusted the rotational speed of the wheel such that water left the buckets with very low velocity than when it entered it. These additions increased the efficiency of the waterwheels to more than 60%.[9] From around the tenth century to the nineteenth century C.E., the number of waterwheels continued to increase in numbers across both Eastern and Western civilizations, such as that they cluttered rivers and obstructed boat traffic. Toward the end of the nineteenth century, the vertical wooden waterwheels underwent several innovations and were replaced by ones in iron which increased its efficiency (80–90%) for low to moderate head chutes and a power range of 10–50 kW per unit.[9]

3.3.2.2 WATER TURBINES

With its large size acting as a major shortcoming, waterwheels began to be replaced by water turbines which were comparatively smaller but more efficient. The major difference between a water turbine and waterwheel was the swirl component of the water which then passed energy to a spinning rotor. This enabled the turbine to be much smaller than a waterwheel of the same power. The migration from waterwheels to turbines took almost 100 years.

Hydroturbines are basically of two types: the impulse turbines and the reaction turbines. The concept of a water turbine was developed independently by Dr. Barker in England in 1744 and in 1750 by Johann Andreas Segner in Germany.[8] In 1832, a French engineer, Benoit Fourneyron, developed the first industrial turbine which was installed for the bellows of a metallurgical propeller-type. In 1843, with some improvements made by Uriah Boyden to the Fourneyron turbine, the Boyden turbine was introduced in the United States. The Fourneyron and Boyden outward-flow turbines remained a prominent technology in the hydraulic industry as well as in other fields of research for more than 70 years even though it was later superseded by other advanced models. In 1849, James B. Francis developed a scientific turbine design method and, after making several improvements to the inward-flow wheel design, installed the first full-scale water turbine at Boott Cotton Mills. As the Fourneyron and Francis turbines were not functional for chutes higher than 120 m, in 1854 in France, Louis-Dominique Girard designed an impulsion turbine to solve this problem. However, the problem of wheel erosion in the case of high chutes still remained. It was in 1866 that Samuel Knight developed the tangential waterwheel which utilized the water from a dam at a high elevation. This was an impulse turbine which made use of a high-pressure nozzle from which water originated was directed slightly off-center to the buckets on the wheel, in such a way that energy was not wasted via water splashing. In 1878, Lester Allan Pelton modified the Knight wheel and developed the Pelton turbine which had a double bucket design with a half-cylindrical profile. This design overcame some of the inefficiencies of the Knight wheel and has to date been used as the reference turbine for high chutes. In 1912, Victor Kaplan created the Kaplan turbine, a propeller-type turbine with an intent to deliver large powers from very low chutes.[8,9]

The use of hydropower was promoted by energy-intensive industries such as aluminum smelters and steelworks. On September 30, 1882, the world's first hydroelectric power plant, Appleton Edison Light Company, began operation on the Fox River in Appleton, Wisconsin. Hydroelectricity is a form of energy that harnesses the power of moving water to generate electricity. Since 1889 onward, all hydropower projects were aimed at electricity generation. The 22.5 GW Three Gorges hydroelectric power plant in China is the largest hydropower generating facility ever built. Though not mentioned here, many other turbines were designed and manufactured, as well as several dams were also established during this period. The world's total installed capacity of hydropower in 2018 was estimated to be 1292 GW with electricity generation reaching 4200 terawatt-hours (TWh). Globally, hydropower is produced in 157 countries (Table 3.1), with 47 countries adding an estimated 21.8 gigawatts (GW) of hydropower capacity into operation in 2018. With Brazil increasing its installed capacity by 3.7 GW in 2018, it thereby reached a total capacity of 104 GW and so has now overtaken the United States (103 GW) as the second-largest country with respect to hydropower capacity (Table 3.2). Currently, hydropower contributes almost two-thirds of the renewable electricity generation, without which the objective of limiting climate change to 1.5 or 2 °C above preindustrial levels would likely be impossible.

TABLE 3.1 Total Installed Capacity and Hydroelectricity Generation of Different Continents in 2018

Continent	Total installed capacity (GW)	Generation (TWh)
Africa	36.2	138
South and Central Asia	148.5	439
East Asia and Pacific	480.4	1534
Europe	251.7	643
South America	170.8	726
North and Central America	204	720
Total	1291.7	4200

GW, gigawatts; TWh, terawatt-hours.

Source: All data in this table are selected based on information provided by International Hydropower Association in 2019.

TABLE 3.2 Top 10 Countries with the Most Hydroelectric Capacity.

Countries	Installed capacity (GW)
China	352
Brazil	104
USA	103
Canada	81
Japan	50
India	50
Russia	49
Norway	32
Turkey	28
France	26

GW, gigawatts.

Source: The information on the countries in this table is based on information provided by International Hydropower Association in 2019.

3.3.2.3 THE TROMPE

A hydraulic air compressor (HAC) or "trompe" as it was originally known is an ancient Italian technology developed in 1588 that utilizes falling water to provide air supply for smelting furnaces. The trompe is a simple device that consists of one or more vertical wooden pipes through which water is channeled via gravity. Upon descent, constriction in the vertical pipe produces a low pressure which causes air to be sucked into the water from an external port, thereby providing a constant air supply. At the bottom of the pipe, the air gets separated from the water and rises to the top of the separation chamber from where it goes to the take-off pipe which can then act as the power source. As the HAC produced compressed air without moving any parts, it was a quite reliable and efficient device. Starting in 1896 almost 18 gigantic HACs were built mostly in the United States, Canada, Germany, and Sweden. Since the air in the trompe undergoes isothermal compression, it does not affect the captured air's temperature, which otherwise would get heated up, thus preventing overheating of any machinery powered by it. The design of HACs was improved further making it all the more efficient and practical. The compressed air produced can be used to power machinery in mining operations, aerate the water, atomize paint, and also serve several other functions.[10]

3.3.3 WIND ENERGY

Keeping in mind the fact that anything that moves has kinetic energy, humans began investigating ways to generate useful forms of energy from the power of the wind. Though wind power is a cost-effective, domestic, and sustainable source of energy, it being an intermittent source of energy that cannot be made or dispatched on demand, cannot be relied on for continuous power supply. The basic design of a wind energy device includes the horizontal axis wind turbine and the vertical axis wind turbine (VAWT); the classification being based on the axis of rotation.

3.3.3.1 WINDMILLS AND WIND PUMPS

While a windmill is a structure used to harness the power of the wind into rotational energy for the purpose of grinding grains, the wind pump uses wind power for pumping water from deep wells back to the surface, for the drainage of water from marshy areas, as well as for automated irrigation of fields in areas surrounded by streams and river. The wind pumps were often coupled to an Archimedean screw, Egyptian noria, or Persian waterwheel; all of them are early pump concepts that could elevate water to a height of 5 m. Prior to the industrial revolution, other applications of wind-powered machines included extraction of oils from oilseeds, nuts, grains, lumber sawing, ventilating of mines, manufacture of gun-powder, and snuff tobacco. Interestingly, it was even used for lopping bee hives into town under siege during warfare. The simplest windmill sails comprised cloth sails attached to the rotating arms or blades. Later, wooden frames covered with cloth began to be used as common sails in order to attain better structural stability. However, when the mills run out of grain and the millstones run dry igniting a spark, there were chances of the common sails catching fire. In cold weather, the cloth sails used to get wet and frozen too. In order to avoid friction-induced fires, a combination of wood and metal was used, and in such instances, the cloth sails were found to last 40–50 years. Modern wind generators utilize wood and glass epoxy, fiberglass, aluminum, and graphite composite materials for their construction.

Although the history of wind-powered devices is quite obscure, it was found that vertical axis wind rotors probably originated around 200 B.C.E. at the Persian-Afghan borders (Seistan), while the horizontal axis windmills of Europe followed much later around the 12th century C.E.[11] While the Seistan rotors were driven by drag forces, the European designs were driven by lift forces. During the 17th century B.C.E., King Hammurabi of Babylon made ambitious plans to irrigate the fertile plains of Tigris and Euphrates Rivers using the vertical axis wind pumps.[12] The Dutch engineers, around 14th century C.E., initiated making improvements to the existing windmill and wind pump designs. The rotors on many of these Dutch mills were twisted and tapered in the same way as modern rotors. The first Dutch marsh mills were started around 1400, and by 1600, there were almost 2000 of them operating to drain almost 2 million acres of land in Holland. It was during the mid-1700s that the Dutch settlers introduced windmills in America. During the American Revolution, wind pumps were used to pump water to make salt on the islands of Bermuda and on Cape Cod. In 1759, John Smeaton through a series of experiments improved the efficiency of windmills and wind pumps, thereby laying the foundation for the aerodynamic theory of wind machines. During the mid-1800s, a need for a distinctive small wind pump for the settlers in American West resulted in the development of American multiblade windmill design. It consisted of a vertical steel structure with a multibladed drag or impulse propeller at the top that caught the wind. The rotational motion of the blade was converted into a linear motion to pump water and carry out functions like irrigation, cattle drinking, and also as a water supply for steam locomotives.[13] Between 1850 and 1970, further modifications of these systems made in the United States resulted in the establishment of over 6 million wind pumps all over Australia, North America, South Africa, and South America, and many were also exported from the United States and Australia to developing countries.[12,13] Wind energy remained a major source of energy prior to Industrial Revolution but later the availability of cheap and plentiful petroleum coal lagged the interest in harnessing wind power which was unreliable at the same time required a high capital cost. However, in different rural areas of the world, vertical axis windmills and wind pumps are still being erected by enterprising farmers for irrigation as well as drainage purposes.

3.3.3.2 WIND TURBINES

A wind turbine is a windmill like structure but specifically built to generate electricity. Professor James Blyth, a Scottish electric engineer, is considered a pioneer in this field as his holiday home in Marykirk was the first known structure in the world to have electricity using wind power in 1887. The American engineer, Charles Brush, has also often been credited as a pioneer in the field. Denmark, in 1891, was the first country to use wind turbines to meet the demand for electrification of rural areas. Poul La Cour, a Danish scientist, was in charge of the experimental station during this period. The wind turbines used were 23 m in diameter and had power outputs between 5 and 25 kW. Being equipped with storage batteries (100–300 A·h capacity), they were capable of meeting energy demands for up to 10 consecutive windless days.[12] In 1903, La Cour founded the Society of Wind Electricians, and in 1904, the society held its first course. He was the first to discover that wind turbines with fewer blades that spin faster are more efficient than those with many blades that spin slower. After World War I, most wind generators transitioned from a drag or impulse system to an airfoil system similar to air propeller thereby significantly improving their power coefficients.

In 1929, the first VAWT, known as the Darrieus turbine after its inventor George Darrieus, was introduced in France and later spread worldwide, and is still in use today.[14] In 1931, the first wind-powered system to be connected to an existing power supply grid was built at Yalta on the Black Sea. It had a rated power of 100 kW in an 11 ms^{-1} wind. The Yalta wind turbine was functional for almost 10 years until it was destroyed during the Second World War. In 1934, Palmer Cosslett Putnam designed the world's first megawatt-size wind turbine that was manufactured by the S. Morgan Smith Company of York, Pennsylvania. It is also considered as the predecessor of the two-bladed turbines built by the United States in the late 1970s and early 1980s.[15] It was connected to the local electrical distribution system in Vermont, United States and was designed to operate at a wind speed of up to 33 ms^{-1}. The Smith–Putnam wind turbine had a rated power of 1000 kW in a wind speed of 13.4 ms^{-1}. The turbine was capable of producing 1250 kW, and it operated for about 1000 hours before an overstressed blade failure in 1945. Fuel shortages and rising demand for electricity, after the Second World War, instilled widespread interest and research in the potential of wind power to provide electricity. However, between the 1950s and 1960s, the increased interest and expectation of producing cheap electricity via nuclear fission diminished the developments of wind turbines.

During the late 1960s, as people became more aware of the environmental consequences of dependence on fossil fuels, a renewed interest in the use of wind energy was ignited. Despite the diverse and technically successful developments and research projects in the field post-1970s, none of them lead to commercial exploitation, as throughout the postwar period (1945–1973), fossil fuels became progressively inexpensive and cheap. Though wind turbines provided a free energy source, the cost of the power so harnessed was determined by the initial cost of the machine and to a certain extent by its maintenance and running cost, and hence, electricity thereby obtained was considered more expensive than that obtained via coal, gas, or oil. However, post-1973, antinuclear protests and sharp rises of unit oil prices spurred some politicians to realize the finite extent of earth's fossil fuel reserves. This initiated research and development programs in wind turbines in some of the more developed countries. In the United Kingdom, the Energy Technical Support Unit for the Department of Energy evaluated the potential of wind energy as a source for generating electricity. During this period, the common misconception that large multi-scale megawatt rotors (MOD series of turbines) offered low energy costs than smaller turbines was corrected. Research in the field suggested the use of more medium-sized wind turbines, at the same time, could produce more energy less expensively. Medium-sized wind turbines of 10–20 m diameter were found to have payback periods of 6–9 years when functioning in parallel to an existing electrical grid system. This led to the widespread development of microturbines of 22 kW.[12,14] The involvement of the US government in wind energy research and development (R&D) after the oil crisis of 1973 was another major milestone in the history of wind turbines. This resulted in the evolution of the commercial wind turbine market from domestic and agricultural (1–25 kW) to utility interconnected wind farm applications (50–600 kW). The US Federal funding and the Public Utility Regulatory Policy Act of 1978, which forced companies to purchase a particular amount of electricity from renewable sources, promoted the first large-scale wind energy penetration in California, resulting in the installment of over 16,000 wind turbines (20–350 kW) between 1981 and 1990.[11] However, with the withdrawal of federal tax credits by the Ronald Reagan administration by the early 1980s, wind rush in the United States collapsed. The 1990s witnessed the demise of the United States' largest manufacturer, Kenetech Windpower. During this period, the focal point of the wind turbine manufacturing industry moved to Europe, particularly Denmark and Germany.[15] Though Denmark attained self-sufficiency in oil a decade later, it still continued its wind development program in order to reduce GHG emissions. During a 15-year period, pioneers in the field made all their inventions and theories available for all, thus enabling the successful development of the wind power industry. It was during this period that some of the large wind turbine manufacturing companies such as Vestas, LM Wind Power, Nordtank, Spanish Gamesa, Micon, and Bonus were founded. Large wind farms were created, and offshore wind farms were born, and as the concern over GHGs grew, the development continued further (Table 3.3). The largest wind farm in the world is the Jiuquan Wind Power Base in China. Also known as the Gansu Wind Farm, it has about 7000 wind turbines with a total installed capacity of 7.96 GW. The farm is set to be expanded to have a total capacity of 20 GW by 2020. In 2018, 50.1 GW of wind power was added thereby increasing the overall capacity of all wind turbines installed worldwide to 596.4 GW. This covered up to 6% of the global electricity demand. While Germany, Spain, Italy, and France showed weak development in the wind market, countries like China, India, and Brazil as well certain African countries showed robust growth (Table 3.4).

TABLE 3.3 World's Biggest Wind Farms.

Wind farm	Country	Installed capacity (GW)
Jiuquan Wind Power Base	China	7.960
Alta Wind Energy Centre	USA	1.548
Muppandal Wind Farm	India	1.500
Jaisalmer Wind Park	India	1.064
Shepherd Flat Wind Farm	USA	0.845
Meadow Lake Wind Farm	USA	0.801
Roscoe Wind Farm	USA	0.781
Fowler Ridge Wind Farm	USA	0.750
Horse Hollow Wind Energy Centre	USA	0.735
Capricorn Ridge Wind Farm	USA	0.662
Walney Extension Offshore Wind Farm	UK	0.659
London Array Offshore Wind Farm	UK	0.630

GW, gigawatts.

TABLE 3.4 Global Wind Installations (2017–2018).

Country/region	Installed capacity (GW)– 2017	Installed capacity (GW)—2018
China	195.7	216.9
USA	88.7	96.3
Germany	56.2	59.3
India	32.9	35.0
Spain	23	23.5
UK	17.8	20.7
France	13.8	15.3
Brazil	12.8	14.5
Canada	12.2	12.8
Rest of the world	93.1	102.1
Total	546.2	596.4

GW, gigawatts.
Source: Pitteloud (2018).

3.3.4 SOLAR ENERGY

The sun plays a vital role in life on earth. With the solar energy received on earth being plentiful, totally renewable, and directly or indirectly being the origin of all energy sources, the potential of the sun's energy to satisfy all our energy needs is immense. Humans have been tinkering with this idea since the dawn of time. They have admired the sun and frequently personified and worshipped it as a deity, and many cultures still continue to do so. Apart from the metaphysical approach, the sun's energy found many practical applications. Solar energy technologies are basically of two types: solar thermal technologies and photovoltaic (PV) technologies. Solar thermal technologies use solar energy to generate heat, and then if needed, electricity is generated from that. PV technologies generate electricity directly from solar energy (Table 3.5).

TABLE 3.5 Largest Solar Power Plants of the World.

Solar power plants	Country	Installed capacity (MW)
Yanchi Solar Park	China	820
Datong "Front Runner"	China	800
Longyangxia Solar-Hydro	China	697
Kamuthi Solar Power Project	India	648
Villanueva	Mexico	640

MW, megawatts.

3.3.4.1 PASSIVE SOLAR HEATING OF BUILDINGS

The basic principle of a passive solar building is that it is built in such a way that it maximizes the exposure of the building to the south, thereby capturing solar energy and further insulating the enclosure to trap the heat within. Neolithic Chinese villagers around 6000 B.C.E. had the sole openings of their houses face south in order to catch the low winter sun rays to warm their interiors. The overhanging thatched roof would keep the high summer sun rays off their houses thereby cooling their interiors. Socrates and Aristotle too advised similar constructions of houses in the third century B.C.E. and often considered as the pioneer of the present passive heating and cooling techniques.[16] The ancient Egyptian Pharaohs solar heated their palaces using black pools of water which captured solar energy by day, and during the night, the hot water was allowed to circulate through pipes on the palace floor. This system helped maintain warmth during the night while lowering the temperature during the daytime to a certain extent. Solar heat assisted the hypocaust system of mechanical heating in the ancient Roman baths built around 212 C.E. Glazing the south-facing windows helped trap heat in the baths, thereby reducing the usage of fuel. Thick walls, hollow tiles, and wooden shutters over the windows helped further to retain heat in the baths. Transparent glass, the Romans discovered, helped in admitting sunlight and also trapping heat in desired spaces. During the sixth century C.E., sunlight was so important to the Romans that a legal precedent for solar rights was actually established and solidified in the Justinian code of law. Around 700–1300 C.E., the south-facing cliff dwellings were chosen by the Anasazi people in the American West to make use of the winter sun to provide warmth. Between the 1500s and 1800s, the south-facing greenhouses, which trap solar heat energy within, were built by the wealthy Europeans who wanted to grow exotic plants in the colder climate. New England "saltbox" houses of the 17th century, the Swiss farmhouses of the 18th century, and the solar houses of the twentieth century are all examples of passive solar buildings.

3.3.4.2 SOLAR FURNACES AND COOKERS

In a solar furnace, high temperature (up to 6330 °F) is obtained for industrial purposes by concentrating the solar radiation onto a substance using a number of heliostats or turnable mirror. Back in the seventh century B.C.E., magnifying glasses were used to concentrate the sun's rays and light a fire. By the third century B.C.E., Greeks and Romans were known to bounce off sunlight of "burning mirrors" to light torches for religious ceremonies. Historians claim that as early as in 212 B.C.E., Archimedes, a Greek inventor, made use of the reflective properties of highly polished bronze shields to concentrate the sun's rays to set fire on the Roman ships attacking Syracuse. Georges-Louis Leclerc, Comte de Buffon, French scientist, and naturalist, in 1695, used a mirror to focus sunlight and achieve a temperature high enough to burn wood and melt lead as well. Antoine Lavoisier, in 1782, focused sunlight using a lens and achieved temperature as high as 3000 °F, capable enough to melt previously unmeltable platinum. The largest solar furnace opened in 1970 at Odeillo in the Pyrénées-Orientales in France employs an array of plane mirrors to gather sunlight, reflecting it onto a larger curved mirror. Asia's largest solar furnace was built in Uzbekistan in 1981 and is also known as the Sun Institute of Uzbekistan. The energy thereby obtained can be used for hydrogen fuel production, foundry applications, and high-temperature testing.

Solar cooker or oven is a device which utilizes the solar energy to heat, cook, or pasteurize food items and drinks. The principle behind the solar furnace and solar cooker remains the same. The world's first documented solar oven was developed in 1767, by Swiss naturalist and physicist Horace Bénédict de Saussure. The so-called "hot box" plate collector consisted of a well-insulated box with three layers of glass that trapped solar heat achieving a cooking temperature of 230 °F. Saussure's invention inspired and informed several others including Augustin Mouchot (French inventor) who saw the great commercial potential of solar appliances in France's sun-rich, fuel poor colonies of North Africa and Asia. In 1877, he devised solar cookers for the French soldiers in Algeria and received the support of the French government to pursue full-time research. Unfortunately, better political relations with England restored France's supply of coal, thereby diminishing interest in solar energy harnessing. The first recorded history of solar cookers use in India dates back to 1876, when William Grylls Adams, a British engineer, developed an octagonal oven in Bombay, India. This helped ease the energy shortfalls and depletion of wood fuel in Colonial India.[17] Baltimore inventor, Clarence Kemp, in 1891 patented a commercial solar water heater for bathing and dishwashing that enjoyed widespread popularity in California. The solar box-type cookers were later commercialized by an Indian pioneer M.K. Ghosh in 1945. Many companies with the help of the Indian government have since then and up to the 1980s harnessed solar energy for cooking purposes. This led to the installations of two of the world's largest solar cookers in India at the Tirupati and the Shirdi Sai Baba temple, where solar energy is used to cook food to feed lakhs of devotees on a daily basis. Though in the 1950s, scientists and researchers devised and constructed solar ovens, they failed due to the availability of lower fuel alternatives. However, the oil crisis of the 1970s again spurred interest in the use of solar energy in China and India. The 1970s in the United States was sound with the several types of concentrating and box-type solar cookers developed by Barbara Kerr using recycling materials and aluminum foil. During 1979, Dr. Bob Metcalf and his student Marshall Longvin performed water pasteurization using box-type solar cookers. India and China during the 1980s expanded the national promotion of box-type solar cookers. In 2000, Paul A. Funk proposed a solar cooker power curve tool that was used to evaluate the heat-storing capacity of solar cookers.[18] Countless styles of solar cookers are continuously being developed by researchers and manufacturers in an intensive effort to enhance the capacity of solar cookers.

3.3.4.3 CONCENTRATED SOLAR POWER

Concentrating solar power (CSP), also known as concentrated solar thermal systems, concentrates a large area of sunlight onto a receiver, which then converts it to high-temperature heat. The generated heat is then used to drive traditional steam turbines or engines that create electricity. Currently, there exist four different CSP technologies, namely, the parabolic trough collector (PTC) technology, linear Fresnel collector (LFC), the Stirling dish collector (SDC) system, and the earliest tower solar power (TSP). The earliest documented use of PTC technology was in 1913, in Maadi Egypt for generating steam that helped drive a 73-kW pump which in turn was used for irrigation purposes. The first LFC prototype was built by Giorgio Francia, an Italian mathematician, at Lacédémone-Marseilles solar station in 1963. The pioneer SDC system was demonstrated in Southern California from 1982 to 1985 by Advanco Corporation. The earliest TSP technology was exhibited by constructing a plant named EURELIOS in a large valley, about 30 km from the sea in Adrano, Sicily, Italy in 1976. The ability of the solar tower technology to generate large-scale electricity (10 MWe) was demonstrated by Solar One plant that was built in California, United States, in 1982. The first commercial-scale CSP plant named SEGS I was built in California in 1984 by using PTC technology to generate 14 MWe; then by 1990, the capacity of the SEGS plants was increased to 354 MWe. Between 1991 and 2005, because of the falling of fuel prices and other policy changes, no CSP plants were built. However, since 2006, the ability of these plants to limit GHG emissions and other environmental impacts of energy generation spurred an interest again in CSP plants. At present, Spain with a total installed capacity of 2.3 GW is considered the largest producer of electricity using CSP technologies. With an estimated 513 MW addition in 2018, the global installed capacity via CSP reached 5.5 GW.[19]

3.3.4.4 PV CELLS

A PV cell or solar cell is a device that converts light into electric current using the PV effect. It was not until the 19th century that the potential of turning sunlight into electricity was discovered. The history of solar PV cells unveils with Alexandre Edmond Becquerel's, a French scientist, discovery of the PV effect in 1839. While experimenting with an electrolytic cell made up of two metal electrodes placed in an electrically conducting solution, electricity generation was found to increase when exposed to light. In 1873, Willoughby Smith, a British electrical engineer, discovered the photoconductive potential of selenium. This discovery was followed by that of William Grylls Adams and his student Richard Evans Day's discovery in 1876 that selenium creates electricity when exposed to light. A few years later, in 1883, an American inventor, Charles Fritts, produced the first solar cells by coating selenium with a thin layer of gold. Though these selenium cells were not as efficient as modern-day PV cells (less than 1% light-to-electrical energy conversion efficiency), for the first time it was proved that light, without heat or moving parts, could be converted to electricity. The paper published by Albert Einstein in 1905 on the photoelectric effect and how light carries energy helped understand the potential of solar energy in generating electricity. By the beginning of the 20th century, Silicon PV cells having five times greater efficiency than selenium PV cells began to be developed. In 1953, Daryl Chapin, Gerald Pearson, and Calvin Fuller hired by Bell Telephone Laboratories developed a solar PV cell by doping strips of silicon with boron and arsenic. This was the first modern PV cell which at first had an efficiency of 4%, and later through modifications was increased up to 11%.

During the early years, when the cost of PV cells was high and the efficiency low, they found applications only in space programs for which cost was not a problem but reliability was vital. For those manufacturing solar cells, the increasing demand for it in space meant a booming industry. The United States' Vanguard satellite, in 1958, was the first satellite to use radios powered by solar energy (less than 1 W). Solar PV cells still remain the accepted energy source for satellites today. The first drivable solar car was a vintage model 1912 Baker Electric car converted to run on PV cells by the International Rectifier Company in 1962. In 1982, Hans Tholstrup and Larry Perkins were the first to cross continents in a solar car, the "BP Solar Trek" making it the world's first practical long-distance solar-powered car.[20] Though during the 1960s and 1970s, the cost of PV cells remained high, the development of large single crystals of silicon for use in integrated circuits moderately brought down the prices of raw materials. The rising oil prices increased the demand for solar power. Starting in the 1970s, the work of Elliot Breman at Exxon resulted in the development of cheaper PV cells. He discovered that using silicon from multiple crystals brought down the prices fivefold than when using silicon from a single crystal, thereby increasing their terrestrial demand. Throughout the 1970s and 1980s, the ready availability of PV cells found a market in navigational buoys and remote telecommunications stations. During the 1970s, the idea of applying solar cells to pump water[21] was put forward by Dominique Campana, a graduate student in Paris. The working model of her idea was later translated by French physicist Jean Alaine Roger who developed the world's first practical PV pump on the island of Corsica.[52] In 1973, the University of Delaware was credited with the construction of the first solar building named "Solar One" which ran on a hybrid supply of solar thermal and PV power. It was the first instance building-integrated PVs was used, that is, the building did not use solar panels; instead, it had solar cells integrated onto its rooftop. In 1982, ARCO Solar develops the first solar power plant in Hesperia, California which generates 1 MW hr^{-1} at full capacity. Two years later, a second solar plant was developed in Carizzo plains, California with 100,000 PV arrays generating 5.2 megawatt (MW) at full capacity. Though the power plants fell into disarray with the return of oil (after the 1973 oil crisis), these plants demonstrate the potential for solar power production.

While single crystal silicon remains the most important and efficient base for solar cells, during the 1990s, polycrystalline silicon being cheaper became popular. PV research and development continued, and cheaper thin-film amorphous silicon PV cells were developed. Thin-film solar cells based on other materials such as cadmium telluride (CdTe) and copper indium gallium diselenide attracted attention and offered as a cheaper alternative to silicon.[22] However, over the years, the reduction in the cost of silicon cells coupled with its high efficiency established its dominance in the solar industry. PV cells began to be used in warning lights, horns on offshore oil rigs, lighthouses, railroad crossings, and even in remote and isolated places where the traditional electricity grid was not available. In 1995, Thomas Faludy filed a patent for a retractable awning with integrated solar cells for use in recreational vehicles (RVs), and it remains one of the popular ways to power RVs to date. By 2005, do-it-yourself solar panels started becoming popular. In 2019, the National Renewable Energy Laboratory in Colorado developed the multi-junction or stacked solar cells with a record-breaking efficiency of up to 45%.[23] The consumer demand for solar panels increased as the price of solar panels decreased from $300 per watt in 1956 to $2.99 per watt in 2019.[24] A solar plant constitutes of a single generating station designed by a single developer or consortium and most often with a single export connection to the grid. Solar parks, on the other hand, may be defined as a group of colocated solar power plants. Many of the largest solar power facilities are installed in China and India (Table 3.4). The total installed capacity of solar PVs reached 480.3 GW (excluding CSP) in 2018 (Table 3.6), with solar PV additions of around 94 GW.

TABLE 3.6 Solar PV Installed Capacity Worldwide.

Region	Solar installed PV capacity (GW)
Africa	8
Asia	280
Europe	121
South America	7
North and Central America	55
Oceania	10
Total	481

GW, gigawatts.
Source: IRENA (2019).

So far, the evolution of the solar industry has been remarkable with several milestones reached in recent years with respect to advancements in technology, installations, cost reductions, and establishment of key solar associations. The use of solar power spans various industries and contributes power to hundreds of different gadgets and technologies all over the world. Over the past 20 years, the solar cell industry has grown dramatically and is often accepted as an effective solution for rising energy needs. In 2018, the world solar power installed total capacity reached 485.8 GW, up slightly by 93.7 GW than last year.[19]

3.3.5 GEOTHERMAL ENERGY

Geothermal energy constitutes the thermal energy generated and stored in the earth's crust; transmitted via conduction/convection through various rocks and natural groundwater reservoirs called aquifers. The energy thus obtained is a sustainable, renewable source of energy as the heat extracted is small compared to the immense magnitude of the earth's heat content. While high and medium temperature resources (approx. > 100°C) are used for power generation, low-temperature resources (approx. < 100°C) are suitable for direct uses such as recreation, heating, and drying. A brief historical outline of different practical applications of geothermal energy is discussed in the following text.

3.3.5.1 DIRECT USE AND DISTRICT HEATING SYSTEMS

Prehistory is abounding with the pieces of evidence pointing to the practical uses of geothermal energy for bathing, cooking, washing, as well as other religious rituals. It was during the beginning of the seventh century B.C.E that the Etruscans considered as the "fathers of geothermal industry" developed their civilizations in the central part of Italy with their settlements and cities built near springs, geothermal manifestations, and products of hydrothermal activity such as alabaster, travertines, iron oxides, sulfur, silica, kaolin, borate, alum, and thermo-mineral muds. Being veritable bartering goods, they were of great economic value to the Etruscans. These minerals were mainly used in pottery, production of enamel, ointments, medicines, and paints. They were also used in the dying of glass, wool, and cloth. A practice still active in the Larderello region in Italy is the production of enamel using borax, which is recovered from boraciferous springs. Many of these hydrothermal products were popularized by the trade of the Etruscans in the Mediterranean Basin. They valued the healing properties of the geothermal waters as well as of its salts and thermo-mineral muds. The practice of geothermal bathing and balneotherapy which was first developed by Etruscans was later perfected by the Romans from the second century B.C.E. The Romans instilled the thermal practice for healing and recreation even in localities lacking geothermal manifestations using artificially heated water. The fall of the Roman Empire in 476 C.E. resulted in a strong decline in thermal bathing as well as extraction and use of most by-products of terrestrial heat everywhere in Italy and in the territories of the old Roman Empire. This lasted for the whole Early and most of the High and Late Middle Ages. From about the 15th century, balneotherapy in the Italian thermal spas underwent a new blooming even though it never reached levels attained in Roman times.[25] The direct utilization of geothermal energy for cooking and therapeutic purposes by ancient people of the world such as the Mexicans, Chinese, Japanese, Greeks, Indians, Turks, Arabs, and Maoris for thousands of years is well documented.

Geothermal heat pumps (GHPs) utilize the relatively constant heat of the earth to provide heating, cooling, and domestic hot water for buildings. Medium-temperature geothermal energy (70–150°C) is used for these purposes. Although the Romans and the Chinese built primitive pipelines to pipe thermal waters and steams for bath, the use of geothermal energy for space heating became common only after the development of metal pipes and radiators. The oldest known geothermal pipeline (a duct of stones, slabs, and clay) was made in Iceland around the 13th century to heat an outdoor bath pool. The earliest documented case of residential heating in the world using geothermal energy was in the 14th century in Chaude Aigues (France). In 1930, the first municipal district heating system using geothermal water was invented in Reykjavik, Iceland, where hot water was piped from a spring 25 km away, and residents used it not only for heating their houses but also for hot tap water. Presently, more than 90% of the population of Iceland live in houses heated by geothermal water. Greenhouse farming using geothermal water was started in Iceland in the 1920s and was later adopted and has been operated for several decades by several countries including the United States, New Zealand, Greece, Hungary, Macedonia, Italy, Romania, Russia, China, and Japan. Large-scale district heating systems have also been installed in many of these countries. The extraction of salt from seawater using geothermal springs is known in Iceland since the 18th century. In 1828, Francesco Larderel replaced the use of firewood in the traditional boric acid industry with the natural terrestrial heat as process heat, thereby solving the problems associated with the wood shortage. This was the first industrial utilization of geothermal energy. The world's first geothermal wells constructed for the boric acid industry were also developed during this period by the manual drilling of small diameter (10–12 cm) shallow wells (6–8 m depth), located near the natural lagoons in Larderello, Italy.[56] It was during the 1950s that the first large-scale industrial application of geothermal steam was utilized in a pulp and paper mill in Kawerau, New Zealand.[26] Geothermal steam-operated air conditioning was first developed in a hotel in New Zealand in the late 1960s. Since the 1960s, in different countries, geothermal steam, heat, and water began to be used in various industries for drying, washing, and dyeing of different products. Nowadays, low-temperature geothermal energy (20–60 °C) finds use in aquaculture (fish farming and algae production), animal husbandry, and even soil heating. There has been a steady increase in the number of countries depending on geothermal energy for direct application with currently almost 73 countries utilizing geothermal energy with a total output of 75.9 TWh per year.[27]

3.3.5.2 ELECTRIC POWER GENERATION

Electricity generation using geothermal steam or hydrocarbon vapor is a much more recent industry dating back to the beginning of the 20th century. On July 4, 1904, at Larderello, Italian, businessman Prince Piero Ginori

Conti powered five bulbs from a dynamo driven by a reciprocating steam engine using geothermal power. In 1905, by using an old Cail steam engine, he increased the power production to 20 kW which he then used to light the Laderel borax factory as well as to drive some small electric engines. Prince Ginori classified the geothermal wells into wet and dry. The wet wells were used in the extraction of boric acid as they produced both water and steam. The dry wells were used to generate electricity as they produced only dry steam and the quantity of steam produced depended on the depth of the well. Despite the shallow depths of the well (average 100 m), the flow rates until the early 1900s were between 6 and 20 tons per hour. Within a decade, in 1913, the first geothermal power station was also built here with an installed capacity of 250 $kW_{(e)}$ which exported power to the local regions from 1913 to 1916. Natural steam was used to heat and evaporate fresh water which was then used in turn to feed the steam turbines constructed by the Italian company Franco Tosi. Using a similar prototype, a much larger and more complex power plant with three Franco Tosi turbo-alternators of 2500 kW each was built in 1916. Following the path set by Italy, in 1919, Japan successfully drilled some wells at Beppu and started producing electricity at a small scale in 1924. It was only in 1966 that geothermal electricity began to be produced here at an industrial scale. In the United States, even though the first well was drilled in California in the Geysers area as early as in 1925, geothermal electricity was generated by Pacific Gas and Electric only in 1960. Since 1958, China and New Zealand also started generating geothermal electricity.[28] Geothermal power plants are basically of three types: "dry" steam, flash, and binary. The Geysers in the United States and the Larderello in Italy are "dry" steam reservoirs as they produce steam with very little water content. The steam produced is directly used to run the turbine. A flash power plant is used in the case of a hot water reservoir like that of Wairakei in New Zealand, wherein hot fluid with temperature greater than ~180 °C is brought up to the surface through a production well, whereupon being released from the pressure of the deep reservoir, some of the water flashes into steam in a "separator." This steam is then used to power the turbines. The binary power plant or the organic Rankine cycle power plant was first demonstrated in 1967 in the Soviet Union and later introduced to the United States in March 1970 by B.C. McCabe, Chairman and CEO of Magma Energy, Inc., a subsidiary of Magma Power Company.[29] Binary power plants make use of low-to-medium enthalpy reservoirs with temperatures (85–150 °C) not hot enough to flash enough steam but instead generate electricity by transferring the heat to a low-boiling point binary liquid. This binary liquid then flashes into vapor which is then used to power the turbine. This technology thus facilitated the use of much lower temperature energy resources than were previously recoverable. At present, 24 countries generate electricity using geothermal energy, and almost 11 other countries including Australia, France, Germany, Switzerland, Japan, the United Kingdom of Great Britain, and Northern Ireland are developing and testing geothermal systems.[27] Geothermal installed capacity increased by 587 MW in 2018, thereby reaching a total capacity of over 14.3 GW by the end of 2018. With geothermal power resources limited to the tectonically active regions of the world, only a few countries use geothermal power, and the United States remains the global leader in generating geothermal power (Table 3.7). Turkey and Indonesia accounted for about two-thirds of the new capacity installed in 2018. The largest geothermal power plant is the Geysers Geothermal complex located in the United States with an installed capacity of 1.5 GW (Table 3.8).

TABLE 3.7 Installed Geothermal Capacity Worldwide in 2018.

Country	Geothermal installed capacity (GW)
USA	3.59
Indonesia	1.94
Philippines	1.86
Turkey	1.20
New Zealand	1.00
Mexico	0.95
Italy	0.94
Iceland	0.75
Kenya	0.67
Japan	0.54
Rest of the world	0.88
Total	14.32

GW, gigawatts.
Source: IRENA (2019).

TABLE 3.8 Some of the Major Geothermal Power Plants in the World.

Geothermal power plant	Country	Installed capacity (MW) (approx.)
The Geysers Geothermal Complex	USA	1517
Larderello Geothermal Complex	Italy	769
Olkaria	Kenya	745
Cerro Prieto Geothermal Power Station	Mexico	720
Makiling-Banahaw (Mak-Ban) Geothermal Power Plant	Philippines	442
Hellisheidi	Iceland	300
Salak Geothermal Power Plant	Indonesia	377
Malitbog Geothermal Power Station	Philippines	232

MW, megawatts.

3.3.6 TIDAL POWER

Tidal power is a type of hydropower that converts the energy of the tide into power or other valuable types of energy. Tidal power draws on energy inherent in the orbital characteristics of the earth–moon system, and to a lesser extent from the earth–sun system. Though being predictable, it is an intermittent source of energy that supplies energy only when the tide surges. While the rise and fall of the tides—in some cases more than 12 m—creates potential energy, the flows due to flood and ebb currents create kinetic energy. Both forms of energy can be harvested by various tidal energy technologies.

3.3.6.1 TIDE MILLS

A tide mill is a water mill driven by the rise and fall of the tide in bays and estuaries. These ancient tide mills relied on the enhanced potential energy within a small basin created with the help of a dike closing off a small bay; sluices installed helped fill up the basin during high tide. The paddlewheel constructed turned when the basin emptied at the ebb, thus transforming the potential energy into mechanical force driven mainly for rasping and dyeing of woods, grinding of cereals, corns, and salt as well as in the manufacturing of tobacco stems. Tide mills were also used to provide fresh water for the city like the mill on the London Bridge. The exploitation of tidal motion for energy has a long history, and tide mills with waterwheels have been known for the best part of a millennium in Europe and elsewhere. The tide mill at Nendrum Monastic site on Stanford Lough on the east coast of Northern Ireland built as early as in 619 c.e. is presently considered as the oldest mill in the world. During the 10th century c.e., the Arab geographer described the Persian Gulf mills he found at Basra in the Tigris-Euphrates delta (Iraq), explaining how the water turned the wheels as it flowed back to the sea. Apart from these mills, tide mills appear to be very much a European, Atlantic Basin technology. Around 1066–1086 c.e., mills in Europe were built at the entrance to the port of Dover on the English Channel. This technology was later exported to Australia. In the Netherlands, where the earliest known mills were built in 1200, almost 20 mills have been reported. The number of mills increased steadily throughout the Middle Ages. By 1300, there were 37 mills in England alone, after which tide mills were built infrequently. Many of the tide mills began to be replaced by windmills as the former were occasionally damaged and destroyed by storms. During the 16th century, in Germany, while tide mills were built in the Hansa City—State of Hamburg, no mills existed along its North Sea Coast. The tidal range was increased and mills were established by using canals as reservoirs and damming certain river branches. Though the perfect location for tide mills is on coasts with a wide tidal range, it does not appear to be the sole determining factor as the Iberian Peninsula with its narrow tidal range had the highest number of tidal mills (some 100 tide mills) in the world. During the 14th and 15th centuries, new mills were built in Britain, near Plymouth and Southampton. From around the 15th century, the construction of tide mills with abutments consisting of strong breakwaters was encouraged as they provided protection against severe storms. The Great voyages of the 15th century and the discovery of America by Columbus in 1492 accelerated the number of the mills built in the following century. In 1613, the French aided by Micmac Indians introduced the first double function tide mills, which made use of both incoming and outgoing tides, at Port Royal (Nova Scotia) in North America. Later in 1635, the first tide mill in the United States was built in Salem (Massachusetts). The growth in the number of mills along the Atlantic coast during the 17th and 18th centuries was attributed to the development of grain crops and the colonization of America. During the 19th and 20th centuries, tide mills were built in Spain, in the North, and around Cadiz. Overall, approximately 800 mills were built on either side of the Atlantic and the North Sea, with over 500 mills on the European littoral. All of these mills were of local importance with the entire industry of the village centered on the mill. The idea of tide mills thus dotted the coasts of England, Wales, France, Portugal,

Spain, Belgium, the Netherlands, Germany, Denmark, Canada, the United States, and China. Despite the mechanical energy produced (30–100 kW) being sufficient for local needs, by the mid-20th century, the use of tide mills declined dramatically as a result of the appearance of the electric motor, need for long-distance transmission of energy, and the advent of power economics. The doom of tide mills started in England and Wales around 1900. Though some of these mills survived World War II, by 1950 they were either abandoned or converted for other uses. The mills in Spain and France also closed down during this period.[30]

3.3.6.2 TIDAL POWER PLANTS (TPPS)

Tide mills may be considered as the forerunners of TPPs that generate electricity. Both the tide mill and TPP comprise a powerhouse, a barrage, and a retaining basin. While the mill transforms the extracted energy into mechanical power, the TPP using tidal turbines transforms kinetic energy into electrical energy. The design of a typical tidal turbine is very similar to that of an offshore wind turbine; however, the former can function independently of weather and climate change. At present, there exist more than 30 different tidal turbine designs under different stages of development: with MCT SEAGEN geared tidal turbine as one of the most advanced ones. Basically, the exploitation of tidal energy involves two approaches. The first one utilizes the rise and fall of the sea level through entrainment which includes either the traditional barrage method or the lagoons. The second approach is analogous to the harnessing of wind power, that is, it utilizes the local tidal currents. The most common and simplest form of power generation is the ebb-flow generation. However, it is also possible to reverse the process and generate power using the flood tide instead of the ebb tide. Another possible mode of operation makes use of both the ebb and flood tide to generate electricity. Due to the high cost for the construction, operation, and maintenance of TPP, only a few commercial projects have been commissioned to date.

Since the 1920s, the idea to transform the ocean's energy into electricity has been formulated with several abortive attempts. A serious proposal for barraging or damming the Severn estuary, with a mean tidal range of 8 m, was made in 1925, and this particular option continued to be examined till 2010. However, following the Severn Tidal Power Feasibility Study (2008–2010), the British government concluded that there was no strategic case for building a barrage but to continue to investigate emerging technologies.

In 1930, the United Kingdom built a TPP with a 16-kW scheme installed at Avonmouth Docks. Apart from small plants built in China during the late 1950s, it was in 1966 that the world's first commercial TPP was built by Electricite de France on the Rance Estuary in Brittany, France. The La Rance tidal power station is a typical example of a tidal energy barrage system. It consists of 24 bulb tidal turbines with a power rating of 10 MW each, thereby having a total power rating of 240 MW. The turbines are placed along with a dam-like structure that directs the energy of the incoming tide directly through the turbines, forcing the blades of the rotor to rotate. The kinetic energy of the rotors is then converted to electricity by the built-in generator. In 1968, Russia began operating a TPP at Kislaya Guba, with a power rating of 0.4 MW. The Kislaya Guba tidal power station too employed a bulb turbine. The Annapolis Royal Generating Station built in 1984 in the Bay of Fundy in Nova Scotia, Canada, is the only tidal electricity generating station in North America. With doubts expressed as to the economic returns while using a bulb turbine, the Annapolis Royal Generating Station utilized the Straflo turbine; a straight-flow turbine with rim type generator. The turbine was, in fact, a modified version of the axial flow turbine patented by Leroy Harz in 1919 and was suggested as a more economic and better efficient system. The Straflo turbine has a generator built into the rim of the turbine runner, enabling the unit to operate in low-head conditions while keeping most of the generator components out of water. It has a power rating capacity of 18 MW. Being installed only in Annapolis power station, the experience with the 18 MW Straflo turbine design is limited.[31] Though since the 1950s China has installed several small-sized power stations, it was only in 1986 that a large-scale power station, the Jiangxia tidal power station, was established with a power rating capacity of 3.9 MW. The Haishan and the Rushankou power stations are other tidal power projects of China.[32] The Uldolmok Tidal Power Station is the first plant commissioned by the South Korean government with a power rating of 1 MW. Later in 2011, another tidal power station of Korea, the Sihwa Lake Tidal Power Station, with an installed capacity of 254 MW became the world's largest power station. It employs the bulb turbine but, unlike most TPPs, power is generated only with tidal inflows. The Swansea Bay tidal lagoon project (United Kingdom) and the MeyGen tidal array project (Scotland) are some of the large-scale tidal projects currently under development. The European Marine Energy Center (EMEC) established in 2007 is the world's first dedicated marine energy testing center to be fully equipped for the testing of tidal and wave energy technology. A similar testing facility, Fundy Ocean Research Centre for Energy, was undertaken by Canada in 2009 in the Minas Passage area of the Bay of Fundy.

In addition to the substantial up-front investment of time and money for its development, tidal barrage systems also cause considerable environmental changes. Artificial lagoons that enclose only an area of coastline with a high tidal range are a proposed alternative to tidal barrages, which often consists of structures spanning the entire estuary. The Swansea Bay in the United Kingdom and the Yalu River in China have both been suggested as potential locations for artificial lagoon-style power plants. However, many planners and engineers favor the development of a tidal current system with respect to a tidal entrainment system, mainly because it is cheaper to implement and has a comparatively lesser impact on the environment. The earliest documented attempts to harness the power of tidal currents began in the early 1990s in the waters of Loch Linnhe in the Scottish West Highlands. The project involved a turbine held mid-water using cables hanged from a seabed anchorage to a floating barge. Planning and development of tidal current powered systems were carried out during the mid-to-late 1990s, and it was only from the beginning of the 21st century that such systems were ready to be tested. In 2001, in the Strait of Messina along the Sicilian coast (Italy), a large floating vertical axis Kobold turbine with an installed capacity of 20 kW was tested as a part of the Enermar project. Bristol-based Marine Current Turbines (MCT) Ltd., in 2002, installed a small-scale tidal power device, Aquanator, that uses rows of hydrofoils to generate electricity from water currents at the Clarence River, New South Wales, Australia. In 2003, they demonstrated a 300-kW horizontal axis tidal turbine, the Seaflow turbine, that they installed off the coast of Devon, the United Kingdom. MCT installed the world's first commercial-scale prototype, SeaGen, with an installed capacity of 1.2 MW in Strangford Narrows in Northern Ireland in 2008. Later in 2018, it was successfully decommissioned after having exported over 11.6 GWh during its lifetime. In 2003, the Hammerfest Strøm system installed a 300-kW pillar-mounted horizontal axis system in a fjord environment in Norway. In December 2011, the company successfully deployed its 1 MW precommercial tidal turbine at EMEC's tidal test site. The device delivered its first energy to the grid in February 2012. Several other models were also developed in Norway, the United Kingdom, the United States, Germany, South Korea, the Netherlands, Australia, Spain, Sweden, France, Japan, Holland, Italy, and Canada.[33]

Placed stationary on the seabed, the most common tidal device typically requires a current of 2.5 m/s or faster to produce electricity cost-effectively.

Though this may be achievable in areas with high marine current inflows, the need to gain energy from watercourses with not too low current velocity remained unmet. This led to the invention of tidal kite turbines by Vienna Austrians Ernst Souczek and Wolfgang Kmentt around 1947.[34] Their invention consists of a holding a turbine, held to a rope anchored at the bottom of the sea, in a floating condition with the help of an underwater carrier connected to the turbine, thereby creating a dynamic buoyancy. In other words, the tidal kite turbines convert tidal energy into electricity by moving through the tidal stream. Even though many others advanced the underwater kite and paravane electric generating systems, the breakthrough in the field happened to be in 2004 with the invention of the Deep Green (then called Enerkite) by Magnus Landberg. In 2007, Minesto, a Swedish marine energy company, develops two models of its Deep Green tidal kite, which is capable to operate in low-velocity (1.2–2.4 m/s) tidal currents. They began installing the infrastructure for its Deep Green pilot project at Holyhead in May 2018. Less than 6 months later, in October 2018, Minesto successfully generated electricity via slow-current water for the first time with its DG5OO model turbine capable of generating 500 kW power which is equivalent to around 1800 solar panels.[35] Though the initial objective was to produce 10 MW to cover the energy needs of 8000 households, it has now been extended to produce 80 MW. Due to the expensive initial cost, tidal power remains a relatively new technology that holds immense potential to satisfy the increasing demand for energy.

3.3.7 WAVE ENERGY

The possibility of converting the wave energy into useful forms of energy has inspired numerous inventors resulting in the registration of more than one thousand patents by 1980. The earliest such patent was filed in Paris by Pierre-Simon Girard and his son in 1799. However, to date, no viable technology exists that has gained merit as a fully functional commercial product capable of harnessing wave energy. In 1910, in France, Bochaux-Praceique constructed a device that utilized the wave energy to light his home at Rayon, near Bordeaux. This was the first oscillating water column (OWC) type of wave energy device to be developed. A former Japanese navy officer, Yoshio Masuda with his experiments on wave energy devices since the 1940s, may be considered as the father of modern wave energy technology. The wave energy-powered navigation buoy installed with air turbines, later named the OWC, developed by him, was commercialized in Japan since 1965. The oil crisis of 1973 created a renewed interest in wave energy as it did in other renewable energies. In 1976, Masuda promoted the construction of a barge (80 m × 12 m), named Kaimei, that was used as a floating testing platform housing several OWCs equipped with different types of air turbines. In 1974, the Salter's duck or the Nodding duck, a laboratory prototype invented by Prof. Stephen Salter, of the University of Edinburgh, Scotland, in a small-scale controlled test could stop 90% of the wave motion with its curved cam-like body and convert 90% of it into electricity. As the oil prices went down in the 1980s, wave energy funding drastically reduced. In 1985, two full-sized (350 kW and 500 kW rated power) shoreline prototypes were installed at Toftestallen, near Bergen, Norway. In 1990, two full-sized OWC prototypes were constructed in Asia at the port of Sakata, Japan (60 kW) and also at Thiruvananthapuram, India (125 kW). In 1991, a small (75 kW) OWC shoreline prototype was deployed at the island of Islay, Scotland. Recently facing the consequences of climate change, there is again a growing interest worldwide for renewable energy, including wave energy. The first commercial wave energy system, the Islay LIMPET (Land Installed Marine Power Energy Transmitter), was installed at Islay island, Scotland in 2000. The 500-kW wave energy collector decommissioned in 2018 was connected to the national grid. The first-ever wave farm was constructed by the Scottish company, the Pelamis Wave Power off the coast of Portugal in 2008. The farm started delivering 2.25 MW using its three Pelamis generators. The current trials, farms, and commercial installations are located in Hawaii (Azura Wave and BOLT Lifesaver), Sweden (Sotenäs), Spain (Mutriku Wave Power Plant), Israel (SDE Sea Wave Power Plant), and Greece (SINN Power). Currently, The European Marine Energy Centre, in Orkney, Scotland established in 2003 is the world's first marine energy test facility and has supported the deployment of more wave and tidal energy devices than at any other site in the world. The worldwide resource potential of coastal wave energy has been estimated to be 2.11 TW; however, at present, the energy harnessed by the wave is still minimal, with an overall capacity under 15 MW. The area with the highest wave energy potential lies in the western seaboard of Europe, the northern coast of the United Kingdom, and the Pacific coastlines of North and South America, Southern Africa, Australia, and New Zealand. Compared to wind and solar energy, government and commercial research and development into wave power have paled.

Ocean Thermal Energy Conversion (OTEC) method is another method of extracting energy from the ocean which utilizes the temperature difference between the deep cold ocean water and the warm tropical surface waters. However, because of the availability of only low-temperature difference (20–25 °C) and the requirement of very large volumes of water needed to be brought to the surface, electricity is generated at low efficiency. Though in 1870 the French novelist Jules Verne introduced the idea of OTEC in one of his books, it was only in 1881 that Jacques Arsene d'Arsonval, a Frenchman physicist, conceptualized the physical development of trapping thermal energy stored in the ocean. In 1930, his student Georges Claude built the first OTEC plant (22 kW capacity) in Matanzas, Cuba. Though research continued throughout the 1940s and 1950s, the availability of cheaper alternatives for power generation halted all work in the field. A more practical compact and economic OTEC power plant was developed by J. Hilbert Anderson and James H. Anderson Jr. in 1962. The mid-1970s witnessed a renewed interest in OTEC research and development due to the Arab oil embargo and consequent skyrocketing of oil prices. The United States, Japan, and India each have conducted and continue to pursue research on small-scale OTEC power plants. In 2015, Makai Ocean Engineering launched the world's largest operational OTEC power plant (100 kW capacity) at the Natural Energy Laboratory of Hawaii Authority. Capable of producing electricity 24 h a day throughout the year, it was connected to the US electrical grid in August 2015.

As most of the tidal and wave energy projects established are relatively small-scale demonstration and pilot projects of less than 1 MW, ocean energy represents the smallest portion of the renewable energy market. Marine energy (Table 3.9) (wave, tide, and ocean energy) generation capacity reached 532 GW in 2018, with only 2 GW additions in the last year.[19] More than 90% of this total is contributed by two large tidal barrage facilities (Table 3.10). By the end of 2018, there were more than 90 tidal power technology developers all around the world, with more than half of them employing a horizontal axis turbine. Other tidal devices included the vertical axis turbines, oscillating hydrofoil, Archimedes screw, and Venturi as well as tidal kites. With respect to wave energy converters, there were at least 200 companies developing wave energy converters of various types including attenuator, point absorber, oscillating wave surge converter, OWC, overtopping/terminator device, rotating mass, bulge wave, and others. Though development activities to harness the enormous potential of the ocean are being carried around the world, the resource remains largely untapped.

TABLE 3.9 Marine Energy Capacity Installed Worldwide (2017–2018).

Region	Installed capacity (GW)—2017	Installed capacity (GW)—2018
Asia	259	259
Eurasia	2	2
Europe	245	247
North America	23	23
Oceania	1	1
Total	530	532

GW, gigawatts.

Source: IRENA (2019).

TABLE 3.10 Biggest Tidal Power Plants Constructed in the World.

Tidal project	Country	Installed capacity (MW)
Sihwa Lake Tidal Power Station	South Korea	254
La Rance Tidal Power Plant	France	240
Annapolis Tidal Generating Station	Canada	20
MeyGen Tidal Power Project	Scotland	6 (planned capacity: 398)
Jiangxia Pilot Tidal Power Plant	China	4.1

MW, megawatts.

3.3.8 BIOENERGY

Bioenergy is considered as a renewable source of energy as it refers to energy obtained from recently derived organic materials such as wood, agricultural crops, or organic waste. On the other hand, although fossil fuels are originally derived from organic matter, they are considered as a nonrenewable source of energy, as they are formed over many millennia through several biological and geological processes. Energy obtained from biomass may be used for the production of transportation fuels, electricity, heat, and other products. Even though accepted as a renewable source of energy, biomass at the point of combustion is not carbon neutral. In fact, biomass is less energy-dense than fossil fuels and contains higher quantities of moisture and less hydrogen, thereby emitting more GHGs per unit of energy produced than fossil fuels. It must, however, be noted that while the burning of fossil fuels releases carbon that has been locked up in the ground for millions of years, burning biomass emits carbon that is part of the biogenic carbon cycle. Proponents of the carbon neutrality of biomass also propose that during combustion, biomass releases back the same amount of CO_2 absorbed by it from the atmosphere during its growth phase, hence resulting in a zero-net balance of CO_2 emission. Though most of the governments, considering biomass as a renewable energy source, promote its use as a green solution to climate change, the scientists, as well as the environmentalists, have been raising concerns for years with respect to the emissions produced by burning biomass.

With the controversy over whether bioenergy is clean energy or not still persisting, a brief history of our dependence on bioenergy is outlined. Biomass-based energy is the oldest source of consumer energy used since the dawn of human civilizations for heating and cooking, and is still the largest source of renewable energy used worldwide.

3.3.8.1 BIOPOWER: HEAT, LIGHTING, AND POWER GENERATION

Solid biofuels: Prior to the 19th century, wood was the primary fuel for the whole world for heating and cooking. Interestingly, prior to the early 20th century, in the United States, wood-burning fireplaces, cookstoves, and heaters were commonly found in households. Interpolation studies conducted by FAO suggest that wood used for bioenergy production increased from 2 billion m^3 in 1970 to 2.6 billion m^3 in 2005. This indicated that by 2030, approximately, 3.8 billion m^3 would be needed. However, the use of energy-efficient woodstoves in developing countries has significantly reduced the consumption of firewood. Even then, today nearly 40% of the world population mainly belonging to rural areas of developing countries in Asia and sub-Saharan Africa rely on firewood to satisfy their energy demands.[36] Cellulosic biomass has also been used traditionally for heat in the past and ash obtained then is used as fertilizers. While the combustion of firewoods is the most energy-efficient way to utilize bioenergy, it being bulky could not be used for small automated heating systems for controlled fuel value. This led to the increased use of wood chips for heating and electricity generation since the beginning of the 21st century. Being expensive than firewood, wood chips are not used in rural areas of developing countries. In the United States, however, wood chips are less expensive than firewoods and so are widely used for heat, hot water, and electricity. Cofiring in which wood chips are combusted along with pulverized coal to drive steam turbines is also being practiced. With incentives being available in the form of Renewables Obligation Certificates for electricity and Renewable Heat Incentive in the

United Kingdom, the use of biomass for energy needs has increased significantly in the last years. In the United Kingdom, wood chips used to feed large power plants are imported from North America and Latin America and constituted about 51 million tons a year in 2013. There exist about 3019 such plants that use wood chips as its fuel source in Europe.[36] In the United States, there exist 222 power plants directly fueled using wood chips with an annual production of 57.6 billion kWh of electricity from wood chips, thereby accounting for 1.42% of the total produced electricity. The world biopower generation capacity was 70 GW in 2010, and it is projected to increase to 145 GW by 2020. With respect to wood chips, wood pellets as well as other biomass pellets are more processed biofuel products hence making them more expensive than wood chips. Thus, the use of wood pellets is limited to residential heating of developed countries. Wood pellets are also used for power generation in countries like China, the Netherlands, the United Kingdom, Japan, and Germany. The global production of wood pellets in 2012 was 19.1 million tons, and the projected increase in 2020 is 45.2 million tons. The rapid increase in wood pellet production in different countries poses severe environmental hazards as more forests and wetlands are being cleared. Woodfire with a typical temperature below 850 °C is unable to melt metals. To overcome these drawbacks, charcoal was produced by the heating of wood in a kiln or retort at about 400 °C in the absence of air until no visible volatiles are emitted. The use of charcoal in metallurgy can be traced back to the Bronze Age (around 3000 B.C.E.). In China's Tang Dynasty (700 C.E.), charcoal was designated as governmental fuel for cooking and heating. With an energy content of 28–33 MJ kg^{-1}, it burns without smoke and flame giving a temperature as high as 2700 °C. Today charcoal is usually made into briquettes for barbecuing and also continues to be used for cooking, heating, air, and water purification, art drawing, as well as steel-making. In 2015, worldwide an estimated 3.7 billion m^3 of wood was extracted from forests, of which 50% (i.e., 1.86 billion m^3) was used as fuel by either directly burning or by converting into other forms of wood fuel. Of this 1.86 billion m^3, 17% (i.e., 315 million m^3) was used as charcoal. More than half of this was produced in Africa (62.1%) followed by America (19.6%) and Asia (17%) with small quantities in Europe (1.2%) and Oceania (0.1%). In 2018, the global production of wood pellets for producing electricity and heat reached an estimated 35 million metric tons. The largest producer and exporter of wood pellets in 2018 was the United States (7.3 million tons). Unsustainable wood harvesting and production of charcoal can cause emission of GHGs along with forest degradation and deforestation. However, the impact can be lessened to a certain extent if charcoal is produced via sustainably managed resources and improved technologies. Though renewable electricity generation sources like wind, solar, and hydro can replace fuelwood and charcoal, the requirement of large up-front investment for them causes the latter to be preferred more.

Wood was the first fuel used for lighting. Resinous pitch oozing from coniferous trees was found to be very flammable and luminous when burned and so were used as torches almost 3000 years ago. By medieval times, the processing of pitch from coniferous trees became a trade governed by guilds. Archaeological evidence of oil being burned in lamps originated almost 4500 years ago in the ancient city of Ur, Mesopotamia. Olive and sesame oils were most commonly used during this period. Since the early Egyptian civilization, tallow (rendered and purified animal fat) has been used for lighting, initially, within lamps, but later began and continued to be used to make candles for almost 2000 years. The pithy stalk of rushes was dipped in animal fat, and a prototype of the candle was made in Egypt. When dried and used, it burnt brightly. These were also used throughout Europe, and in some places, they continued to be used until the 19th century. Evidence of modern candles consisting of a small wick and a thick, hand-formed layer of tallow emerged from Rome during the first century C.E. Animal oils, especially from fish and whales, were used in Northern Europe. Around the 18th century, sperm whale oil was found to be an excellent illuminant and became widely used in colonial America. Far more superior candles were produced when in 1825 French chemists Michel Chevreul along with Joseph Gay patented stearin, a tallow derivative. Ethanol became a common fuel, as feedstock was abundant, and it could be produced by anyone with a still. In the early 19th century, lamps burned many different fuels including all kinds of vegetable oils (castor, rapeseed, peanut, and olives), animal oils (especially whale oil and tallow from beef and pork), refined turpentine from pine trees, as well as alcohol (especially wood alcohol and grain alcohol). By the 1850s, the US patent office recorded almost 250 different patents for different fuels, lamps, wicks, and burners used for illumination. In the 1820s, the most common fuel used was a dangerous mixture called the burning fluid composed of turpentine and alcohol, and was popularized by Isaiah Jennings in New York in 1830. In 1835, a mixture of one part turpentine with four parts alcohol and a small amount of camphor oil extracted from the camphor tree (used for aroma) was patented by Henry Porter of Bangor, Maine and came to be popularly known as the "Porter's burning fluid." Despite the risks associated with the mixture, its demand increased. In 1856, Henry Porter's business in Boston was taken over by Rufus H. Spalding, who then became the sole manufacturer of Porter's Patent Composition. And so alcohol blends replaced the expensive whale oil by the late 1830s in most of the country. By 1860, thousands of distilleries churned out at least 90 million gallons of alcohol per year for lighting. However, with the isolation of paraffin wax from petroleum products in 1830, high-quality candles began to be made from it cheaply. The candle industry thus made a transition from biomass-based to petroleum-based industry around the 1860s. With the advent of kerosene, coal gas, and incandescent light bulb, the candle industry declined.[37]

Gaseous biofuels: Anecdotal pieces of evidence suggest that biogas was used in Assyria back in the 10th century B.C.E. for heating bathwater. However, it was only in 1630 that Jan Baptista van Helmont, a Belgium scientist, found that the decaying organic material produced flammable gases. Later in 1776, Count Alessandro Volta concluded that there was a direct correlation between the amount of organic matter decaying and the amount of corresponding gas produced. During the period between 1804 and 1808, John Dalton and Humphrey Davy working independently established that the combustible gas produced was methane. In 1859, the first recorded anaerobic digester plant was established in Bombay, India. A Frenchman in 1881 developed a crude version of the septic tank, named "automatic scavenger," to treat wastewater via anaerobic digestion. The success of the anaerobic digester resulted in the introduction of this system in the United States in 1883. In 1895, the term "septic tank" was coined by Donald Cameron, an Englishman who constructed a watertight covered basin to treat sewage by anaerobic decomposition. The increased efficiency and successful treatments of septic tanks inspired the local government of Exeter, England, to make use of the methane evolved from septic tanks for lighting its streets in 1895. Commercial use of biogas in China has been attributed to Guorui Luo who in 1921 constructed an 8 m^3 biogas tank fed with household wastewater. Microbiological research work in an attempt to identify best anaerobic bacteria and digestion conditions for promoting methane production was conducted by US scientist A.M. Buswell and his colleagues in the 1930s. The sharp rise in fossil fuel prices in the early 1970s stimulated interest in the use of anaerobic digestion systems for energy as well as industrial wastewater depollution.[38] Currently, worldwide there exist approximately 50 million microscale digesters with almost 42 million of those operating in China and another 4.9 million in India. Almost 700,000 biogas plants operate in the rest of Asia, Africa, and South America. These microscale digesters may be operated as stoves for heating and cooking, thereby replacing the high emission solid fuels like firewood and charcoal. The generation of heat and electricity using a combined heat and power plant is an established technology used worldwide. Electricity generation via biogas plants increased from 46,108 GWh in 2010 to 87,500 GWh in 2016 (a growth of almost 90% in 6 years). In 2018, there were more than 10,000 digesters in Europe as well as almost 2200 sites in all 50 US states producing biogas for electricity and heating.

Upgradation of biogas to biomethane plants, though new, is now a proven technology. Globally, there exist almost 700 plants that upgrade biogas to methane. There are over 540 upgrading plants in Europe, followed by 50 in the United States, 25 in China, 20 in Canada, and the remaining few in Japan, South Korea, Brazil, and India. While certain plants upgrade biogas to be used as vehicle fuel, others inject it to the local and national grid. Many plants are also beginning to capture CO_2 to be used in greenhouses and the food and drink industry.[39] Syngas produced via the gasification or pyrolysis of plant materials is another gaseous biofuel that can also be used directly to generate electricity. The raw material most commonly used for syngas production is woody biomass low in nitrogen and ash content. Global biopower capacity increased by almost 6 GW in 2018 reaching an estimated 115.7 GW.[19] The total electricity generated via biomass increased by 9% from 532 TWh in 2017 to 581 TWh in 2018. While the European Union remained the largest generator of bioelectricity by region, the country which produced the largest amount of electricity was China. Other major producers of bioelectricity were the United States (69 TWh), Brazil (54 TWh), Germany (51 TWh), India (50 TWh), the United Kingdom (35.6 TWh), and Japan (29 TWh).

3.3.8.2 BIOFUEL

Currently, biomass is the only renewable source of feedstock material to produce liquid fuel. The history of biofuels roots back into the earliest automotive days during which the first cars were built to run on biofuels rather than fossil fuels. In 1826 in the United States, Samuel Morey developed the first internal combustion engine (ICE) which used a mixture of turpentine and alcohol to power a small boat up the Connecticut River at 7 and 8 mph (miles per hour). German engineer Nikolaus August Otto in 1860 developed another ICE that ran on ethanol fuel blend. During the 1860s, the main obstacle which prevented ethanol from being used as an engine fuel in the United States was the imposition of tax on alcohol to fund the civil wars.

Henry Ford, in the period between 1908 and 1925, mentioned ethanol as the fuel of the future and used it to power tractors and his model T cars. The United States witnessed an increased demand for ethanol as necessitated by the shortage of raw materials and natural resources during the years of World War. The 1973 oil crisis further spurred interest in the use of ethanol as a fuel. The first pilot bioethanol plant with a distillation column was established at South Dakota University in 1979. During the 1980s, the US Environmental Protection Agency (EPA) headed gasoline phaseout resulted in increased demand for ethanol as an octane booster and volume extender. Throughout the 1990s, methyl tert-butyl ether (MTBE) derived from fossil fuels was used in most of the oxygenated gasoline markets. However, restrictions on the use of MTBE as a fuel oxygenate led to the rapid growth of ethanol production in the United States since 2002. Fermentation of vegetable biomass results in the production of ethanol. Starch/sugar-based crops such as sugar cane, sugar beet, sweet sorghum, corn, wheat, barley, potato, yam, and cassava are used for the commercial production of bioethanol. Bioethanol produced from such food crops is called "first generation" biofuels, while that produced from non-food crop lignocellulosic plant materials is called "second-generation" biofuel. World's first commercial-scale cellulosic ethanol plant, the Crescentino Biorefinery, Crescentino, Vercelli in Italy started operating in 2013. Run by the Italian company Beta Renewables, it has an annual production of 20 million gallons of ethanol. Commercial production of second-generation ethanol is expected to grow in the future, as an alternative to gasoline. "Third-generation" biofuels based on improvements in the production of biomass takes advantage of energy crops such as algae. The world's largest ethanol producers, POET and Archer Daniels Midland, are from the United States. The global commercial bioethanol production has increased from 15.1 billion liters in 1990 to 106 billion liters in 2018.[40]

In the 1890s, Rudolf Diesel invented the diesel engine which ran on different fuels including vegetable oil. A French Otto Company at the World's Fair in Paris at 1900 demonstrated a peanut oil fuel-based diesel engine. However, widespread availability and low price of fossil fuels discouraged further research for developing vegetable-based fuel. Still, experiments conducted in various countries identified that the main problem with using vegetable-based fuel to run diesel engines was its high viscosity. During World War II, fuel shortages necessitated research to find ways to convert vegetable oil to diesel. Several attempts were made to overcome the problem associated with the high viscosity of vegetable-based fuels. In 1937, George Chavanne, a Belgian scientist, presented and patented the "Procedure for the transformation of vegetable oils for their uses as fuels." It was the Brazilian scientist, Expedito Parente, who developed the first industrial-scale biodiesel production process in 1977. Later in 1989, the world's first commercial-scale biodiesel plant which produced biodiesel from rapeseed began operation in Asperhofen, Austria. The world production of biodiesel has steadily increased from 7.4 million tons in 2000 to 45 million tons in 2019.

Both biodiesel and bioethanol display comparatively higher oxygen content and greater dissolution capacity with respect to fossil fuels and hence are corrosive to the engine, fuel storage, as well as to the distribution system. In order to overcome these problems, drop-in biofuels are suggested as an alternative. As the name suggests, they can be readily dropped into the existing fuel system without any significant modification in engine or infrastructure. Drop-in biofuels, currently at the research and development stage, are biomass-derived liquid hydrocarbons that are functionally equivalent to the existing petrol distillate fuel. Currently, the dominant pathway to making drop-in biofuels is the oleochemical/lipid pathway wherein animal and vegetable oils are hydrotreated to produce biofuels. To date, this remains the only pathway to produce significant amounts of drop-in biofuel. Currently, the commercialization of this technology is still in progress.

In an attempt to produce fuels similar to fossil fuels, human beings began treating vegetative biomass in a simulated high-temperature (300–900 °C), high-pressure, oxygen-free environment. However, the final product of such pyrolysis: biochar (the black, solid residue), bio-oil (the brown vapor condensate), and syngas (the uncondensable vapor) was little like fossil fuels. Although about 3000–5000 years ago, a similar technique was used in ancient China and by indigenous Amazonians to produce charcoal and biochar, respectively, the pyrolysis of biomass to produce bio-oil for use as fuel is quite recent. There are basically two types of pyrolysis with respect to the heating rate: fast pyrolysis and slow pyrolysis. Since the development of fast pyrolysis, various types of reactors (bubbling fluidized bed, circulating fluidized bed, ablative pyrolysis, vortex, rotating cone pyrolysis, vacuum, etc.) have been developed since the 20th century for the production of bio-oil.[41] Though containing combustible components, crude pyrolysis bio-oil due to its high moisture and acid content is instable, corrosive, viscous, and difficult to ignite thus requiring significant upgradation before it can be used as a petrol distillate fuel alternative. The past few decades have investigated a number of techniques (e.g., hydrogenation, esterification, emulsification, etc.) to upgrade crude bio-oil to use as a substitute for heavy fuel oils to power static appliances such as electric generators, turbines, furnaces, engines, and boilers. The bio-oil produced is used as heating fuel and sometimes as potential industrial feedstock material for chemicals, lubricants, thickeners, paints, and so on. Although there are a number of pilot plants that convert woody biomass to bio-oil, techniques that are economically feasible and industrially applicable are still under development. Hence, commercial production, as well as utilization of biofuel as a petrol fuel alternative, faces many technological challenges.

At present, the volume of renewable diesel in the world is about 5 billion liters, and production is mainly dominated by freestanding facilities based on hydrotreating.[42] In 2018, with increased production of 10 billion liters made in the year, the global production of all biofuels (ethanol, biodiesel) reached a total of 154 billion liters. The United States and Brazil have dominated the biofuels market together producing nearly 70% of all biofuels in 2018, followed by China (3.4%), Germany (2.9%), and Indonesia (2.7%).[43]

3.4 DEVELOPMENT OF GREEN CHEMISTRY AND GREEN ENGINEERING AS ACADEMIC DISCIPLINES

Chemistry via the medical revolution has greatly improved the quality of human life raising the life expectancy from 47 years in the 1900s to over 80 years in 2019 thereby increasing the human population to an estimated 7.8 billion. The consequent increased demand for food and energy has led to massive global industrialization and materialization. As initially the manufacturers were concerned with only the successful production, the growing use and complexity of chemicals, as well as the application of synthetic amendments in agriculture since the industrial revolution, began causing environmental pollution which grew into global transboundary problems affecting human health and well-being. Eventually, humans developed concerns regarding the control of the waste stream as well as the impact of chemicals on health and the environment. Despite major growth in renewable energy over the past decade, global GHG emissions too kept rising, as human demand for fossil fuels also simultaneously increased (Figure 3.2). All of this demonstrates the urgency in relying completely on all clean energy solutions as well as improving the efficiency, investment, and innovations meant for curbing emissions and capturing carbon.

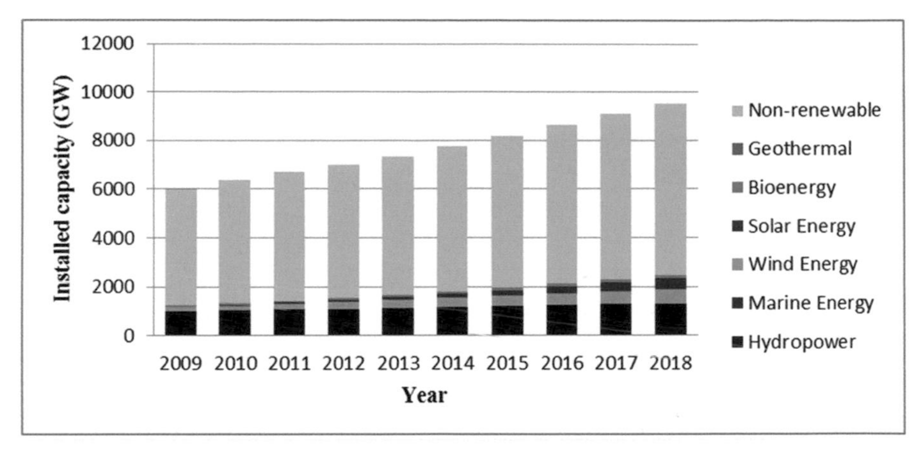

FIGURE 3.2 Global power capacity trend by source between 2009 and 2018.
Source: This figure is drawn based on selected information provided by IRENA in 2019 and REN21 in 2019.

3.4.1 GREEN CHEMISTRY

The Pollution Prevention Act (also called the P2 Act) of 1990, established as a national policy of the United States, was developed with an intention to limit the creation of pollution at the source through improved design (including cost-effective changes in products and processes as well as in the use of raw materials) rather than monitoring and clean up. As per the Act, the pollution that cannot be prevented should be either recycled or else treated in an environmentally safe manner whenever possible, and that disposal or release into the environment must be sought only as a last resort and that too must be conducted in an environmentally safe manner. The idea of green chemistry was initially developed as a response to this Act authored by the EPA. It applies fundamental principles from all chemical disciplines including organic, inorganic, analytical, biochemistry, and even physical chemistry to develop chemical products that are inherently less toxic than currently existing products. A research grant program launched by the EPA Office of Pollution Prevention and Toxics by 1991 encouraged the redesigning of chemical products and designs in a way that was less toxic to human health as well as the environment. During the early 1990s, EPA along with United States National Science Foundation began to fund basic research in green chemistry. Another key driver for the development of green chemistry was the annual US Presidential Green Chemistry Challenge

Awards Program. Established in 1995, it drew attention to fundamental breakthroughs in green chemistry success stories at both the academic and industrial levels. In 1997, the Green Chemistry Institute (GCI) was formed to advance the growth of green chemistry and green engineering. Later in 2001, the GCI became a part of the American Chemical Society. The mid-to-late 1990s also witnessed an increase in the number of international meetings devoted to green chemistry such as the Gordon Research Conferences on Green Chemistry, and green chemistry networks developed in the United States, the United Kingdom, Spain, and Italy. Green chemistry was first defined by Paul Anastas and John Warner in 1998 as, "the utilization of a set of principles that reduces or eliminates the use or generation of hazardous substances in the design, manufacture, and application of chemical products." Although seemingly intuitive, they clearly explained the 12 principles of green chemistry (Box 3.1) that helped the chemists and chemical engineers to apply the principles of sustainability to their research.[44] The rate of green chemistry patent applications and the new green technologies emerging have been increasing. In 1999, the Royal Society of Chemistry launched the journal Green Chemistry which provides a platform for the publication of innovative research on the development of alternative green technologies. Green chemistry concepts have continued to gain popularity, and in 2005, the Nobel Prize for Chemistry was awarded jointly to Yves Chauvin, Robert H. Grubbs, and Richard R. Schrock, "for the development of the metathesis method in organic synthesis." This represented a great step forward for green chemistry, by reducing potentially hazardous wastes through smarter production. In 2018, Frances H. Arnold won the Nobel Prize in chemistry for the directed evolution of enzymes, a technique that she has pioneered to pursue new avenues within green chemistry. The application of her results included the greener manufacturing of chemical substances such as pharmaceuticals as well as the production of renewable fuels. Though green chemistry has been tremendously successful so far in producing chemical products that are inherently less toxic, it still holds plenty of room for improvement.

BOX 3.1 Principles of Green Chemistry

Principles of Green Chemistry

1. It is better to prevent waste than to treat or clean up waste after it has been created
2. Synthetic methods should be designed to maximize the incorporation of all materials used in the process into the final product
3. Wherever practicable, less hazardous chemical synthesis must be designed

BOX 3.1 *(Continued)*

Principles of Green Chemistry

4. Designing safer chemical
5. The use of auxiliary substances should be made unnecessary wherever possible and innocuous when used
6. Energy requirements of chemical processes should be recognized for their environmental and economic impacts and should be minimized
7. Use of renewable feedstock
8. Reduce unnecessary derivatization
9. Catalytic reagents are superior to stoichiometric reagents
10. Chemical products at the end of their function should be designed for degradation
11. Real-time analysis for pollution prevention
12. Inherently safer chemistry for accident prevention

3.4.2 GREEN ENGINEERING

Green engineering involves the design, development, and commercialization of industrial processes and products that are economically feasible. It minimizes the generation of pollution at the source as well as poses minimal risk to human health and the environment. The 12 principles of green engineering (Box 3.2) which provide a framework for scientists and engineers to engage in while designing products, processes, and systems were proposed by Paul Anastas and Julie Zimmerman in 2003.[45] Engineers apply risk assessment concepts and systems analysis to processes and products, with the intent to consider the life cycle impacts of that particular product or process. These analyses allow one to appropriately consider and thereby control all the environmental impacts as well as risks associated at various stages of the process. Green engineering and chemistry together have made a significant sustainable contribution in the fields of agriculture, food processing, potable water, sustainable energy, consumer products, automobiles, construction, industrial automation, education, health, aircraft, and space travel as well as in the communication sector. The tracing back of the history and current status of each of these sectors is beyond the scope of this book.

BOX 3.2 Principles of Green Engineering

Principles of Green Engineering

1. Designing all material and energy inputs and outputs as inherently nonhazardous as possible
2. Prevention of waste is better than to treat or cleanup of waste after it is formed
3. Minimization of energy consumption and material use during separation and purification processes
4. Maximize mass, energy, space, and time efficiency of all operations and products
5. All operations and products should be output-pulled rather than input-pushed via the use of energy and material
6. Embedded entropy and complexity must be viewed as an investment when making design choices
7. Targeted durability, not immortality, should be a design goal
8. Meet need, minimize excess
9. Material diversity in multicomponent products should be minimized
10. Integrate local material and energy flows in the design of products, processes, and systems
11. Design for commercial "afterlife"
12. Material and energy inputs should be from renewable resources to the maximum extent possible

3.5 FUTURE OUTLOOK AND OPPORTUNITIES

Dependence on green technology is deeply rooted in the need for sustainable solutions for the survival of our planet facing the dare consequences of climate change and global warming. The rising population and increasing demand for food and energy are other key drivers. Though still a dominant global energy source, the future of fossil fuels is quite unpromising due to the lack of its availability at affordable prices as well as due to the knowledge its impact has on the environment and global climate change. Thus, the future of the energy industry lies outside of the traditional fossil fuel-based centralized power station models which cause serious environmental problems. With solar and wind energy available almost anywhere, a fall in price and advancement of renewable technologies is likely to diminish our dependence on fossil fuels. Though output from wind turbines is found to decrease with the age of installation, by 2050, wind power is expected to supply 14.3 exajoules (EJ) globally. With the primary energy then expected to be as high as 1000 EJ, wind energy will remain a marginal source of energy if it follows the linear growth rate of the past decade. Electricity generation via solar energy followed an exponential growth rate recently, and given the possibility of technical outbreaks such as perovskite solar panels, solar spray, and multi-junction cells, any predictions of future output are unlikely to be useful. However, it must be noted that wind energy also had experienced an exponential growth phase between the mid-1990s up until 2008 after which output has only risen linearly. At present, hydropower is a mature technology with only little in the way of technical progress to be expected in the future. Still annual electric output via hydropower by 2050 is estimated to be well below 30 EJ. Geothermal electricity outgrowth has been linear for the past three to four decades, and if this trend is to continue, it is estimated to reach 20 GW by 2050; a very minor fraction of global electricity output. Enhanced geothermal system is a looming technology which offers to dramatically expand the use of geothermal energy as presently only 10% of the world's area with hydrothermal convection systems is fit for geothermal power production. Though there are pitfalls in this technology, with concerns regarding whether such drilling would cause seismic activity, there is plenty of research and development activity going on. Currently, ocean energy is the world's greatest remaining source of untapped renewable energy. In spite of the halting progress the industry has made, it is quite evident that it remains decades behind other forms of renewable energy. Several promising technologies (wave energy converters, wave carpet) are being developed, and many have undergone ocean trials; however, no commercial-scale wave power exists now. The renewable energy sector output will steadily expand in the future, given that the remaining fossil fuels will be much more costly to extract than at present. The impact of ongoing climate change on the output of renewable energy will vary depending on the source. For example, while geothermal and tidal energy will not be directly affected, as they are independent of climate change, bioenergy and hydropower will be most affected.[46]

Over the last few years, global investment in green technology has been increasing by almost 20% in sectors like energy, innovation, and manufacturing. As green technology continues to emerge as a growing force, several strong industry clusters with varying levels of investment have emerged. With some of the major concerns of modern society being rising plastic as well as electronics consumption and consequent waste production, the development of innovative biopolymer materials that are fully biodegradable for use in packaging, agriculture, medicine, electronics, and other areas through the application of green chemistry and engineering is quite promising. The complete transition from synthetic nonbiodegradable plastics to edible as well as biodegradable plastics such as the E6PR (Eco Six Pack Ring) produced from the by-products of brewing beer could make a noticeable dent in the plastic pollution of oceans. "Green" electronics is an emerging area of research aimed at tackling the consequences of the colossal demand of electronics by delivering low-cost and energy-efficient materials and devices. Though organic (carbon-based) electronics entered the research field in the mid-1970s, the performance, as well as stability of organic semiconductors, remains a major obstacle in its development as solid competitors for its inorganic counterparts. In the not too distant future, the widespread use of roadside energy harvesting technologies like piezoelectric devices and triboelectric devices might be able to generate enough electricity for smart cities.

Even though electric vehicles (EVs) were highly successful when compared to the ICE cars around the 19th century, the cheaper availability of fossil fuel-powered vehicles and the wide accessibility of fuel stations by 1912 led to the rapid decline of EVs. The transportation sector became responsible for 24% of direct CO_2 emissions from fossil fuel combustion in 2018. Road vehicles such as cars, buses, and trucks as well as two- and three-wheelers account for nearly three-quarters of this transmission. In the past as well as in the current scenario, the cleaning up of these emissions is a pressing issue. A lot of problems that put electric cars out of favor in the early 1900s are still prevalent today such as the batteries being too heavy and taking too long to charge. The low mileage, at the same time great expense, and the short life span of EVs is what makes it unattractive to buyers. In spite of the drawbacks, the number of electric light-duty vehicles on road has exceeded 5 million. With a great deal of progress made in battery technology, EVs are expected to reenter the market on a large scale within a couple of years. By 2050, EVs could represent more than 60% of new sales and may constitute up to 25% of the global car fleet. Breakthrough research into the production of drop-in biofuels, chargers such as Sonnen EV chargers and electrified roads such as eRoadArlanda are promising, sustainable, low-carbon, green-powered alternatives to petroleum-based fuels.

In this modern world, now more than ever before, consumers are willing to buy environmentally friendly products and are also concerned with the impact of their decisions on the environment. Governments can help build a green future through measures such as green industrial policies, research, and development activities as well as implementing incentives and subsidies for green technology. Companies that green their supply chains by optimizing operations with environmentally friendly power generation and storage methods, taking into consideration both energy and material efficiency, as well as waste management and recycling, are now at the forefront of the business world. By lowering costs and minimizing waste, they are capable of enhancing their competitive position by increasing productivity while maintaining sustainability. This "Green" revolution will surely wipe off companies and industries that fail to offer environmentally sustainable practices and technologies as current economic theories postulate that the increased energy consumption will eventually exhaust the fossil resources and consequently raise the cost to extract them. While at present, green technology products are among the many options available for us, in the very near future, it is going to be the only option toward satisfying the growing demand for energy in a sustainable way. The present generation is in dire need of a new green deal, which integrates green technologies with economics and policy thereby delivering more service to the entire population, reducing and eradicating extreme poverty, and generating more energy without compromising the development of the economy. All this must be achieved within the overall framework of low carbon emissions in order to protect future generations.

KEYWORDS

- **green technology**
- **products**
- **sustainable development**
- **living standards**
- **green gas emissions**

REFERENCES

1. *Green Paper on Integrated Product Policy*; Commission of the European Communities, Brussels, 2001.
2. *Green Technology and Sustainability Market by Technology—Global Forecast to 2024*; MarketsandMarkets, Pune, India, 2019.
3. Valoch, K. Archaelogy of the Neanderthalers and Their Contemporaries. In *History of Humanity Volume I Prehistory and the Beginnings of Civilization*; De Laet, S. J., Ed.; Routledge: New York, 1996; pp 312–333.

4. Hardy, K. Prehistoric String Theory. How Twisted Fibres Helped to Shape the World. *Antiquity* **2008,** *82* (316), 271–280. https://doi.org/10.1017/S0003598X00096794.
5. De Laet, S. J. From the Beginnings of Food Production to the First States. In *History of Humanity Volume I Prehistory and the Beginnings of Civilization*; De Laet, S. J., Ed.; Routledge: New York, 1996; pp 877–890.
6. Yannopoulos, S.; Giannopoulou, I.; Kaiafa-Saropoulou, M. Investigation of the Current Situation and Prospects for the Development of Rainwater Harvesting as a Tool to Confront Water Scarcity Worldwide. *Water (Switz.)* **2019,** *11* (10), 1–16. https://doi.org/10.3390/w11102168.
7. Guo, M.; Song, W.; Buhain, J. Bioenergy and Biofuels: History, Status, and Perspective. *Renewable Sustainable Energy Rev.* **2015,** *42*, 712–725. https://doi.org/10.1016/j.rser.2014.10.013.
8. Viollet, P. L. From the Water Wheel to Turbines and Hydroelectricity. Technological Evolution and Revolutions. *C. R. Mec.* **2017,** *345* (8), 570–580. https://doi.org/10.1016/j.crme.2017.05.016.
9. Lewis, B. J.; Cimbala, J. M.; Wouden, A. M. Major Historical Developments in the Design of Water Wheels and Francis Hydroturbines. *IOP Conf. Ser. Earth Environ. Sci.* **2014,** *22*, 1–11. https://doi.org/10.1088/1755-1315/22/1/012020.
10. Taylor, C. H. *Illustrated Description of the Taylor Hydraulic Air Compressor and Transmission of Power by Compressed Air*; Montreal, Quebec, 1897.
11. Kaldellis, J. K.; Zafirakis, D. The Wind Energy (R)evolution : A Short Review of a Long History. *Renew. Energy* **2011,** *36*, 1887–1901. https://doi.org/10.1016/j.renene.2011.01.002.
12. Johnson, G. L. *Wind Energy Systems* (Electronic Edition); Kansas University Reprints: Manhattan, KS, 2006.
13. Ragheb, M. History of Harnessing Wind Power. In *Wind Energy Engineering: A Handbook for Onshore and Offshore Wind Turbines*; Letcher, T. M., Ed.; Elsevier Inc.: London Wall, United Kingdom, 2017; pp 127–143. https://doi.org/10.1016/B978-0-12-809451-8.00007-2.
14. Fleming, P. D.; Probert, S. D. The Evolution of Wind-Turbines: An Historical Review. *Appl. Energy* **1984,** *18* (3), 163–177. https://doi.org/10.1016/0306-2619(84)90007-2.
15. Manwell, J. F.; Mcgowan, J. G.; Rogers, A. L. *Wind Energy Explained Theory, Design and Application*, 2nd ed.; John Wiley and Sons Ltd: West Sussex, UK, 2009.
16. Yellott, J. I. Solar Heating and Cooling of Homes. In *Solar Energy Engineering*; Sayigh, A. A. M., Ed.; Academic Press Limited: London, 1977; pp 365–382.
17. Bradford, T. *Solar Revolution: The Economic Transformation of the Global Energy Industry*; MIT Press: London, England, 2006.
18. Funk, P. A. Evaluating the International Standard Procedure for Testing Solar Cookers and Reporting Performance. *Sol. Energy* **2000,** *68* (1), 1–7. https://doi.org/10.1016/S0038-092X(99)00059-6.
19. *Renewable Energy Statistics 2019*; IRENA, Abu Dhabi, 2019.
20. Jonokuchi, H.; Maeda, S. History of Solar Car and Its Electric Components Advancement and Its Future. In *Smart Innovation, Systems and Technologies*, Vol. 68; Springer Science and Business Media Deutschland GmbH, Berlin, Germany, 2017; pp 812–819. https://doi.org/10.1007/978-3-319-57078-5_76.
21. Perlin, J. The French Connection: The Rise of the PV Water Pump. *Refocus* **2001,** *2* (1), 46–47. https://doi.org/10.1016/s1471-0846(01)80096-2.

22. Righini, G. C.; Enrichi, F. Solar Cells' Evolution and Perspectives: A Short Review. In *Solar Cells and Light Management*; Enrichi, F., Righini, G. C., Eds.; Elsevier: Amsterdam, The Netherlands, 2020; pp 1–32. https://doi.org/10.1016/b978-0-08-102762-2.00001-x.
23. Geisz, J. F.; Steiner, M. A.; Jain, N.; Schulte, K. L.; France, R. M.; McMahon, W. E.; Perl, E. E.; Friedman, D. J. Building a Six-Junction Inverted Metamorphic Concentrator Solar Cell. *IEEE J. Photovoltaics* **2018,** *8* (2), 626–632. https://doi.org/10.1109/JPHOTOV.2017.2778567.
24. Matasci, S. How Much Does a Solar Panel Installation Cost? https://news.energysage.com/how-much-does-the-average-solar-panel-installation-cost-in-the-u-s/ (accessed Dec 6, 2019).
25. Parri, R.; Lazzeri, F.; Cataldi, R. Larderello: 100 Years of Geothermal Power Plant Evolution in Italy. In *Geothermal Power Generation: Developments and Innovation*; DiPippo, R., Ed.; Elsevier Inc.: Cambridge, United States, 2016; pp 537–590. https://doi.org/10.1016/B978-0-08-100337-4.00019-X.
26. Fridleifsson, I. B.; Freeston, D. H. Geothermal Energy Research and Development. *Geothermics* **1994,** *23* (2), 175–214. https://doi.org/10.1016/0375-6505(94)90037-X.
27. Van, M.; Sigurjón, N.; Margeir, A.; Páll, G.; Pálsson, G. *Uses of Geothermal Energy in Food and Agriculture Opportunities for Developing Countries*; FAO: Rome, 2015.
28. Barbier, E. Geothermal Energy Technology and Current Status: An Overview. *Renew. Sustain. Energy Rev.* **2002,** *6* (1–2), 3–65. https://doi.org/10.1016/S1364-0321(02)00002-3.
29. DiPippo, R. Magmamax Binary Power Plant, East Mesa, Imperial Valley, California, USA. In *Geothermal Power Plants*; DiPippo, R., Ed.; Elsevier: Oxford, UK, 2012; pp 361–374. https://doi.org/10.1016/b978-0-08-098206-9.00018-x.
30. Charlier, R. H.; Menanteau, L. The Saga of Tide Mills. *Renew. Sustain. Energy Rev.* **1997,** *1* (3), 171–207. https://doi.org/10.1016/S1364-0321(97)00005-1.
31. Charlier, R. H. Re-invention or Aggorniamento? Tidal Power at 30 Years. *Renew. Sustain. Energy Rev.* **1997,** *1* (4), 271–289. https://doi.org/10.1016/S1364-0321(97)00003-8.
32. Zhou, Q.; Bai, Y.; Li, Y.; Wang, X.; Wang, H.; Du, M.; Zhang, S.; Meng, J.; Duan, L.; Shi, Y.; Cai, X.; Zhang, R.; Wu, Y. Reviews of Development and Utilization of Tidal Energy over Chinese Offshore. In *OCEANS 2016—Shanghai*; Institute of Electrical and Electronics Engineers Inc.: Piscataway, NJ, 2016. https://doi.org/10.1109/OCEANSAP.2016.7485635.
33. Bryden, I. G. Tidal Energy. In *2010 Survey of Energy Resources*; Gadonneix, P., Pacific, A., Asia, S., Frei, C., Eds.; World Energy Council: London, United Kingdom, 2010; pp 1–608.
34. Souczek, E. Stream Turbine. U.S. Patent 2,501,696, March 28, 1950.
35. Banerjee, R. "Deep Green": How a Swedish Startup Is Making Waves Underwater https://www.newstatesman.com/spotlight-america/energy/2019/05/deep-green-how-swedish-startup-making-waves-underwater-1 (accessed Dec 13, 2019).
36. *European Wood Chips Plants—Country Analysis BASIS – Biomass Availability and Sustainability Information System*; European Biomass Association, Belgium, 2015.
37. DiLaura, D. A Brief History of Lighting. https://www.osa-opn.org/home/articles/volume_19/issue_9/features/a_brief_history_of_lighting/ (accessed Dec 16, 2019).
38. Abbasi, T.; Tauseef, S. M.; Abbasi, S. A.; Abbasi, T.; Tauseef, S. M.; Abbasi, S. A. A Brief History of Anaerobic Digestion and "Biogas." In *Biogas Energy*; Springer: New York, 2012; pp 11–23. https://doi.org/10.1007/978-1-4614-1040-9_2.

39. Jain, S.; Newman, D.; Nzihou, A.; Dekker, H.; Le Feuvre, P.; Richter, H.; Gobe, F.; Morton, C.; Thompson, R. *Global Potential of Biogas*; World Biogas Association: London, United Kingdom, 2019.
40. Garside, M. Fuel Ethanol Production in Major Countries 2018. https://www.statista.com/statistics/281606/ethanol-production-in-selected-countries/ (accessed Dec 17, 2019).
41. Steele, P. H.; Yu, F.; Gajjela, S. Past, Present, and Future Production of Bio-Oil. In *Woody Biomass Utilization : 2009 FPS Conference Proceedings,* Mississippi State University, MS, 2009; pp 17–22.
42. van Dyk, S.; Su, J.; Mcmillan, J. D.; Saddler, J. N. *'Drop-in' Biofuels : The Key Role That Co-Processing Will Play in Its Production*; Society of Chemical Industry and John Wiley & Sons, Ltd, 2019.
43. *Renewables 2019 Global Status Report*; REN21, Paris, 2019.
44. Anastas, P. T.; Warner, J. C. *Green Chemistry: Theory and Practice*; Oxford University Press: New York, 1998.
45. Anastas, P. T.; Zimmerman, J. B. Through the 12 Principles of Green Engineering. *Environ. Sci. Technol.* **2003,** *37* (5), 94A–101A. https://doi.org/10.1021/es032373g.
46. Moriarty, P.; Honnery, D. Global Renewable Energy Resources and Use in 2050. In *Managing Global Warming*; Letcher, T. M., Ed.; Elsevier: Cambridge, United States, 2019; pp 221–235. https://doi.org/10.1016/b978-0-12-814104-5.00006-5.

CHAPTER 4

THERMODYNAMIC AND TRANSPORT PROPERTIES OF AMINO ACIDS IN AQUEOUS SOLUTION

CARMEN M. ROMERO[1*] and MIGUEL A. ESTESO[2]

[1]*Universidad Nacional de Colombia, Bogotá, Colombia*

[2]*Universidad de Alcalá, Alcalá de Henares, Spain*

[*]*Corresponding author. E-mail: cmromeroi@unal.edu.co*

ABSTRACT

Amino acids can be considered as mixed solutes that contain hydrophilic groups (amine and carboxyl) and the hydrophobic aliphatic chain. These facts, combined with their biological importance, make the study of the physicochemical properties of these compounds in aqueous solvents very interesting and relevant. The information about structural changes derived from their physicochemical properties contributes to the interpretation of solute-solvent and solute-solute interactions in aqueous solutions.

The amino acids considered in this work are α-amino acids and, α,ω-amino acids. They are: glycine, DL-α-alanine (2-aminopropanoic acid), DL-α-aminobutyric acid (2-aminobutanoic acid), DL-α-norvaline (2-aminopentanoic acid), DL-α-norleucine (2-aminohexanoic acid), β-alanine (3-aminopropanoic acid), γ-aminobutyric acid (4-aminobutanoic acid), 5-aminovaleric acid (5-aminopentanoic acid), and 6-aminocaproic acid (6-aminohexanoic acid).

The compounds selected have a linear hydrocarbon chain that increases its length by one methylene group allowing the application of additivity approaches to the properties at infinite dilution of these systems.

4.1 INTRODUCTION

Amino acids are organic compounds that contain a basic (-NH_2) and an acid carboxyl (-COOH) functional groups, along with a generic side chain (R group) characteristic of each amino acid; they can contain other functional groups.

Several classifications of amino acids have been proposed according to their characteristics such as polarity, side chain group, hydrophobicity and the locations of the principal functional groups as alpha- (α-), beta- (β-), gamma- (γ-) or delta- (δ-) amino acids. When amino acids have the amine group bound to the first (alpha-) carbon atom adjacent to the carboxylic group, they are called 2-, or α-amino acids. When the amine group is attached to the other terminal carbon atom, called the omega (ω) carbon atom, they are named α,ω-amino acids. The alpha carbon is a chiral carbon atom. The alpha-amino acids are the most common form found in nature, but only in the L-isomer with the exception of glycine that does not has a chiral carbon atom.[1–3] According to the side chain, the amino acids can be divided into different categories, classified as aliphatic, hydrophobic, aromatic, polar or charged.

Aliphatic amino acids in aqueous solution have amphiprotic properties. The pH at which amino group is protonated and the carboxyl deprotonated is known as the isoelectric point; at this pH, the net charge is zero. When the amino acids contain only one amine and one carboxylic acid functional groups, the isoelectric point is close to the pH of water, so these compounds exist in aqueous solution in the form of zwitterions or dipolar ions. As a result, these solutes exhibit in water negative carboxylate and positive α-ammonium groups; this fact has important consequences in their behavior.

Amino acids are the structural building units of proteins. They are attached to one another by peptide bonds, forming a chain. More than 500 amino acids occur naturally but only 22 of them are common constituents of proteins essential for life; for this reason, they are considered as essential amino acids; of these, 20 are encoded by the universal genetic code. Other known amino acids are not found in proteins or are not produced by regular cellular processes.

All of the protein amino acids are available commercially. Due to their characteristic properties, their applications include diverse fields such as additives to foods and animal feeds, medical products, cosmetics, synthetic polymers, detergents, etc.[2,4] They are also used as precursors to obtain chemicals used in various industries that include the production of drugs, biodegradable plastics, chiral catalysts and they are also used in agricultural processes involving pesticides and herbicides.

In the food industries, a number of amino acids have been widely used as flavor enhancers and flavor modifiers and to improve the nutritional value of a protein by the addition of amino acids which are scarce in a particular protein.[1] They are used to chelate metal cations in order to improve the absorption of minerals from supplements, which may be required to improve the health or production of animals.[2–5]

The study of thermodynamic properties for substances of biological interest in aqueous solutions has received increased attention in the last years. The information about structural changes derived from the properties of amino acids in aqueous solution contributes to the understanding of their behavior in aqueous solution and the elucidation of the nature of interactions between nonpolar and polar groups with water. Besides, they are of special importance to improve its practical applications and for the information that can be obtained about the factors affecting protein stability in aqueous solutions.[6–10]

The aqueous solutions of aliphatic amino acids have a special interest. Their behavior is not only determined by the electrostatic interactions between the ionic groups and water molecules. The presence of the hydrocarbon chain is responsible for the hydrophobic interactions between the nonpolar chain and water. The behavior of their solutions clearly depends on the number of methylene groups of the hydrocarbon chain, as well as in the position of the ionic groups.

The aim of writing this chapter was to contribute to the understanding of the molecular interactions governing the behavior of biomolecules in aqueous environments. This review surveys the partial molar volumes, the partial molar compressibilities, the viscosity *B* coefficients and diffusion coefficients of aliphatic amino acids in aqueous solution at several temperatures.

4.1.1 AMINO ACIDS: MODEL COMPOUNDS

Several studies have been carried out in relation to the noncovalent interactions that occur between solute molecules and the solvent and the role they play in biological systems. They seek to understand the energy balance of the forces involved in stabilization of the native protein

However, the characterization of the hydration thermodynamic properties of biomolecules is complex. For this reason, the use of small model solutes in aqueous solution provides elements to understand the interactions involved in aqueous solutions.[8–10]

Hydrophobic interactions have been considered to be the most important driving force for many biochemical processes. However, hydrophilic interactions also play a fundamental role in processes, such as protein folding. For this reason, the use of model compounds that contain both polar and nonpolar groups in water such as amino acids, amides, and alcohols have provided important information in relation to the interactions involved in protein stability.[11,12]

The dependence with the concentration of a property is a measure of the solute–solute interactions, whereas the extrapolation to infinite dilution of the same property provides information about the solute-solvent interactions.

In solution, at low solute concentrations, the dominant interactions are the solute–solvent interactions. The most relevant forces that determine the strength and specificity of noncovalent interactions in aqueous solutions are electrostatic interactions, London dispersion forces, and hydrophobic interactions. Electrostatic interactions are strong, long-range interactions and comprise ion–ion, ion–dipole, dipole–dipole interactions, as well as hydrogen bonds.[11,12]

When a solute dissolves in the water this solute is surrounded by a hydration shell or sphere formed by water molecules. Water in the hydration sphere has different properties from bulk water and these properties depend on the solute nature.

The interaction between the nonpolar solutes and water is called hydrophobic hydration and the interaction between the polar or ionic groups with water is referred to as hydrophilic hydration. The addition of nonpolar surfaces, strengthen the structure of the water molecules in the hydration shell due to the hydrophobic character of the solute. This structuring effect is called a forming effect of water structure. In the case of mixed solutes, two different hydration shells are formed: One hydrophilic shell around the polar group or groups and a hydrophobic shell around the nonpolar region.[11–13]

When the concentration of solute molecules becomes larger, the interaction between solute molecules becomes important. The interaction between solvated nonpolar groups or molecules in the presence of water is called hydrophobic interaction.

The hydrophobic interaction refers to the tendency to aggregate that the nonpolar species or molecules of a solute with important apolar regions present in aqueous solution, giving rise to the apolar–solute/apolar–solute interaction. This type of interaction is entropically determined.[11–12]

The hydrophobic interaction is an effective force that acts to minimize the surface area that apolar molecules expose to water and is a consequence of the strong interactions by hydrogen bonds between water molecules, which are disturbed by the insertion of a nonpolar solute into the solvent.

Proteins are macromolecules composed of amino acid residues bonded by peptide bonds. Protein hydration plays a fundamental role to maintain their three-dimensional structure. The interpretation of solute–solvent and solute–solute interactions are fundamental in the study of protein stability. However, this work is limited by the structural complexity of macromolecules, such as proteins and nucleic acids, as they have many and diverse functional groups.

The structure of amino acids and, in particular, the functional groups they have, make these compounds ideal model systems for the study of larger molecules such as biomolecules in aqueous solution.[13–16] Consequently, the thermodynamic properties of aqueous solutions of model compounds such as amino acids provide elements to understand the behavior of proteins in water and the factors affecting protein stability.

For this reason, the physicochemical properties of amino acids in aqueous solutions have been extensively studied using different techniques. Many studies have been done on naturally occurring α-amino acids but the information related to α,ω-amino acids is not so abundant and, moreover, a few works have been reported at temperatures different from 298.15 K. Despite the information reported in the literature, the behavior of this type of solutes in water is not well understood due to the complexity of solute–water and solute–solute interactions that take place in aqueous solutions. The presence of hydrophilic and hydrophobic hydration regions around polyfunctional molecules, such as amino acids, makes a complex issue to study the hydration of these molecules.

Some properties are related to structural changes in aqueous solutions so they have special relevance for probing the hydration of the solutes in water and in aqueous solutions. Important information is reported on the calorimetric properties,[13–16] on the partial molar volumes[17–35] and compressibilities[17–19,21–24,27] of amino acids in aqueous solution. The group additivity approaches used to describe the behavior of thermodynamic properties has been considered by several authors.[13–16,30–32] Other properties that are important in the description of the behavior of amino acids in aqueous solutions are the transport properties: Viscosity[21–23,26–29,36–44] diffusion.[45–56] The information has been used to provide an insight into hydrophobicity, hydration properties, and solute–solvent interactions in aqueous solutions. Most of these studies refer to naturally occurring small α-amino acids at 298.15 K and very few data are reported in the literature for α,ω-amino acids especially at temperatures different from 298.15 K.

Besides the practical interest in the behavior of amino acids in water and in aqueous mixed solvents mentioned in the introduction, the physicochemical properties of its aqueous solutions are of considerable interest for the study of hydrophobic and hydrophilic interactions between these solutes and water and their effect on the water structure.[8–14] In particular, volumetric, molar compressibility, viscometric, and diffusion properties of aqueous solutions of amino acids are important because they are sensitive to small changes in solution behavior giving information on solute hydration and on the nature of interactions in aqueous solution. This information is also fundamental to understand which type of interaction is dominant and determines the native conformation of a protein.[6–10]

For this review the amino acids selected were: Glycine, DL-α-alanine (2-aminopropanoic acid), DL-α-aminobutyric acid (2-aminobutanoic acid), DL-α-norvaline(2-aminopentanoicacid),DL-α-norleucine(2-aminohexanoic acid); the α,ω-amino acids were: β-alanine (3-aminopropanoic acid), aminobutyric acid (4-aminobutanoic acid), aminovaleric acid (5-aminopentanoic acid), and aminocaproic acid (6-aminohexanoic acid). In the two series, the amino acids differ in the length of the alkyl chain and the comparison between the behavior of both series will allow the analyses of the effect of the position of the polar groups.

4.1.2 AMINO ACID HYDRATION

Solute hydration of polar molecules that have alkyl groups is characterized by the strong tendency of water molecules to organize themselves around nonpolar surfaces, and the interaction between the polar groups with water, usually by hydrogen bonds. The resulting behavior is a consequence of the balance between the two interactions: Hydrophobic hydration and hydrophilic hydration. This fact is evidenced by the properties of the infinitely diluted solution.

When the interaction between the polar group and water dominates over the interaction between the apolar chain and water, the enthalpic term ΔH dominates over the entropic term $T\Delta S$. If the alkyl chain is long enough, the interaction between the apolar chain and water can dominate over the polar group–water interaction and the entropic term TΔS can compensate and even can become larger than the enthalpic term. This is the more general criteria to characterize the hydrophobia hydration $|T\Delta S| > |H|$.[11,12]

Amino acids can be considered as mixed solutes containing highly hydrophilic groups (amine and carboxyl) and the hydrophobic aliphatic chain. The physicochemical properties of α-amino acids and α,ω-amino acids exhibit very different behavior due to the different hydration of the molecules which is a consequence of the effect of the position of the polar groups.

α-Amino acids in aqueous solution have a polar region and a hydrocarbon nonpolar residue exposed to water. The vicinal charged hydrophilic groups are well separated from the hydrophobic domain. As a consequence, there are two clear hydration zones: A hydrophilic zone and a hydrophobic zone that can overlap. The areas of overlap are not large and for this reason, the action of various functional groups is sufficiently localized. This allows considering the action of the different factors on the effects of solvation as a superposition of them in the zones of polar and nonpolar hydration.

α,ω-Amino acids have two polar regions located at the extremes of the molecule and a hydrocarbon nonpolar residue between them; the separation of the ionic groups increases the polarity of the molecule and causes a decrease of its hydrophobic character.

When the concentration of the amino acid increases, the effect among the solvated solute molecules can be interpreted as the result of hydrophobic interactions involving solvated nonpolar chains, hydrophilic interactions between hydrated polar groups and the partial dehydration of the solvation layers.[6–8,13–15]

4.1.3 ADDITIVITY OF GROUP CONTRIBUTIONS

Several studies have suggested that the thermodynamic properties of aqueous solutions of nonpolar and mixed organic compounds can be expressed in terms of the contribution of the groups that form the solute molecule.

The group contribution models assume that a molecule can be reduced to the sum of its structural elements, these elements being independent of each other. They are used to predict thermodynamic properties that are determined from an appropriate combination of the contributions of the different groups that make up a molecule.[11–14,67–73]

The simplest models to describe the additivity of the solute–solvent interactions assume that the contribution of the methylene groups is independent of the group or polar constituent groups and, consequently, it must be the same in the different organic compounds. The additive character of the properties at infinite dilution with the increase of the hydrocarbon chain is based on the assumption that each methylene group interacts with the solvent independently of the other methylene groups. This method has been used specially to determine volumetric and heat capacity contributions of methylene and functional groups.[25, 30–32]

Other approaches have been used to describe the interactions between solvated solute molecules in terms of group contributions. The Savage–Wood (SWAG) additivity principle assumes that each group can interact independently with the other functional groups and the interaction is independent of the positions of the functional groups in the two solute molecules, and their relative position or stereochemistry in each molecule. This allows the calculation of the properties as the sum of the contributions of all possible interactions. This model neglects intramolecular interactions with neighboring groups. The Savage-Wood (SWAG) approach has been used to predict different properties such as enthalpy, entropies and Gibbs energy.[13–15,57,58]

4.2 VOLUMETRIC PROPERTIES OF AQUEOUS SOLUTIONS OF AMINO ACIDS

Volumetric properties are very sensitive to solute–solvent interactions. The limiting partial molar volumes at infinite dilution give information about solute hydration and solute–solvent interactions while the dependence of the partial molar volumes on concentration provides information about solute–solute interactions. The influence of temperature on the behavior of volumetric properties of aqueous solutions has been used to obtain information about solute hydration and the balance between hydrophobic and hydrophilic interactions between solute and water and has particular significance for industrial and practical applications.

Most studies on volumetric properties of amino acids have been done on naturally occurring amino acids at 298.15 K. However, fewer data are reported in the literature for amino acids such as norvaline and norleucine in spite of their linear hydrocarbon chain, and for α,ω-amino acids at temperatures different from 298.15 K.[17–28]

The partial and the apparent molar volumes of the amino acids in solution can be determined from density measurements using different techniques such as picnometry, magnetic float densimetry, and vibrating tube densimetry. The last method is one of the more precise methods and it is frequently used.

The apparent molar volume, φ_v, can be calculated from density data using eq (4.1):

$$\varphi_v = \frac{M}{\rho} + \frac{(\rho_0 - \rho)}{m\rho\rho_0} \tag{4.1}$$

where M is the molar mass of the solute, m is the molal concentration of the aqueous solution, ρ_0 is the density of the solvent, and ρ is the solution density.

The dependence of the apparent molar volumes of the amino acids with the concentration at the working temperatures is fitted by least squares method to obtain an equation that describes the behavior of the apparent molar volumes of the amino acids. Concentration is expressed in molality, m, and the experimental data usually are fitted to second-order polynomial equations as it is shown by eq (4.2). In this equation, a and b are empirical constants and $\overline{V}_2^{\infty}$ is the partial molar volume at infinite dilution.

$$\varphi_v = am^2 + bm + \overline{V}_2^{\infty} \tag{4.2}$$

For both series of amino acids at each temperature, it is observed that the partial molar volume at infinite dilution is a linear function of the number of carbon atoms of the hydrocarbon chain. As a result, the partial molar volume at infinite dilution of one of these amino acids can be described in terms of group contributions according to a simple methylene additivity approach based on the dependence of the property on the number of carbon atoms or the number of methylene groups. According to Hakin, the number of carbon atoms is approximately equivalent to the number of methylene groups,[25] and thus the partial molar volume at infinite dilution can be represented by the following eq (4.3):

$$\overline{V}_2^0 = \overline{V}_2^0(NH_3^+, COO^-) + n\overline{V}_2^0(CH_2) \tag{4.3}$$

where n represents the number of methylene groups, $\overline{V}_2^0(CH_2)$ and $\overline{V}_2^0(NH_3^+, COO^-)$ are the volumetric contributions of the amino and carboxylic groups to the partial molar volume of the amino acid.[17–28]

The behavior of the partial molar volumes at infinite dilution of α-amino acids and α,ω-amino acids as a function of the number of methylene groups is presented in Figure 4.1. Both series of compounds follow a similar behavior. The dependence of the partial molar volumes at infinite dilution of α-amino acids and α,ω-amino acids with the number of methylene groups is well represented by linear equations indicating that the volumetric contribution of the methylene group is additive.

The volumetric properties of amino acids in aqueous solutions show that the partial molar volumes at infinite dilution of α,ω-amino acids in aqueous solutions are smaller than the corresponding values for α-amino acids at the same temperatures. This suggests that solute–solvent polar interactions are higher for α,ω-amino acids.[17–28]

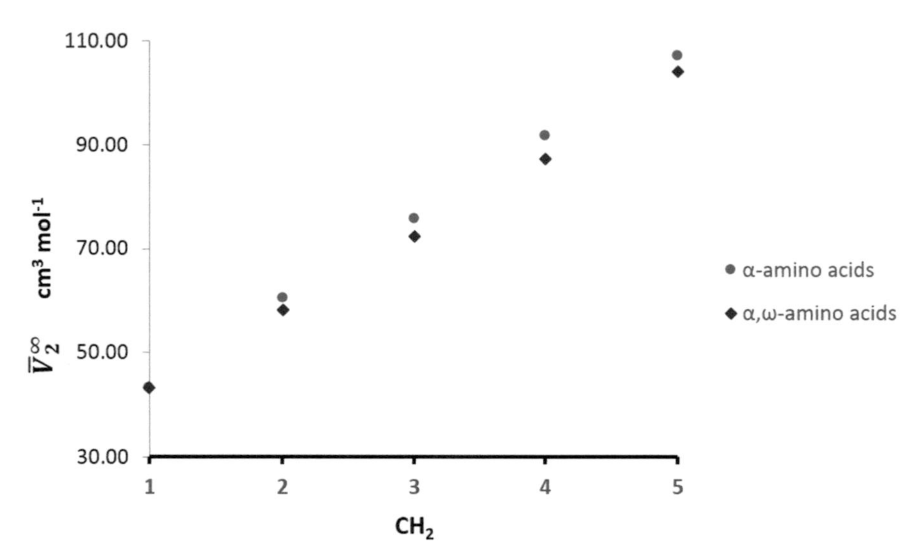

FIGURE 4.1 Comparison of the partial molar volumes at infinite dilution of α-amino acids and α,ω-amino acids as a function of the number of methylene groups.[20–24]

Table 4.1 presents a comparison of the volumetric contributions of the methylene group and the polar carboxylic and amino groups in water determined in previous works at several temperatures.

TABLE 4.1 Volumetric Contributions of the Methylene, Amino, and Carboxylic Groups to the Partial Molar Volume of α-Amino Acids and α,ω-Amino Acids.

T / K	$\bar{V}_2^0$ (CH_2) / ($cm^3 mol^{-1}$)		$\bar{V}_2^0$ (NH_3^+, COO^-) / (cm^3 mol^{-1})	
	α-Amino acids[a]	α,ω-Amino acids[b]	α-Amino acids[a]	α,ω-Amino acids[a]
293.15	15.93	15.39	27.47	26.50
298.15	16.00	15.34	27.76	26.95
303.15	16.03	15.41	28.11	27.01
308.15	16.08	15.41	28.41	27.18

[a]Ref. [22].
[b]Ref. [21].

The group contributions for α-amino acids at 298.15 K are similar to the volumetric contributions corresponding to other homologous series such as alcohols, amides, and other model solutes.[30–32] The group contributions for α,ω-amino acids are considerably smaller than the contributions reported for the same groups in the case of α-amino acids.[14–17,19] This has been considered as a consequence of a larger electrostrictive effect due to charge separation.[33] The results suggest that the assumption of the group contribution approaches that the properties of a molecule can be represented by the sum of its structural elements, which are independent of each other is valid for homologous series of compounds. The approach cannot be generalized and the group contribution depends on other factors such as the group position.

For the two series of solutes, the value of the group volumetric contributions increases slightly with temperature in the range considered in this work.[17–28]

The electrostrictive effect or the solute–solvent interaction associated with electrostatic interactions between the charged groups and water has been used to evaluate the electrostriction volume and the number of hydrated water molecules based on the simple model proposed by Millero.[17,28] According to it, the infinite dilution partial molar volume can be described as the sum of two contributions: the intrinsic volume and the electrostriction volume. For α-amino acids, electrostriction volumes, and hydration numbers become smaller as the temperature increases and with increasing the chain length. This shows that the electrostrictive effect decreases with both temperature and the chain length and so the number of water molecules hydrated predominantly by polar groups.

Another way to evaluate electrostriction is based in the scaled particle theory, by calculating the interaction volumes, using the approach presented by Chalikian.[18] The results obtained for the α,ω-amino acids 3-aminopropanoic, 4-aminobutanoic, and 5-aminopentanoic acid show that the interaction volumes become more negative with the increase in surface area and the number of methylene groups, suggesting that solute–solvent interactions become stronger.[20] The trend changes for 6-aminohexanoic acid and the interaction volume becomes less negative suggesting that the surface area is smaller. The same behavior has been reported by Chalikian for α,ω-amino acids[18] and could be explained as a result of polar–polar interactions between amino and carboxylic groups strongly hydrated that can participate in intramolecular hydrogen bond interactions.

The influence of the temperature on the volumetric properties of aqueous solutions has been used to obtain information about solute hydration and the balance between hydrophobic and hydrophilic interactions between solute and water. For both series of amino acids, the effect of temperature is small and, in all cases, partial molar volumes become slightly larger as the temperature increases.

The dependence of the partial molar volumes of both series of amino acids in water is linear, which is probably a result of the narrow range of temperatures considered. As a result of this, the partial molar expansibilities obtained are constant. The dependence of the partial molar volumes at infinite dilution with the temperature is represented by the partial molar expansibility $\overline{E}_2^{\infty}$ that can be determined using the eq (4.4).

$$\overline{E}_2^{\infty} = \frac{d\overline{V}_2^{\infty}}{dT} \tag{4.4}$$

Table 4.2 shows the partial molar expansibilities for amino acids in water.

TABLE 4.2 Partial Molar Expansibilities of α-Amino Acids and α,ω-Amino Acids in Aqueous Solution.

Amino acid	$\overline{E}_2^{\infty}$ **cm^3·mol^{-1}K^{-1}**
Glycine[1]	0.088
2-Aminopropanoic acid[a]	0.065
3-Aminopropanoic acid[b]	0.061
2-Aminobutanoic acid[a]	0.086
4-Aminobutanoic acid[b]	0.045
2-Aminopentanoic acid[a]	0.098
5-Aminopentanoic acid[b]	0.075
2-Aminohexanoic acid[a]	0.130
6-Aminohexanoic acid[b]	0.053

[a]Ref. [22].

[b]Ref. [21].

These results agree well with the values reported in the literature. For α-amino acids in water, the expansibilities are very small and positive.[17,19,22,24–26] For α,ω-amino acids the expansibilities are also positive and their values are smaller than those found for α-amino acids in water.[18,20,21,23]

4.3 ADIABATIC COMPRESSIBILITIES OF AQUEOUS SOLUTIONS OF AMINO ACIDS

Compressibility is a physical property that is very sensitive to solute hydration and thus reflects intermolecular interactions in solution. The apparent molar adiabatic compressibilities, K_ϕ, of the amino acids in solution, are determined from both the density, ρ, and the adiabatic compressibility, β_S, of the corresponding solution by using the eq (4.5):

$$K_\varphi = \frac{M\beta_S}{\rho} + \frac{\left(\beta_S\rho_0 - \beta_S^0\rho\right)}{m\rho\rho_0} \tag{4.5}$$

where β_S^{o} is the adiabatic compressibility of the solvent. The adiabatic compressibility, β_S, of the aqueous solution, is calculated from both the sound speed, u, and the density measurements using the Newton-Laplace relationship represented by equation (4.6):

$$\beta_S = 1/\rho u^2 \tag{4.6}$$

The dependence of the apparent molar compressibilities of the amino acids with molality is fitted, by a least-squares method, usually to a second-order polynomial equation. In eq (4.7), a and b are empirical coefficients and $\overline{K}_2^\infty$ is the partial molar compressibility at infinite dilution

$$K_\varphi = am^2 + bm + \overline{K}_2^\infty \tag{4.7}$$

For the two series of amino acids considered, the partial molar compressibilities of the solutes in water are negative at all temperatures and become larger as the number of methylene groups of the nonpolar chain of the amino acid increases.[21–24] The results presented agree well with other literature values.[17,18]

The partial molar compressibilities at infinite dilution, $\bar{K}_2^\infty$, of α-amino acids and α,ω-amino acids in aqueous solutions at 298.15 K are shown in Table 4.3.

TABLE 4.3 Partial Molar Compressibilities at Infinite Dilution of α-Amino Acids and α,ω-Amino Acids in Aqueous Solutions at 298.15 K.

Amino acid	$-\bar{K}_2^0$ 10^4 $cm^3 \cdot mol^1 \cdot GPa^{-1}$
Glycine	25.10
2-Aminopropanoic acid[a]	25.16
3-Aminopropanoic acid[b]	26.04
2-Aminobutanoic acid[a]	26.56
4-Aminobutanoic acid[b]	30.80
2-Aminopentanoic acid[a]	30.03
5-Aminopentanoico acid[b]	33.76
2-Aminohexanoico acid[a]	34.52
6-Aminohexanoic acid[b]	36.78

[a]Ref. [24].
[b]Ref. [21].

The compressibilities at infinite dilution become less negative as the temperature increases. The dependence of the partial molar compressibilities of the amino acids with the temperature adjusts well to second-order polynomial equations.

Figure 4.2 shows the behavior of the partial molar compressibilities of the amino acids in water at 298.15 K.

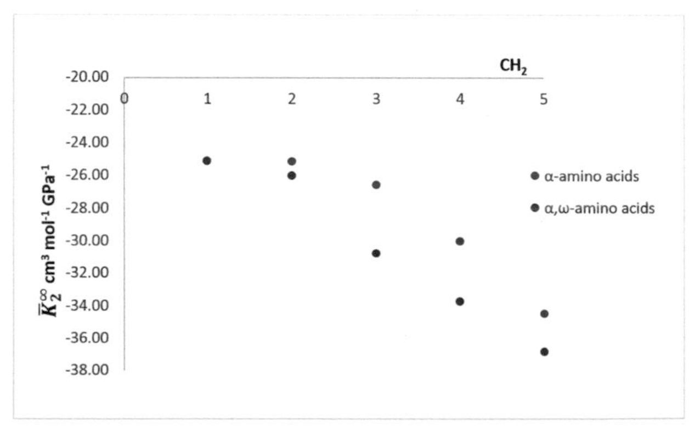

FIGURE 4.2 Comparison of the partial molar compressibilities at infinite dilution of α-amino acids and α,ω-amino acids as a function of the number of methylene groups at 298.15 K.

The behavior of the thermodynamic properties: Partial molar volumes and partial molar compressibilities at infinite dilution of the α-amino acids and α,ω-amino acids in aqueous solution, agree with the results reported in the literature.[17–19,25–27] They confirm that glycine and alanine behave as hydrophilic solutes, whereas interactions between aminobutyric acid, norvaline, and norleucine and water are mainly hydrophobic and become enhanced as the number of methylene groups of the hydrocarbon chain becomes larger. Similar behavior is observed for the series of the α,ω-amino acid in water.

4.4 VISCOSITY OF AMINO ACIDS IN AQUEOUS SOLUTIONS

Viscometric properties of aqueous solutions provide valuable information for the hydrophobic interactions of nonpolar side chains with water and the hydrophilic interactions between polar groups and water.

Several studies on the viscosities of α-amino acids in aqueous solutions have been performed.[21–23,28,30,41–48] However, less information is reported in the literature for α,ω-amino acids in aqueous solutions, in particular at temperatures different from 298.15 K.[21–23,,41–48]

Viscosity, η, can be determined by using the Hagen–Poiseuille equation which is applicable to Newtonian incompressible fluids flowing in laminar regime:

$$\eta = \frac{\pi r^4 \rho g h t}{8Vl} \tag{4.8}$$

where h is the height of the capillary, g is the gravity constant, ρ is the density of the liquid, r and l are the radii and the length of the capillary tube, respectively, V is the volume of the liquid that flows through the capillary tube and t is the efflux time.

The viscosity of aqueous solutions can be determined by using several techniques, but one of the most accurate methods uses the suspended-level Ubbelohde capillary viscometers. These instruments have the advantage that the values obtained are independent of the total volume. They measure the efflux time of the liquid through a capillary tube at a given temperature.

The relative viscosities $\eta_r = \eta/\eta_o$ are calculated from the viscosity values of both the solution, η, and the solvent, η_o. The values obtained as a function of the molality are adjusted by a least-squares method to a second-order polynomial equation as has been proposed by Tsangaris–Martin for mixed solutes[39]:

$$\eta_r = 1 + Bm + Dm^2 \tag{4.9}$$

In this equation, *B* and *D* are empirical coefficients. It has been suggested by several authors that the *B* coefficient depends on the size, shape, and charge of the solute molecule and is sensitive to the solute–solvent interactions while the *D* coefficient has been related to the solute-solute interactions.[36–40,43] According to them, large and positive *B* values are related to structure-making effects of strongly hydrated solutes on water structure, whereas small *B* positive or negative values are considered to reflect structure-breaking effects of hydrophilic solutes on water structure.

The viscosity of α-amino acids and α,ω-amino acids solutions at all temperatures adjust well to the Tsangaris–Martin equation and its values increase with the amino acid concentration.

Table 4.4 presents the comparison between the *B* coefficients for α-amino acids and α,ω-amino acids aqueous solutions at 298.15 K.[28,21,43]

TABLE 4.4 Viscosity *B* Coefficients of α-Amino Acids and α,ω-Amino Acids in Aqueous Solutions at 298.15 K.

n_{CH2}		**B/ kg mol^{-1}**
	α-amino acids	**α,ω-amino acids**
1	0.143	0.143
2	0.232	0.190
3	0.299	0.270
4	0.371	0.327
5	0.454	0.414

The amino acids have small positive *B* coefficients. The coefficients for the α-amino acids are slightly larger than the *B* coefficients for the α,ω-amino acids in aqueous solutions and this can be a consequence of the larger hydrophobic surface exposed to the solvent in the case of the α-amino acids and the hydrophilic character of the α,ω-amino acids.[21–23,26–29,37–43] The coefficients of the amino acids in water increase as the number of methylene becomes larger, indicating that the magnitude of the coefficients is a consequence of the increase in the hydrophobic character of the amino acid as the hydrocarbon chain length increases.

The viscosity *B* coefficients present a linear dependence on the number of methylene groups so they can be analyzed in terms of group additivity expressing the values as the contribution of polar groups and number of CH_2 units as shown in Figure 4.3.

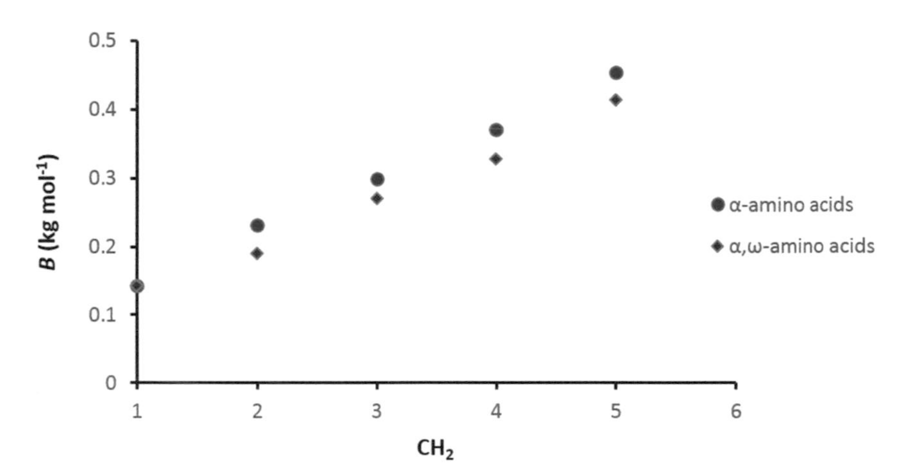

FIGURE 4.3 Comparison of the viscosity *B* coefficients of α-amino acids and α,ω-amino acids as a function of the number of methylene groups at 298.15 K.[21–23,28,41]

For all the amino acids with the exception of glycine, the viscosity decreases as the temperature increases, following the behavior previously reported for most systems, including amino acids.[26–29,37–44] The smallest amino acid, glycine, shows a different behavior with temperature and the *B* coefficient shows a slight increase with temperature as has been reported by other authors.[37–43] This result has been considered a consequence of its hydrophilic behavior.

The first derivative of the *B* coefficient with respect to the temperature, dB/dT, has been used as an indication of the effect of the solute on water structure.[43] The negative sign for the derivative has been attributed to a structure-making effect, while the positive sign to a structure-breaking effect on the water structure. In this study, all the solutes considered show negative values for dB/dT except glycine, confirming its high hydrophilic character and the breaking-structure effect that it has on the water structure.

4.5 DIFFUSION STUDIES IN AQUEOUS SOLUTIONS OF AMINO ACIDS

The mutual diffusion coefficient describes transport based on diffusion, that is, the flow of molecules or atoms from a region of high concentration to a region of low concentration, through an irreversible process. The diffusion process in liquids can be described using the Fick laws, being the diffusion coefficient of each species present in the system a quantitative measurement of this process.[45,46]

One of the most used methods to determine the diffusion coefficients in liquids is the Taylor dispersion one which has been described in several studies. The method is based on the dispersion of a small amount of solution injected into a laminar carrier stream of solvent or solution flowing through a capillary tube.[47,56] The experimental mutual diffusion coefficients, D, of the solutes in aqueous solution are fitted by using a least-squares method to a second-order polynomial relationship to obtain, by extrapolation, the value for the limiting diffusion coefficient at infinite dilution.

The information about the diffusion of amino acids in water is very scarce especially for the diffusion coefficients of α,ω-amino acids in water.[48,51]

TABLE 4.5 Limiting Diffusion Coefficients and Hydrodynamic Radius for α- and α,ω-Amino Acids in Water at 298.15 K.

Amino acid	$D^0_W/10^{-9}$ m²s⁻¹	R_H/nm
Glycine	1.061[a]	0.231
2-Aminopropanoic acid	0.930[a]	0.264
2-Aminobutanoic acid	0.839[a]	0.292
4-Aminobutanoic acid	0.838[b]	0.293
2-Aminopentanoic acid	0.774[a]	0.317
5-Aminopentanoic acid	0.768[b]	0.319
2-Aminohexanoic acid	0.737[a]	0.333
6-Aminohexanoic acid	0.715[b]	0.343

[a]Ref. [48].
[b] Ref. [21].

The diffusion coefficients D for α-amino acids in water increase with the concentration. The coefficients at infinite dilution, D^o, are strongly dependent on the length of the hydrocarbon chain. The coefficients decrease as the number of methylene groups of the hydrocarbon chain increases as it is shown in Table 4.5.

The coefficients are very sensitive to temperature changes. They become larger as the temperature increases.[48,49] The same behavior is observed for aqueous solutions of α,ω-amino acids. The values obtained for α,ω-amino acids in the water at 298.15 K are slightly lower than those for α-amino acids.

These results agree with the data presented by Umecky et al.[48] at 298.25 K; the limiting diffusion coefficients at infinite dilution depend on the number of carbon atoms of the hydrocarbon chain. The hydrodynamic radius is not very sensitive to the carbon chain length neither to the position of the charged groups. However, a small increase in the number of methylene groups is observed.

4.6 CONCLUSIONS

A comparative study is presented of the partial molar volumes, the partial molar compressibilities at infinite dilutions, the viscosities and the diffusion coefficients for the α-amino acids: Glycine, 2-aminopropanoic acid (alanine), 2-aminobutanoic acid (aminobutyric acid), 2-aminopentanoic acid (norvaline), and 2-aminohexanoic acid (norleucine) and for the α,ω-amino acids: 3-aminopropanoic acid (β-alanine), 4-aminobutanoic acid (γ-aminobutyric acid, GABA), 5-aminopentanoic acid, and 6-aminohexanoic acid in water.

The results from the partial molar volumes and compressibilities at infinite dilution for the α-amino acids in water show that glycine and alanine behave as hydrophilic solutes, whereas the character of the interaction between aminobutyric acid, norvaline and norleucine and water is mainly hydrophobic. The experimental results for α,ω-amino acids confirm they exhibit a hydrophilic behavior.

The group contributions for α-amino acids are similar to the volumetric contributions corresponding to other homologous series; however, the volumetric contributions for α,ω-amino acids are considerably smaller. This is a consequence of a larger electrostrictive effect due to the increase in the distance between charges in the case of α,ω-amino acids. The results suggest that the assumption of the group contribution methods that the properties of a molecule can be represented by the sum of its structural elements, which are independent of each other, is valid for homologous series of compounds. The approach cannot be generalized and the group contribution depends on other factors such as the group position.

The partial molar compressibilities, $\bar{K}_2^{\infty}$, of the amino acids in water at all temperatures are negative and become larger as the number of methylene groups of the nonpolar chain of the amino acid increases. The dependence of the partial molar compressibilities of the α-amino acids with temperature is well represented by a second-order polynomial equation, indicating that $\bar{K}_2^{\infty}$ is more sensitive to the changes of temperature than $\bar{V}_2^{\infty}$, since $\partial\bar{K}_2^{\infty}/\partial T$ depends on the temperature, while $\bar{E}_2^{\infty}$ is a constant in the temperature range considered.

The amino acids have small positive viscosity *B* coefficients. The coefficients for the -amino acids are slightly larger than the *B* coefficients for the α,ω-amino acids in aqueous solutions and this can be a consequence of the larger hydrophobic surface exposed to the solvent in the case of the α-amino acids and the hydrophilic character of the α,ω-amino acids. The coefficients of the amino acids in water increase with the length of the hydrocarbon chain and present a linear dependence on the number of methylene groups.

The values of the diffusion coefficients at infinite dilution obtained for α,ω-amino acids in water at 298.15 K are slightly lower than those for α-amino acids. It is important to highlight that the limiting diffusion coefficients at infinite dilution depend on the number of carbon atoms of the hydrocarbon chain following a similar behavior of the trend exhibited by α-amino acids. They decrease as the number of methylene groups of the hydrocarbon chain increases.

For both series of solutes, the diffusion coefficients at infinite dilution, D^{o} are strongly dependent on the length of the hydrocarbon chain. The coefficients decrease as the number of methylene groups of the hydrocarbon chain increases.

The most important effect of the interaction between ions is the elimination of water molecules from the hydration sphere of the amino acids toward the dissolution, reducing the electrostriction

KEYWORDS

- **α-amino acids**
- **α**
- **ω-amino acids**
- **thermodynamic properties**
- **transport properties**
- **aqueous solutions**

REFERENCES

1. Hardy, P. *Chemistry and Biochemistry of the Amino Acids;*. Barret Chapman, G. C. Ed.; Hall-Methuen: London, 1985.
2. Araki, K.; Ozeki, T. Amino Acids. *Kirk-Othmer Encyclopedia of Chemical Technology*, 4th ed; Wiley-Interscience: Hoboken, New Jersey, 2003.
3. Cohn, E. J.; Edsall, J. T. *Proteins, Amino Acids, and Peptides*, Reinhold Publishing Corporation, New York, 1943.
4. Izumi, Y.; Chibata, I.; Itoh, T. Production, and Utilization of Amino Acids. *Angew. Chem. Int. Ed.* **1978,** *17*, 176–183.
5. Davis, J. Amino Acid Uses in Industry. News Medical Life Sciences. https://www.news-medical.net/life-sciences/Amino-Acid-Uses-in-Industry.aspx. (Accessed 22 April 2019)
6. Cohn, E. J.; Edsall, J. T. *Proteins, Amino Acids, and Peptides as Ions and Dipolar Ions*. Hafner Publishing Company: New York; 1965.
7. Pace, C. N.; Fu, H.; Fryar, K. L.; Landua, J.; Saul, R. T.; Shirley, B. A.; Hendricks, M. M.; Limura, S. K.; Scholtz, J. M. Contribution of Hydrophobic Interactions to Protein Stability. *J. Mol. Biol.* **2011,** *408*, 514–528.
8. Scharnagl, C.; Reif, M.; Friedrich, J. Stability of Proteins: Temperature, Pressure and the Role of the Solvent. *Biochim. Biophys. Acta*. **2005,** *1749*, 187–213.
9. Zweifel, M. E.; Barrick, D. Relationships between the Temperature Dependence of Solvent Denaturation and the Denaturant Dependence of Protein Stability Curves. *Biophys. Chem.* **2002,** *101*, 221–237.
10. Cooper, A. Thermodynamics of Protein Folding and Stability. In *Protein: A Comprehensive Treatise;* Allen, G., Ed.,. JAI Press Inc: Stanford; 1999, Vol 2, pp. 217–270.
11. Franks, F. Water: A Matrix of Life. *Royal Society of Chemistry*, 2nd ed; Cambridge, 2000.
12. Ben-Naim, A. *Hydrophobic Interactions*, Plenum Press: New York, 1980.
13. Lilley, T. H. Interactions in Solutions: The Interplay Between Solute Solvation and Solute-Solute Interactions. *Pure Appl. Chem.* **1994,** *66*, 429–434.
14. Lilley, T. H. *The Chemistry and Biochemistry of Amino Acids*; Barret Chapman, G. C., Ed.; Hall-Methuen: London, 1985.
15. Palecz, B. Enthalpic Homogeneous Pair Interaction Coefficients of L-α-Amino Acids as a Hydrophobicity Parameter of Amino Acid Side Chains. *J. Am. Chem. Soc.* **2002,** *124* (21), 6003–6008.
16. Romero, C. M.; Cadena, J. C.; Lamprecht, I. Effect of Temperature on the Dilution Enthalpies of α,ω-Amino Acids in Aqueous Solutions. *J. Chem. Thermodyn.* **2011,** *43*, 1441–1445.
17. Millero, F. J.; Lo Surdo, A.; Shin, C. The Apparent Molal Volumes and Adiabatic Compressibilities of Aqueous Amino Acids at 25 Degree. *J. Phys. Chem.* **1978,** *82*, 784–792.
18. Chalikian, T. V.; Sarvazyan, A. P.; Breslauer, K. J. Partial Molar Volumes, Expansibilities, and Compressibilities of Alpha, Omega-Aminocarboxylic Acids in Aqueous Solutions Between 18 and 55 Degrees C. *J. Phys. Chem.* **1997,** *97*, 13017–13026.
19. Kharakoz, D. P. Volumetric Properties of Proteins and their Analogs in Diluted Water Solutions: 1. Partial Volumes of Amino Acids at 15–55°C. *Biophys. Chem.* **1989,** *34*,115–125.

20. Romero, C. M.; Cadena, J. C. Effect of Temperature on the Volumetric Properties of α,ω-Amino Acids in Dilute Aqueous Solutions. *J. Sol. Chem.* **2010,** *39*, 1474–1483.
21. Romero, C. M.; Esteso, M. A.; Trujillo, G. P. Effect of Temperature on the Partial Molar Volumes, the Partial Molar Compressibilities and the Viscosities of α,ω-Amino Acids in Water and in Aqueous Solutions of Sodium Chloride. *J. Chem. Eng. Data* **2018,** *63*, 4012–4019.
22. Romero, C. M.; Rodríguez, D. M.; Ribeiro, A. C. F.; Esteso, M. A. Effect of Temperature on the Partial Molar Volume, Isentropic Compressibility and Viscosity of DL-2-Aminobutyric Acid in Water and in Aqueous Sodium Chloride Solutions. *J. Chem. Thermodyn.* **2017,** *104*, 274–280.
23. Romero, C. M.; Rodríguez, D. M.; Ribeiro, A. C. F.; Esteso, M. A. Effect of Temperature on the Partial Molar Volumes, Partial Molar Compressibilities and Viscosity B-Coefficients of DL-4-Aminobutyric Acid in Water and in Aqueous Sodium Chloride Solutions. *J. Chem. Thermodyn.* **2017,** *115*, 98–105.
24. Rodríguez, D. M.; Romero, C. M. Effect of Temperature on the Partial Molar Volumes and the Partial Molar Compressibilities of α-Amino Acids in Water and in Aqueous Solutions of Strong Electrolytes. *J. Molec. Liq.* **2017,** *233*, 487–498.
25. Hakin, A. W.; Duke, M. M.; Marty, J. L.; Preuss, K.E. Some Thermodynamic Properties of Aqueous Amino Acid Systems at 288.15, 298.15, 313.15, and 328.15 K: Group Additivity Analyses of Standard-state Volumes and Heat Capacities. *J. Chem. Soc. Faraday Trans.* **1994,** *90*, 2027–2035.
26. Yan, Z.; Wang, J.; Liu, W.; Lu, J. Apparent Molar Volumes and Viscosity B-Coefficients of Some α-Amino Acids in Aqueous Solutions From 278.15 to 308.15 K. *Thermochim. Acta.* **1999,** *334*, 17–27.
27. Ogawa, T.; Mizutani, K.; Yasuda, M. The Volume, Adiabatic Compressibility, and Viscosity of Amino Acids in Aqueous Alkali-Chloride Solutions. *Bull. Chem. Soc. Jpn.* **1984,** *57*, 2064–2068.
28. Romero, C. M.; Negrete, F. Effect of the Temperature on Partial Molar Volumes and Viscosities of Aqueous Solutions of a-DL-Aminobutanoic Acid, DL-Norvaline, and DL-Norleucine. *Phys. Chem. Liq.* **2004,** *42*, 261–267.
29. Ellerton, H. D.; Reinfelds, G.; Mulcahy, D. E.; Dunlop, P. J. Activity, Density, and Relative Viscosity Data for Several Amino Acids, Lactamide, and Raffinose in Aqueous Solution at 25°. *J. Phys. Chem.* **1964,** *68*, 398–402.
30. Cabani, S.; Gianni, P.; Mollica, V.; Lepori, L. Group Contributions to the Thermodynamic Properties of Non-Ionic Organic Solutes in Dilute Aqueous Solution. *J. Solution Chem.* **1981,** *10*, 563–595.
31. Gianni, P.; Lepori, L. Group Contributions to the Partial Molar Volume of Ionic Organic Solutes in Aqueous Solution. *J. Solution Chem.* **1996,** *25*, 1–42.
32. Sedlbauer, J.; Jakubu, P. Application of Group Additivity Approach to Polar and Polyfunctional Aqueous Solutes. *Ind. Eng. Chem. Res.* **2008,** *47*, 5048–5062.
33. Marcus, Y. Electrostriction in Electrolyte Solutions. *Chem. Rev.* **2011,** *111*, 2761–2783.
34. Marcus, Y. Viscosity B-Coefficients, Structural Entropies and heat Capacities, and the Effects of Ions on the Structure of Water. *J. Sol. Chem.* **1994,** *23*, 831–848.
35. Pierotti, R. A. A Scaled Particle Theory of Aqueous and Nonaqueous Solutions. *Chem. Rev.* **1976,** *76,* 717–726.

36. Devine, W.; Lowe, B. M. Viscosity B-Coefficients at 15 and 25°C for Glycine, B-Alanine, 4-Amino-N-butyric Acid, and 6-Amino-n-hexanoic Acid in Aqueous Solution. *J. Chem. Soc. A: Inorg. Phys. Theor.* **1971,** *1971*, 2113–2116.
37. Jenkins, H. D. B.; Marcus, Y. Viscosity B-Coefficients of Ions in Solution. *Chem. Rev.* **1995,** *95*, 2695–2724.
38. Yan, Z.; Wang, J.; Liu, W. Lu, J. Apparent Molar Volumes and Viscosity B Coefficients of Some A-Amino Acids in Aqueous Solutions from 278.15 to 308.15 K, *Thermochim. Acta* **1999,** *334*, 17–27.
39. Tsangaris, J. M.; Martin, R. B. Viscosities of Aqueous Solutions of Dipolar Ions. *Arch. Biochem. Biophys.* **1965,** *112*, 267–272.
40. Mason, L. S.; Kampmeyer, P. M.; Robinson, A. L. The Viscosities of Aqueous Solutions of Amino Acids at 25 and 35°. *J. Am. Chem. Soc.* **1952,** *74*, 1287–1290.
41. Romero, C. M.; Beltrán, A. Effect of Temperature and Concentration on the Viscosity of Aqueous Solutions of 3-Aminopropanoic Acid, 4-Aminobutanoic Acid, 5-Aminopentanoic Acid, 6-Aminohexanoic Acid. *Rev. Colomb. Quim.* **2012,** *41*, 123–131.
42. Daniel, J.; Cohn, E. J. Studies in the Physical Chemistry of Amino Acids, Peptides and Related Substances: VI. The Densities and Viscosities of Aqueous Solutions of Amino Acids. *J. Am. Chem. Soc.* **1936,** *58*, 415–423.
43. Zhao, H.; Viscosity B-Coefficients and Standard Partial Molar Volumes of Amino Acids, and Their Roles in Interpreting the Protein (enzyme) Stabilization. *Biophys. Chem.* **2006,** *122*, 157–183.
44. Mirikar, S. A.; Pawar, P. P.; Bichile, G. K. Ultrasonic Velocity, Density, and Viscosity Measurement of Amino Acid in Aqueous Electrolytic Solutions at 308.15K. **2015,** *2*, 19–25.
45. Tyrrel, H. J. V.; Harris, K. R. Diffusion in Liquids, Butterworths, London, 1984.
46. Tyrrel, H. J. V. The Origin and Present Status of Fick's Diffusion Law. *J. Chem. Ed.* **1964,** *41*, 397–400.
47. Taylor, G. Dispersion of Soluble Matter in Solvent Flowing Slowly Through a Tube. *Proc. R. Soc. London A* **1953,** *219*, 186–203.
48. Umecky, T.; Kuga, T.; Funazukuri, T. Infinite Dilution Binary Diffusion Coefficients of Several α-Amino Acids in Water over a Temperature Range from (293.2 to 333.2) K with the Taylor Dispersion Technique. *J. Chem. Eng. Data* **2006,** *51*, 1705–1710.
49. Ma, Y.; Zhu, C.; Ma, P.; Yu, K. T. Infinite Dilution Binary Diffusion Coefficients of Several α-Amino Acids in Water over a Temperature Range from (293.2 to 333.2) K with the Taylor Dispersion Technique. *J. Chem. Eng. Data* **2005,** *50*, 1192–1196.
50. Longsworth, L. G. Diffusion Measurements, at 25° of Aqueous Solutions of Amino Acids, Peptides and Sugars. *J. Am. Chem. Soc.* **1953,** *75*, 5705–5709.
51. Ellerton, H. D.; Reinfelds, G.; Mulcahy, D. E.; Dunlop, P. J. The Mutual Frictional Coefficients of Several Amino Acids in Aqueous Solution at 25°. *J. Phys. Chem.* **1964,** *68*, 403–408.
52. Ribeiro, A. C. F.; Rodrigo, M. M.; Barros, M. C. F.; Veríssimo, L. M. P.; Romero, C. M.; Valente, A. J. M.; Esteso, M. A. Mutual Diffusion Coefficients of L-Glutamic Acid and Monosodium L-Glutamate in Aqueous Solutions at T = 298.15 K. J. *Chem. Thermodyn.* **2014,** *74*, 133–137.

53. Rodrigo, M. M.; Valente, A. J. M.; Barros, M. C. F.; Veríssimo, L. M. P.; Romero, C. M.; Esteso, M. A.; Ribeiro, A. C. F. Mutual Diffusion Coefficients of L-lysine in Aqueous Solutions. *J. Chem. Thermodyn.* **2014,** *74*, 227–230.
54. Rodrigo, M. M.; Esteso, M. A.; Barros, M. C. F.; Verissimo, L. M. P.; Romero, C. M.; Suárez, A. F.; Ramos, M. L.; Valente, A. J. M.; Burrows, H. D.; Ribeiro, A. C. F. The Structure and Diffusion Behavior of the Neurotransmitter Γ-Aminobutyric Acid (GABA) in Neutral Aqueous Solutions. *J. Chem. Thermodyn.* **2017,** *104*, 110–117.
55. Rodriguez, D. M.; Verissimo, L. M. P.; Barros, M. C. F.; Rodrigues, D. F. S. L.; Rodrigo, M. M.; Esteso, M. A.; Romero, C. M.; Ribeiro, A. C. F. Limiting Values of Diffusion Coefficients of Glycine, Alanine, -Aminobutyric Acid, Norvaline and Norleucine in a Relevant Physiological Aqueous Medium. *Eur. Phys. J. E.* **2017,** *40*, 21–25.
56. Rodrigo, M. M.; Esteso, M. A.; Veríssimo, L. M. P.; Romero, C. M.; Ramos, M. L.; Justino, L. L. G.; Burrows, H. D.; Ribeiro, A. C. F. Diffusion and Structural Behavior of the DL-2-Aminobutyric Acid. *J. Chem. Thermodyn.* **2019,** *135* (2019), 60–67.
57. Savage, J. J.; Wood, R. H. Enthalpy of Dilution of Aqueous Mixtures of Amides, Sugars, Urea, Ethylene Glycol, and Pentaerythritol at 25°C: Enthalpy of Interaction of the Hydrocarbon, Amide, and Hydroxyl Functional Groups in Dilute Aqueous Solutions. *J. Sol. Chem.* **1976,** *5*, 733–750.
58. Desnoyers, J. E.; Perron, G.; Avédikian, L.; Morel, J.-P. Enthalpies of the Urea-Tert-Butanol-Water System at 25°C. *J. Sol. Chem.* **1976,** *5* (1976), 631–644.

CHAPTER 5

NATURE'S GREEN CATALYST FOR ENVIRONMENTAL REMEDIATION, CLEAN ENERGY PRODUCTION, AND SUSTAINABLE DEVELOPMENT

BENNY THOMAS[1*], DIVYA MATHEW[2], and K. S. DEVAKY[3]

[1]*Assistant Professor, Department of Chemistry, St. Berchmans College, Changanassery, Kerala, India*

[2]*FDP Substitute, Department of Chemistry, St. Berchmans College, Changanassery, Kerala, India*

[3]*Professor, School of Chemical Sciences, Mahatma Gandhi University, Kottayam, Kerala, India*

[*]*Corresponding author. E-mail: thomasbennyk@gmail.com*

ABSTRACT

Green chemistry is the design and application of chemical processes to reduce the practice and generation of materials hazardous to human health and the environment. The exclusion of widely dispersed anthropogenic pollutants is one of the main concerns for a sustainable improvement for our planet. Comparing to traditional physicochemical methods, bioremediation is the safest, least troublesome, and most economic treatment. Enzymatic bioremediation has risen as an attractive alternative to traditional methods. Nowadays exciting new opportunities for biocatalysis toward the production of renewable and clean energy sources are rapidly emerging. Based on the premise that these alternatives can contribute to a cleaner environment, especially when using renewable agricultural products, the demand for these energies is increasing.

5.1 AN INTRODUCTION TO BIOREMEDIATION

The overall quality of the environment inextricably determines the quality of life on earth. Collective awareness of the environment where we live will lead to an intensive search for designing greener technologies. Biotechnology can offer new platforms for providing proper awareness of the environment to the mankind. Further, biotechnology provides opportunities for the transformation of the pollutants into benign substances, generation of biodegradable materials from renewable sources, and emerging ecologically safe industrial and discarding procedures for the welfare of the population. Bioremediation technology is an effective eco-friendly approach for removing toxic pollutants from the soil and aquatic environment through appropriate organization, conservation, and renovation of the environment.[1] In this process, organic wastes are biologically degraded under controlled manner to a harmless state below established concentration limits.

5.2 PERSPECTIVE FOR ENZYMATIC BIOREMEDIATION

The prime goal of green chemistry is the reduction or elimination of the practice and generation of hazardous materials through the design, fabrication, and application of chemical procedures and products. Biocatalysts—enzymes and whole cells—offer a greener alternative to conventional organic synthesis under mild reaction conditions in the framework of low energy necessities and minimal problems of isomerization and rearrangement. In addition, biocatalysts are biodegradable. They may exhibit chemo-, regio-, and stereoselectivity ensuring huge reduction in the formation of by-products and evading the necessity for functional group activation, protection, or deprotection.[2]

5.3 ENVIRONMENTAL APPLICATIONS AND BENEFITS OF DIFFERENT ENZYMES

Removal of widely dispersed anthropogenic organic pollutants is one of the main challenges for the sustainable improvement for our planet and the survival of mankind.[3] Compared to the conventional physicochemical methods, bioremediation offers enzymatic bioremediation green, minimum troublesome, and most cost-effective treatment using whole microorganisms, either naturally occurring or introduced or isolated enzymes. They degrade the persistent hazardous contaminants into nontoxic or less toxic compounds. Furthermore, enzymatic bioremediation offers more simple systems than whole organisms. Most xenobiotics—the chemical compounds like drugs, pesticides, or carcinogens that are foreign to a living organism—can be submitted to enzymatic bioremediation. For instance, polycyclic aromatic hydrocarbons (PAHs), polynitrated aromatic compounds, pesticides such as organochlorine insecticides, bleach plant effluents, synthetic dyes, polymers, and wood preservatives like creosote and pentachlorophenol can be succumbed to enzymatic bioremediation. New developments in the design and application of enzymatic "cocktails" for biotreatment of wastewaters have recently emerged. From a green environmental point of view, the application of enzymes instead of chemicals or microorganisms undoubtedly has advantages like lack of toxic side-products, enhanced bioavailability, and higher scale production.[4] For enzymatic bioremediation, the enzyme should be kept at optimal reaction conditions to display high substrate affinity with K_m in the micromolar range, supporting thousands of product turnovers. Altogether, the enzymes should exhibit high robustness under the selection of external factors and low dependency on expensive redox cofactors like NAD(P)H, which would be prohibitive in a commercial setting. The disposal and management of the sludge produced from biomass during living cell mediated bioremediation is a problematic concern in developing countries where the demand for water treatment is high. Depending on the choice, enzymes can either function at mild conditions replacing harsh conditions and harsh chemicals or work in extreme conditions and hence save energy and prevent pollution. The high specificity of enzymes results in fewer unwanted side effects and by-products. Besides, enzymes can readily be absorbed back into nature owing to their biodegradability. In short, for a sustainable future, enzymatic bioremediation will definitely open an eco-friendly strategy in a wide range of industries. In literature, there are lots of reports on enzymatic environmental remediation satisfying the principles of green chemistry.

5.3.1 OXIDOREDUCTASES

Oxidoreductases are a class of enzymes used to detoxify hazardous compounds by oxidative coupling.[5] The oxidation is carried out through the transfer of electrons from reductants to oxidants and results in the release of chloride ions, CO_2, and methanol. These enzymes cleave the chemical bonds of organic pollutants yielding energy and transferring electron from the reduced organic matter (donor) to another compound (acceptor). During the redox reactions, the contaminants are finally oxidized to harmless compounds. Various species of bacteria, fungi, and higher plants are responsible for the production and secretion of oxidoreductases. The energy generated by oxidoreductase during degradation and oxidation of pollutants to harmless compounds is utilized by microorganisms for their metabolic process. Oxygenases, monooxygenases, and dioxygenases are the main members of the oxidoreductase enzyme family.[4] Oxidoreductases have been used in the degradation of many natural and man-made pollutants, especially from textile industries, for example, phenolics, anilinics, and dyes.[6] The Gram-positive bacteria *Bacillus safensis* produces oxidoreductase to degrade the petroleum compounds. Chlorinated phenolic compounds are the major harmful component in the effuents of paper and pulp industry. Many fungal species are rich in extracellular oxidoreductase enzymes and are well-thought-out to be apt for the exclusion of these chlorinated phenolic compounds. The filamentous structure of fungi helps them to reach the soil pollutants more effectively than bacteria. White-rot fungi, *Panus tigrinus*, and its extracellular oxidoreductase can remove the phenols, color, and organic load released from olive mill wastewater. Petroleum-based hydrocarbons and chlorinated compounds like DDT, BHC, and so on can be effectively degraded by the oxidoreductases secreted from the plant families of *Fabaceae*, *Gramineae*, and *Solanaceae*. Further, chlorinated solvents, explosives, and petroleum hydrocarbons can be degraded by phytoremediation.[7] Oxidoreductase enzymes released by bacterial species are helpful for the reduction of radioactive metals. The electrons released by the oxidation of organic pollutants are used for the reduction of radioactive elements.[8]

5.3.1.1 OXYGENASES

Enzyme oxygenase belongs to the family of oxidoreductases. They transfer oxygen from molecular oxygen utilizing FAD (flavin adenine dinucleotide)/NADH (nicotinamide adenine dinucleotide)/NADPH (nicotinamide adenine dinucleotide phosphate) as a co-substrate. Oxygenases are responsible for the aerobic degradation of aromatic compounds by increasing their reactivity or water solubility or bringing about cleavage of the aromatic ring.[9,10] The addition of molecular oxygen is an essential step for their degradation. Thus, they catalyze the cleavage of the aromatic ring by adding one or two molecules of oxygen. Additionally, halogenated organic compounds especially herbicides, insecticides, fungicides, plasticizers, and intermediates for chemical synthesis can be degraded by specific oxygenases. Furthermore, they assist dehalogenation reactions of halogenated methanes, ethanes, and ethylenes in association with multifunctional enzymes.[4,11] On the basis of the number of O_2 molecules involved, oxygenases are classified into two subclasses.

5.3.1.2 MONOOXYGENASES

Monooxygenases catalyze the oxidation of simple alkanes, complex steroids, and fatty acids.[12] Monooxygenases act as biocatalysts in bioremediation process and synthetic chemistry due to their highly region selectivity and stereoselectivity on a wide range of substrates.[4] These enzymes require only molecular oxygen for their activities and utilize the substrate as reducing agent.[13] They utilize the substrate as reducing agent and necessitate only molecular oxygen for their activities. The main reactions catalyzed by monooxygenases include desulfurization, dehalogenation, denitrification, ammonification, hydroxylation, biotransformation, and biodegradation of various aromatic and aliphatic compounds.[14] Majority of monooxygenases comprise cofactors,[4] for example, flavin-dependent monooxygenases and P450 monooxygenases. Methane monooxygenase is the best one for the degradation of substituted aliphatic, aromatic, and heterocyclic hydrocarbons. Monooxygenase catalyzes oxidative dehalogenation reactions under oxygen-rich conditions whereas they consequence reductive dechlorination affording the formation of labile products under low oxygen levels. Methane monooxygenases are found in cytoplasmic membrane and cytoplasm.

5.3.1.3 DIOXYGENASES

Dioxygenases are multicomponent enzyme systems responsible for the degradation of aromatic pollutants into nontoxic materials. On the basis of their mode of action, aromatic dioxygenases have the element of both aromatic ring hydroxylation and aromatic ring cleavage. Aromatic ring hydroxylation dioxygenases degrade the chemical compounds by the addition of two molecules of oxygen into the ring while aromatic ring cleavage dioxygenases cleave the aromatic rings of compounds.[4] They have Rieske (2Fe–2S) cluster and mononuclear iron in their alpha subunit, and hence, they thought to belong to the family of Rieske nonheme iron oxygenases.[15]

Aromatic hydrocarbon dioxygenase includes toluene dioxygenase, catechol dioxygenase, and so on. They introduce molecular oxygen into their substrate.[16] They are effectual for environmental remediation owing to their enantiospecific oxygenation property of a wide range of substrates

5.3.2 LACCASES

Laccases are an interesting group of ubiquitous, oxidoreductase enzymes offering great potential for bioremediation applications.[17] They are produced by certain plants, fungi, insects, and bacteria. They are multi-copper oxidases that catalyze the oxidation of phenolic and aromatic compounds present in the soil and water. They are always produced in the cell, but can be secreted extracellularly. They are able to degrade the ortho- and paradiphenols, aminophenols, polyphenols, polyamines, lignins, and aryl diamines as well as some inorganic ions into less hazardous or nontoxic materials.[18–20] In addition, they can oxidize, decarboxylate, and demethylate the methoxy substituted phenolic acids.

Depolymerization of lignin is catalyzed by these enzymes resulting in a variety of phenols. The generated phenol derivatives can be utilized as nutrients for microorganisms or they can be repolymerized to humic materials by laccase itself.[21] Laccases can effectively decolorize azo dyes by oxidizing their bonds and finally transform them into less harmful substances. The decolorization activity is mainly due to two laccase isozymes purified from fungus *Trametes hispida*.[22] Laccase immobilization on solid support increases their stability, half-life, and resistance to protease enzymes. Laccase (isolated from fungus *Trametes versicolor*) immobilized on porous glass beads offers higher thermal, pH, and storage stabilities in the remediation platform.[23] Laccase can reduce the dioxygen molecules of pollutants into water by removal of electrons from the organic substrate.[24] Synthetic laccase is also reported in paper and pulp industry to boost bleaching of pulp and textiles.[4] The substrate specificity of laccases displays strong dependence on pH, and their specificity can be subdued by azides, cyanides, and halides excluding iodide.[25] Different laccases display different tolerance power toward the inhibition by halides. The nitrogen concentration in fungi is responsible for laccase production and shows a direct proportionality. Both homologous and heterologous methods can be adopted for the production of recombinant laccases.

5.3.3 PEROXIDASES

Peroxidases are universal enzymes useful for the oxidation of lignin and other phenolic compounds. Hydrogen peroxide and mediators are essential for their oxidation and degradation activities.[4] For instance, phenolic radicals produced by oxidation of phenolic compounds, aggregates and become less soluble and hence precipitated quickly. These peroxidases may be haem and non-haem proteins.[26] In mammals, they strengthen the immune system and assist hormone regulation. Auxin metabolism, lignin and suberin formation, cross-linking of cell wall components, defense against pathogens, or cell elongation are their major functions in plants.[26,27] The haem peroxidases are of two types. They are found in animals, plants, fungi, and prokaryotes. In mammals, they strengthen the immune system or help in hormone regulation. In plants, they assist auxin metabolism, lignin and suberin formation, and defense against pathogens or cell elongation. Non-haem peroxidases comprise thiol peroxidase, alkylhydroperoxidase, haloperoxidase, manganese catalase, and NADH peroxidase. The thiol peroxidase is the largest one and comprises glutathione peroxidases and peroxy redoxins as subfamilies.[26] The lignin peroxidase (LiP) and manganese peroxidase (MnP) are the most studied enzymes because of their greater potential for the degradation of toxic substances. Horseradish peroxidases can be effectively immobilized as cross-linked enzyme aggregates (HRP-CLEAs) using ethylene glycol-bis [succinic acid N-hydroxysuccinimide] as the cross-linker. HRP-CLEAs are found to be effective in the oxidative para-dechlorination of toxic contaminants and carcinogens like 2,4,6-trichlorophenol with high efficiency.[28] Soybean peroxidase (SBP) and chloroperoxidase are well studied for the degradation of thiazole compounds. Peroxidases are further classified into LiP, MnP, and versatile peroxidase (VP).[4,29]

5.3.3.1 LIPS

LiPs are monomeric haem-containing proteins and secondary metabolites of white-rot fungus.[4,30] In LiPs, Fe (III) is penta-coordinated with histidine residue and four tetrapyrrole nitrogens. They necessitate hydrogen peroxide as co-substrate and veratryl alcohol as mediator for catalyzing the oxidation of toxic pollutants.[31,32] The degradation of pollutants proceeds through two-electron oxidation of the native ferric enzyme by H_2O_2.[33] Additionally, LiP degrades lignin and other phenolic compounds. The presence of H_2O_2 and veratryl alcohol (mediator) is essential for the degradation activity of LiP.

During the course of the reaction, H_2O_2 gets reduced to H_2O, and the enzyme LiP gets oxidized. The oxidized LiP further gains an electron from veratryl alcohol and returns to its native reduced state forming veratryl aldehyde. But veratryl aldehyde gets reduced back to veratryl alcohol by taking an electron from the substrate. LiPs display a great potential for the treatment of wastewater in the field of bioremediation. Lignin degradation by bacterial peroxidases is more efficient as compared to fungal peroxidases regarding their specificity and thermostability. LiPs generally degrade the plant cell wall constituent lignin. For LiP-catalyzed degradation, the aromatic contaminants should have a minimum redox potential higher than 1.4 V.[34]

5.3.3.2 MnPs

MnPs are haem-containing extracellular enzymes. They are produced by lignin-degrading basidiomycetes fungi. They can effectively oxidize Mn^{2+} into Mn^{3+} by a sequence of reactions.[4,32] Further, they catalyze the degradation of several phenols, amine-containing aromatic compounds, and dyes.[31] Several acidic amino acid residues and one haem group containing manganese binding site are present in enzyme MnP. Mn^{2+} ion can stimulate the production of the enzyme MnP and serve as the substrate for the enzyme MnP. The Mn^{3+} ions generated act as a mediator for the oxidation of phenolics and afford Mn^{3+}-chelates. Xenobiotic pollutants may be buried deep within the soil, and hence, they are not primarily accessible to the enzymes. But the Mn^{3+}-chelates are small enough to diffuse into areas inaccessible even to the enzyme.[31] For instance, MnP immobilized on chitosan beads activated by glutaraldehyde shows a greater potential for decolorization of dye effluent from the textile industry. In addition, the stability and half-life of MnP can be enhanced by immobilization.

5.3.3.3 MICROBIAL VPs

VPs are capable of oxidizing Mn^{2+}, methoxybenzenes, and phenolic aromatic compounds directly. In comparison with peroxidases, VP exhibits extraordinary substrate specificity to oxidize the phenolic and nonphenolics even in the absence of manganese.[35] Consequently, a highly efficient VP overproduction is essential for the bioremediation of recalcitrant pollutants.[36,37]

5.3.4 HYDROLYTIC ENZYMES

The soil and water pollution by industrial effluents is a serious problem of the modern world.[4] Bioremediation offers a safe and economic alternative to commonly used remedies. Extracellular hydrolases secreted by microbes play a key role in degradation of organic polymers and toxic compounds with molecular weights less than 600 Da that can diffuse through the pores of the cell.[38] Hydrolytic enzymes are most commonly used for the bioremediation of pesticides and insecticides by disrupting chemical bonds like ester, peptide, and carbon-halide in the toxic molecules. Extracellular hydrolytic enzymes like amylases, proteases, lipases, DNases, pullulanases, and xylanases have diverse potential applications in the areas of food industry, feed additive, biomedical sciences, and chemical industries.[39] The hemicellulase, cellulase, and glycosidase are of much application in biomass degradation.[40]

5.3.4.1 LIPASES

Lipases are specific for the breakdown of lipids.[41] They are produced by bacteria, plants, actinomycetes, and animal cells. They are helpful in the drastic reduction of total hydrocarbon content in the contaminated soil through processes like hydrolysis, interesterification, alcoholics, and aminolysis. They are beneficial for the hydrolysis of triglyceride, the main component of natural oil or fat, into diacylglycerol, monoacylglycerol, glycerol, and fatty acids.[4] Lipolytic reactions occur at the lipid–water interface, where lipolytic substrates usually form equilibrium between monomeric, micellar, and emulsified states. In the biphasic oil–water system, the enzyme lipase gets adsorbed on to the oil–water interface in the bulk of the water phase and then breaks the ester bonds of triolein to diolein, monoolein, and glycerol, and oleic acid is formed at each consecutive reaction stage. The glycerol formed is hydrophilic in nature and gets dissolved into the water phase. Monoacylglycerol is effectively used as an emulsifying agent in the food, cosmetic, and pharmaceutical industries. Lipase activity is one of the most useful parameters for testing the degradation of hydrocarbon in soil.[42]

5.3.4.2 CELLULASES

Cellulases are the key enzymes for the degradation of cellulose, the most abundant biopolymer found on the earth.[43] Cellulases are generally produced by microorganisms.[44] They may be cell-bound, associated with cell envelope, and extracellular. Usually, cellulases are composed of a mixture of several enzymes such as endoglucanase and exoglucanase or cellobiohydrolase.[45] The endoglucanase attacks regions of low crystallinity in the cellulose fiber and creates free chain ends. But the exoglucanase removes cellobiose units from the free chain ends of cellulose and degrades it. The enzyme action of β-glucosidase is helpful for the hydrolysis of cellobiose to glucose units. Cellulases are useful in the detergent and washing powders manufacturing industries, where cellulose microfibrils produced during processes are removed by these enzymes.[46] Alkaline cellulases can be employed for the bioremediation of ink in paper and pulp industry during the recycling of paper and waste management.[47,48] Cellulase can adapt to harsh environmental conditions like extreme pH and temperature.

5.3.4.3 PHOSPHOTRIESTERASES

Phosphotriesterases have application in the degradation of chemical wastes released from industrial fields and pesticides (e.g., parathion, malathion) used in crop fields.[49] Parathion is an organophosphatic component in herbicides and insecticides.[50] Organophosphate is an ester of phosphoric acid and is degraded by phosphotriesterases. There are aryldialkylphosphatase, organophosphorus hydrolase, and recombinant thermostable phosphotriesterases too.

5.3.4.4 HALOALKANE DEHALOGENASES

Halogenated compounds produced by both natural activities and man-made efforts are present everywhere in the soil. These compounds may be hazardous, toxic, mutagenic, or carcinogenic. Haloalkane dehalogenases are useful for the hydrolysis of carbon halogen bonds present in the various halogens containing contaminants and produce alcohol and halides.[51] The active site of haloalkane dehalogenase is present between the main domains of an eight-stranded β-sheet helices. First haloalkane dehalogenase discovered from the bacterium *Xanthobacter autotrophicus* has the ability to degrade 1, 2-dichloroethane. Several dehalogenases have been cloned and characterized from Gram-positive and Gram-negative haloalkane degrading bacteria.

5.3.4.5 PROTEASES

Proteases are class of enzymes responsible for the hydrolysis and reverse synthesis of peptide bonds in aqueous environment and nonaqueous environment, respectively.[52] They hydrolyze the proteinaceous substance produced by shedding and molting of appendages, death of animals, and as the by-product of poultry, fishery, leather, detergent, and pharmaceutical industries.[53] Based on the mode of catalysis of peptide chain, there are endopeptidase like serine endopeptidase, cysteine peptidase, aspartic endopeptidases, metallopeptidases, and so on, and exopeptidases such as aminopeptidase, carboxypeptidase, and so on. The endopeptidase acts on the inner regions of peptide chain whereas the exopeptidases are sensitive to the terminal amino or carboxylic position of chain only.[54] Proteases are beneficial for the manufacture of cheese, detergent, non-calorific artificial sweetener, and effective therapeutic agents. The alkaline proteases are useful for removing hairs in the leather industry. Subtilisin in combination with antibiotics is helpful in the treatment of burns and wounds.[55]

5.4 ENZYMATIC REMEDIATION OF INDUSTRIAL WASTES

Enzymes are increasingly being used in process industries to create a "cleaner" technology and decrease the use of harsh chemicals that are not environment friendly.[56] Enzymes are powerful tools for sustaining a clean environment in several ways. They are effectually utilized in a number of industries like agro-food, oil, animal feed, detergent, pulp and paper, textile, leather, petroleum, chemical, and biochemical.

5.4.1 FOOD INDUSTRY

Enzymes have diverse potentials in the industries like manufacturing of cheese, vinegar and wine, leavening of bread, brewing of beer, and so on.[56] Enzymes support food processes by saving energy and resources, and hence improving the efficiency of the overall process. The use of enzymes decreases the number, variety, and toxicity of by-products and effluents. In the food industry, amylases are an effective alternative to strong acids and high temperatures required for the breakdown of starch. The industrial xylanase is a good enzyme for wheat separation because it shows a high level of activity toward soluble arabinoxylans and a low level of activity toward insoluble arabinoxylans and thus releasing previously bound starch molecules and proteins. Only a minimum amount of water is needed to wash out the starch. Hence, apart from the higher yield, water saving is one of the most important advantages of using xylanase. Besides, less water is used in the process, and hence, less wastewater is disposed of into the environment and leads to a substantial economic benefit. Moreover, the reduced viscosity of the soluble fraction makes the removal of liquid easier in the subsequent starch concentration step. Thus, less energy is required for the final evaporation stages.

5.4.2 EDIBLE OIL INDUSTRY

Even though enzymes are extensively used in carbohydrate processing, they play little role in edible oil and fat processing because of the high expense of the necessary enzymes like phospholipases for the degumming of oils.[57] The introduction of enzymes has opened up a new avenue for vegetable oil extraction. Hexane was the traditionally used solvent to dissolve oil from crushed seeds. It is not only dangerous to breathe in but also highly explosive. Enzymatic extraction process in aqueous medium replaces hexane-based technology.[56]

5.4.3 ANIMAL FEED INDUSTRY

It is estimated that billion tons of feces and urine are produced each year by livestock. The way of treatment of these hazardous materials is one of the serious pollution threats in the United States. The animal feed industry has gone through many changes in the past few years. The consumers and the industry itself have looked more closely than before into the production of animal feeds, rearing of animals, and the impact of animal husbandry systems on the environment. In the animal feed industry, enzymes offer a relatively new development.[58] The enzymatic treatment of catfish and the inclusion of fiber degrading enzymes in the diets of grower pigs result in the reduction of polluted wastes. Enzymes in the feed industry are mostly used to either enhance the feed digestibility or improve the utilization of raw material. They increase the amounts of the dry feed matter, particularly nitrogen and soluble carbohydrates, which can be retained by the animal, thus reducing excretions and hence reducing the pollution load. Commonly used enzymes for this purpose are amylases, pectinases, proteases, β-glucanases, xylanases, and hemicellulases.[56,59]

5.4.4 DETERGENT INDUSTRY

Eutrophication of water bodies caused by the presence of phosphorous in detergents paved the way for the introduction of enzymes in detergent formulations. Enzymes have been used in laundry detergents since the 1960s.[56,60] Proteases were the first enzymes to appear, followed by amylases, lipases, and cellulases. The detergent industry has remained the largest market for industrial enzymes, and new enzyme products are constantly being developed for such use. The efficiency of the enzymes is reflected in the necessity of trace amounts, typically less than 1% by weight, for remediation. Washing at lower temperatures in comparison with conventional high-temperature washing cycles is one of the main environmental benefits of using enzymes. Moreover, enzymes are consistent with the principal necessities in the detergent industry like cost-effectiveness, acceptable environmental profile, biodegradability, no negative impact on sewage treatment processes, and saving of energy by lower temperature washing. They are also multifunctional, providing an advantage in stain removal, anti-redeposition, whiteness/brightness retention, color maintenance, and fabric softening. Furthermore, the addition of enzymes to detergent formulations reduces the amounts of bleach, caustic, and phosphate.

5.4.5 PULP AND PAPER INDUSTRY

The paper–pulp industry uses harsh chemicals and bleaching compounds at many stages of the processes and discharges considerable amounts of deleterious by-products like lignin into the environment. Hence, paper–pulp industry is one of the most eminent sectors for environmental pollution.[61] Even though the introduction of enzymes has not fully substituted any manufacturing task to date, they are very helpful in either improving the processes or in reducing pollution. During the manufacture of paper, in the pulping process, lignocellulosic raw material is digested with calcium sulfite, and the digested raw material is then dewatered and refined. Subsequently, the refined pulp is bleached with Cl_2 to make the pulp completely white. In this bleaching process, the residual lignin is virtually removed completely, and enzymes like xylanases and ligninases are introduced for effective bleaching of the pulp reducing the use of oxidizing chemicals to about 15–20%. Cellulases are beneficial in deinking, an important part of secondary fiber processing.[56] Cellulases release ink particles, thereby enhancing the removal of ink by flotation and consequently substantial decrease in the usage of deinking chemicals.[59,61]

5.4.6 TEXTILE INDUSTRY

Enzymes can replace harsh organic or inorganic chemicals currently used in the textile industry.[56] Consequently, the utilization of highly specific enzymes for various textile processing applications is popular. Chemicals such as acids widely used in desizing of cloths, bleaching chemicals, and the production of coloring agents and dyes are responsible for the pollution associated with the textile industry.[62] In denim manufacturing, even though the wastewater streams can be readily disposed of by maximum dilution with cleaner waters, the stone washing is associated with a problem of disposal of the sand produced by eroding of stones. The pumice grit may clog up drains, and hence, their exclusion from the wastewater is essential. An enzyme substitute for pumice stones can alleviate this environmental problem. With the appropriate enzymes, stones are no longer needed. The cost of enzymes is comparable to the cost of stones yet when other savings are taken into account; the stone-free process turns out as being more affordable. The major advantages of a stone-free process include (1) some people rejected jeans as the stones damage the denim, (2) less wear on laundry machines, (3) no storage and handling of bulky stones, (4) no destoning and no need to wash off pumice grit, (5) more room for garments in the wash loads, (6) no disposal problems due to solid pumice waste, and (7) less variation in wash results. In the textile industry, desizing is essential for the removal of all the starch paste from the fabric. The starch strengthens the warp thread without breaking during the weaving process. Up to that time, textiles soaked in water for several days or treated with acid, alkali, or oxidizing agents to break down the starch. However, both of these methods are difficult to control and sometimes may damage or decolorize the material. Therefore, crude enzyme extracts in the form of malt extract or pancreas extract are used to carry out desizing.

Fabric softening is another area in the textile industry where enzymes are replacing the traditional practices. For years, chemicals are used by the manufacturers in order to make fabrics softer. However, the softness is removed after continuous washing, and the chemicals used are deleterious to the environment. Biopolishing or enzymatic softening offers long-lasting softness to the fabric. Moreover, it offers a green softening procedure. Recent enzymatic development in the textile industry moved away from the use of chemical bleaches, particularly chlorine- and hydrogen peroxide-based bleaches. In the degradation of hydrogen peroxide by the enzyme catalase, there is a significant reduction in the amount of water consumed in the overall process, since the series of rinses between primary bleaching and dyeing processes can be excluded. Enzymes can also be utilized in the dyeing step by lowering their amounts in the effluents. Cellulases, amylases, and proteases are useful in the prewashing step of dyeing to improve the dyestuff uptake. Enzymes can improve the degree of whiteness prior to dyeing, color shade, and felting behavior. Further, LiP, MnP, and laccase have the potential to degrade common dyes found in the textile effluents.

5.4.7 LEATHER INDUSTRY

Animal hides and skins are excessively used in the leather industry. Leather manufacturing is a multistep process including curing, soaking, dehairing, dewooling, bating, and tanning.[56] Traditionally, dehairing and dewooling requires an extremely alkaline condition in the soaked and swollen epidermis and corium of the skin, followed by the reaction of hair proteins with sulfides to break the bonding of hair protein fibrils and dissolve the proteins of hair root. The presence of these chemicals in the effluents poses a severe environmental problem. The incorporation of enzymes, especially proteases, is beneficial for improving the leather quality and reducing the pollution. In the presence of enzymes, the hair is not dissolved and can be filtered out from the liming float. In this manner, chemical oxygen demand (COD) and biological oxygen demand (BOD) of the waste discharges can be reduced to a considerable extent. Bating makes the leather soft, supple, and able to accept an even dye and demonstrate the grain in an adequate fashion, and encompasses deliming and deswelling and degrading of the collagen of the skins and the protein fibers. Bating is helpful for the control of the quality of leather: Stiff leather is used for soles and is only lightly bated, while the soft leather used in gloves results from intense bating. The introduction of trypsin leads to a reliable process reducing the BOD content in the effluent. Degreasing or the removal of fat from the leather before tanning is significant since once most of the natural fat has been removed, and subsequent chemical treatments such as tanning, retanning, and dyeing have a better effect. In addition to the final quality of the product, lipases offer an environmentally better way of removing the fat. The practice of lipases in the degreasing step permits the partial or full replacement of the solvents or surfactants that may be harmful to the environment.

5.4.8 PETROLEUM INDUSTRY

Enzymes are progressively piercing the sectors of petroleum industry, where their use was once considered not viable because of either technical or economic reasons.[56,63] Enzymes can remove sulfur from petroleum at mild operating conditions like temperature and pressure. In addition to the cost advantages compared to the existing methods, biodesulfurization technology offers the benefit of operating under milder conditions, consuming less energy, and emitting fewer greenhouse gases. The future may realize the practice of enzymes to catalyze cracking, viscosity reduction, and demetalization in a petroleum refinery.[63]

5.5 ENZYMES AND WASTE MANAGEMENT

Enzymes constitute an alternative to traditional, nonbiological cleanup methodologies. They catalyze the transformation of single or complex mixtures of pollutants. Furthermore, enzymes are useful to detect and quantify the amount of pollutants in the environment before, during, and after the remediation process. The use of enzymes for waste management can be classified into four categories: (1) effluent treatment and detoxification, (2) renewable energy resources, (3) bioindicators for pollution monitoring, and (4) biosensors.

5.5.1 EFFLUENT TREATMENT AND DETOXIFICATION

An imperative aspect of satisfactory industrial operations is the effective breakdown of solids and the clearing and prevention of fat blockage or filming in waste systems. Most of the bioremedial procedures explore the biotransformation ability of living organisms, mainly bacteria including *Pseudomonas*, *Flavobacterium*, *Arthrobacter*, and *Azotobacter*. Tolerance of these organisms to extremely toxic molecules and to changing loadings of less aggressive substances can be increased by enzymatic pretreatment of the wastes, because enzymes increase the available nutrients without additional demand on enzyme production and secretion mechanisms. This in turn increases the available energy for growth and detoxification activity. Organic solvents, polyaromatic hydrocarbon containing wastes, oily wastes, halogenated aromatic hydrocarbons, pesticides, and munition wastes are subjected to enzymatic detoxification.[64,65]

5.5.1.1 PAHs

PAHs are mutagenic or carcinogenic compounds composed of two or more fused aromatic rings. They are widely distributed in the natural environment and cause severe health concerns. Coal and petroleum are two major natural sources of PAHs and are produced by coal gasification, coking, and wood preservation. Low molecular weight PAHs are readily biodegradable by many microbes like *Pseudomonas*, *Micrococcus*, and *Flavobacterium*. But high molecular weight PAHs having more than four fused aromatic rings are quite recalcitrant and toxic.[66,67] LiPs, MnPs, and P450 monooxygenases are responsible for the degradation of PAHs. The addition of surfactants along with some of these microorganisms can stimulate the biodegradation of PAH.[68]

5.5.1.2 TNT

Life on our planet earth is based on the continuous cycling of elements. Inadequate incorporation of natural and synthesized molecules into biological cycles may cause immense mobilization of natural resources and severe environmental problems. Nitroaromatic compounds are relatively recalcitrant to biological degradation. But microorganisms are capable of mineralizing them at a relatively slow rate.[69] TNT is the most explosive nitroaromatic pollutant in the world. In TNT, the nitro groups are symmetrically located on the aromatic ring, and hence, it is more recalcitrant than mono- and dinitrotoluenes. The electronegative nitro groups make the nucleus electrophilic by withdrawing the π electrons from the aromatic ring. The ease of reduction of nitro group is due to the high electronegativity of N atom and the accumulation of partial positive charge of the nitrogen atom owing to the polarity of N–O bond. The toxicity of nitroaromatic compounds is mainly due to the presence of nitroso and hydroxylamino groups. These functionalities react with biological molecules and cause chemical mutagenesis and carcinogenesis. The mutagenic effect can be alleviated by the complete reduction of nitro group to amino group. MnP and LiP are useful for the degradation of TNT and 4-amino-2,6-dinitrotoluene. Further, mineralization is more desirable for the degradation of TNT than the reduction of nitro group into amino group. MnP is capable of mineralizing TNT and its intermediates in the field. The degradation of TNT through mineralization by MnP holds promise for future field application. The rate of mineralization can be enhanced by changing the concentration of the enzymes used or by the addition of a co-substrate like catechol.

5.5.1.3 PHENOL

Water pollution threatens the ecological advancement of human civilization. Phenols are aromatic compounds created during coal refining, oil refining, plastic production, and resin production. The majority of the phenolic compounds are harmful, lethal, endocrine disrupting, mutagenic, teratogenic, and carcinogenic in nature. They also cause severe damages to the marine ecosystem.[70] Consequently, thc exclusion of phenolics from polluted wastewaters prior to their discharge to the environment is essential. Enzymatic remediation is progressively becoming a gorgeous, environmentally friendly, and sustainable alternative to wastewater treatment technology. The exclusion of diverse phenolic pollutants from wastewaters can be carried out by both homogenous and heterogeneous enzymatic processes. Additionally, recombinant DNA technology encourages the utilization of enzymes in wastewater remediation. This technology offers large-scale production of enzymes, with enhanced stability and activity and at a lower cost. Enzymes are more attractive than the living microorganisms because of several reasons such as (1) they do not require nutrients, do not require biomass acclimation, and do not generate sludge; (2) lower mass transfer to contaminants compared to the living microorganisms; (3) form simpler systems great ease of control; and (4) highly effective systems even at very low concentrations. Peroxidases and laccases are the most prevalent enzymes useful for the removal of phenolic pollutants. HRP is the most widely accepted peroxidase for the removal of phenolic compounds from wastewaters which can remove over 99% of phenols within 35 min of treatment. However, the potential use of HRP on industrial scale is strictly limited due to high production cost, high susceptibility to deactivation, laborious cultivation and extraction processes, and hence limited availability. SBP can be used as an alternative to HRP. SBP is more abundant than HRP because of its low production cost. Furthermore, it exhibits lower vulnerability to irreversible deactivation by H_2O_2. Laccases are another example which can be extensively used to treat wastewaters contaminated with phenols. Laccases are nonspecific and they can remove more than 90% phenol from a refinery wastewater sample within a shorter time.

5.5.1.4 POLYCHLORINATED BIPHENYLS (PCBs)

PCBs are threat to both the natural ecosystem and human health. PCBs are synthetic chemical compounds that carry one to ten chlorine atoms on a biphenyl carbon skeleton. Among the 209 theoretically possible congeners, 20–60 are present in commercial products. This versatile composition of PCBs, with their high hydrophobicity and chemical stability, makes them persist in the environment.[71] Various microorganisms have been isolated for the degradation of PCBs. The degradation of PCB occurs by co-metabolism of biphenyl, which is not so foreign to nature. Fungal derived extracellular MnP, LiP, and laccase are capable of degrading PCBs. However, the ability of the enzymes to degrade PCBs is determined by their parent fungus. For example, laccase from *Trametes versicolor* is capable of degrading over 65% of the 4-hydroxybiphenyl in a solution in 3 h and at a pH of 4.0; that from *Pleurotus ostreatus* has the ability to degrade over 50% of the same compound under the same conditions. In addition, for PCBs, the amount of chlorination is also a decisive factor for its biodegradability. Even though the enzymes get more inhibited as the number of chlorine atoms increases, the ionization potential increases with the extent of chlorination. There is an inverse linear correlation between the ionization potential and the enzymatic removal rate coefficients for PCBs. Enzymatic biodegradability of a compound depends on the ionization potential, type, and structure of the compound.

5.5.1.5 DYES

Synthetic organic colorants are widely accepted candidates in food, textile, cosmetic, and pharmaceutical industries. The majority of them are highly recalcitrant. The discharge of these carcinogenic dyes and dye precursors constitutes water pollution. Further, the colored wastewaters cause a serious environmental problem and public health concern.[72] The stability and xenobiotic nature of reactive azo dyes make them recalcitrant. Complete degradation of azo dyes cannot be accomplished by conventional wastewater treatments. Consequently, they are released into the environment in the form of colored wastewater resulting in acute effects on exposed organisms. Phytoplanktons absorb light that enters the water and display abnormal coloration and reduction in photosynthesis. This also alters the pH, BOD, and COD levels, and causes intense coloration of river water. Biotreatment offers a cheaper and eco-friendly alternative for the exclusion of color in textile effluents. Enzymes have been used in the textile industry as detergents for decades. But, only recently, extracellular enzymes have been scrutinized for their ability to decolor, degrade, and mineralize dyes. The chromophores in the dyes must be oxidized and cleaved during the degradation process in wastewater treatment plant. Azo reductases, laccases, and MnPs are the most promising enzymes for the exclusion of color in textile effluents.

5.5.1.6 ORGANOPHOSPHATES

Organophosphorus compounds (OPs) are highly toxic molecules. They are one of the major components of herbicides, pesticides, and nerve gas.[73] In modern agriculture, pesticides maintain the quality and productivity of agricultural crops through controlling the weeds and pests. OPs are among the most commonly used pesticides due to their less expense and ease of application. The nerve gases were used as chemical warfare agents in the past. OP pesticides and nerve gases collectively referred to as nerve agents. Chemically, OPs are phosphotriesters. The widespread practice of pesticides resulted in global issues concerning both environment and human health. The potential lethality of these compounds developed different types of enzymes capable of hydrolyzing OPs that have been identified and characterized. Some of them are phosphotriesterases, organophosphorus hydrolases, serum paraoxonases, methyl parathion hydrolases, and organophosphate acid anhydrolases. In 2003, Qiao et al. developed a genetically engineered form of carboxylesterase responsible for the degradation and detoxification of malathion, parathion, and monocrotophos. The enzyme is capable of degrading about 80% of the malathion in 90 min, about 85% of the parathion in 6 h, and about 20% of the monocrotophos in 9 h. The presence of carboxylester bonds specific to the enzyme carboxylesterase is the reason for faster and more complete degradation of malathion and parathion than monocrotophos. Furthermore, these contact times appear to be quite feasible for its field implementation.

5.5.1.7 WASTEWATER

Both the inadequate availability of fresh water and the emergent consumption of fresh water by anthropogenic activities are global crisis. Unfortunately, our water bodies are used as sinks for wastewater from domestic and industrial sources. Hence, it is necessary to take cumulative attention to replenish our water resources. The strategies taken to purify water and return it to its source in the least toxic form possible in order to reutilize it are collectively called "wastewater treatment." Moreover, dwindling water resources is a global problem.[74] Effective effluent treatment is an imperative step in the conservation of our water resources. Pollutants like dyestuffs, PCBs, pesticides, and so on are resistant to degradation by conventional treatment methods and persist in the environment. It emphasizes the necessity for current and future research on developing economically achievable and ecologically viable wastewater treatments. Both intracellular and extracellular enzymes are explored as biochemical means of wastewater treatment with the intention of selectively degrading a target pollutant without affecting the other components in the effluent. Further, they can operate under mild reaction conditions. From the eco-friendly perception, the acceptability of enzymes mainly relies on their biodegradability. Oxidoreductases are most widely accepted for the remediation of dyestuffs. For example, the azo groups in dyes are converted to amines by reductive cleavage using azo reductases. Peroxidases like HRP, MnP, and LiP require peroxides like H_2O_2 for their functioning. The administration of the enzyme to the target effluent is carried out by simple addition of the cells or tissues that produce the enzyme into the effluent directly. The direct administration of enzymes can be adopted to co-metabolize target contaminants using suitably adapted strains of microorganisms. Moreover, immobilization methods are effectual for enhancing the reusability of enzymes by preventing the loss of enzyme during the course of the reaction and minimizing the loss of activity of enzymes under harsh treatment conditions.

Enzymatic treatment can reduce many of the harmful effects of wastewater on the environment and human health. Enzymes are able to kill pathogens through the degradation of their cell walls. Peptidoglycan layer composed of amino acids, various sugars, and teichoic acid is responsible for the rigidity of the cell wall. The outer membrane must be degraded by a lysozyme or similar enzyme or EDTA before the enzymatic degradation of the peptidoglycan layer. Enzymes are able to break apart large sludge particles consist of polysaccharides and proteins. They are held together by extracellular polymeric substances that come from cell autolysis, bacterial metabolic reactions, and wastewater itself. Their cleavage facilitates more surface area for microbes to attack and allows for a more complete and more efficient degradation of the sludge particles. In the presence of free β-glucosidase, cellulase, and protease, these bonds break apart and the particles deflocculate. The enzyme mixture makes the wastewater more biodegradable, allowing the microbes to work more effectively. Indeed, the increased rate of oxygen consumption in wastewater signposts greater microbial activity and higher removal rates. The demand for more oxygen means that the treatment system must be aerated to prevent the wastewater from becoming anoxic or anaerobic. This cost may act as a deterrent from using enzyme mixtures on a large scale, so further studies must be done to reduce the ensuing oxygen consumption.

5.5.2 RENEWABLE ENERGY RESOURCES

5.5.2.1 FORMATION OF BY-PRODUCTS

If the concentration of one or more components in a waste or effluent stream exceeds a specific level, that material is termed as by-product. Industries are the major contributors to unlimited varieties of wastes. The recovery of the by-products can save energy. Enzymes are widely accepted for this purpose. These wastes can be broadly classified into four categories: starches and sugars, nonstarch carbohydrates, proteins, and fats and oils.[75]

5.5.2.2 ENZYMES FOR CLEAN ENERGY PRODUCTION

Nowadays, the production of renewable and clean energy sources like biodiesel, bioethanol, and biohydrogen using biocatalysts is promptly emerging. These alternatives can contribute to a cleaner environment, and the demand for these energies is increasing especially when using renewable agricultural products.[76,77] Indeed, nowadays, bioenergy could meet about 15% of the world's energy consumption. We trust, possibly, that in the next 20–30 years, we will be able to convert biomass for our transportation fuels.

Based on the biomass and enzymatic platform employed, we can differentiate diverse groups of clean energies.

5.5.2.2.1 Biodiesel

Recently, the conversion of vegetable oils to methyl or other short-chain esters by lipase-catalyzed transesterification reaction has led to high-grade production of biodiesel. The enzyme-catalyzed transesterification could overcome the disadvantages of acid/base catalyzed chemical transformations to a great extent. It reduces the consumption of energy and costly and chemically wasteful separation of the catalyst from the reaction mixture.[76] Biodiesel is a renewable source of energy. Besides, it has low emissions per volume. Moreover, biodiesel is exempt from diesel tax through special legislation in several European countries. Thus, the processes involving biocatalysis become more competitive. Likewise, efficient solvent-free syntheses have been achieved with immobilized lipases derived from *Pseudomonas cepacia*, *Rhizomucor miehei*, and *Candida antarctica*. However, the relatively high production cost, moderate reaction yields, and difficulties found during purification of the unreacted substrates necessitate new future advances.

5.5.2.2.2 Bioethanol

Carbohydrates were the major raw materials used for the production of food, clothing, and energy, prior to the discovery of petroleum. Ethanol fuels can be derived from renewable resources like agricultural crops (corn, sugar cane, and sugar beet) or from agricultural by-products such as whey.[76] Ethanol can replace the toxic oxygenate, methyl tert-butyl ether. Hence, it can be used as an effective supplement for petroleum fuels or as an extender. Acid hydrolysis of biomass into sugars is the best available technology for bioethanol production. The enzymatic substitutes like α-amylases, invertases, lactases, cellulases, and so on to hydrolyze carbohydrates into fermentable sugars are surprisingly growing up. In the second step, bacteria, yeasts, and fungi are found to be effectual instead of strong acids, for the fermentation of sugars to produce ethanol-avoiding by-products. Utilization of natural renewable resources under safer factory working conditions, significant reduction in harmful automobile emissions and consumer benefit are the expected environmental benefits from bioethanol as a safer alternative to the existing supply of liquid fuel, gasoline. Bioethanol derived from "lignocellulosic biomass" is commercialized as an alternative to fossil fuels in road transport in many countries. The industry is looking for low-cost, renewable, abundant, and accessible feedstock from food crops to woody biomass and agricultural waste for producing bioethanol. Simultaneously, cumulative waste generation linked to growing population and living standards is a worldwide challenge. Waste papers and cardboard are rich in cellulosic fractions, and hence, they are considered as potential feedstock for bioethanol production.[77]

5.5.2.2.3 Biohydrogen

Recently, molecular hydrogen is considered as a renewable, efficient, and pollution-free energy source for fuel. On burning, it yields only water; this makes it different from every other common fuel we use today. Hydrogen derived from biomass has the potential to compete with that produced from natural gas by catalytic conversion of hydrocarbons or photochemical water splitting. Even if certain microbes are effective in the production of hydrogen from fruit and vegetable wastes, the process is still at the laboratory stage. Consequently, most research has focused on the utilization of hydrogenases for the production of hydrogen, for instance, by the fermentation of sugars or more attractively from waste. On the other hand, right now, typical production ranges are only 0.37–3.32 moles hydrogen per one mole of glucose. Because of both the cost and low production, hydrogen is not our primary fuel. Hence, the search for new hydrogenases using genome database mining and metagenomes is a growing field of curiosity.[78]

5.5.3 BIOINDICATORS FOR POLLUTION MONITORING

Environmental monitoring involves the chemical analysis of defined pollutants. The availability of advanced technologies for environmental monitoring is one of the inherent drawbacks of chemical analysis. The induction of hepatic cytochromes in wild rodents can be used as a bioindicator for monitoring terrestrial environmental pollution.[79] Induction of specific families of hepatic cytochrome enzymes in various species of animals can be associated with environmental contaminants like polychlorinated dibenzodioxins, polychlorinated dibenzodifurans, polyaromatic hydrocarbons, and PCBs. Hydrolytic enzymes in activated sludge can be used as indicators of biodegradation activity of biomass. Enzyme activity has been shown to be a microbial population indicator, a monitor of active biomass, or an indicator of specific engineering parameter such as COD or phosphorous removal.

5.5.4 BIOSENSORS

The safety of living organisms from harmful pollutants relies on environmental security. Biosensors are ideal for the detection of hazardous materials and the measurement of the extent of pollution.[80–82] A biosensor is a self-sufficient integral tool that provides precise, quantitative, and analytical information. The prime elements of a biosensor are biological recognition element, transducer, and signal processing system. The biochemical receptor is in direct contact with transduction element. The accuracy and precision during the measurement, speed, sensitivity, specificity and range of measurement, reliability to testing, calibration and long-term stability, robustness, size, safety and portability, cost of analysis, and acceptability by users are the key properties for designing an enzyme-based biosensor. Enzymes can detect the consumption or production of compounds like CO_2, NH_3, H_2O_2, H^+, or O_2. Then the transducers identify the pollutants and correlate their existence into the substrates.[83] The key advantages offered by biosensors over conventional investigative methods for environmental applications include their portability, compactness; work on site; and capacity to analyze contaminants in composite matrices with the least sample preparation. Laccase, tyrosinase, and peroxidases are exploited for the development of biosensors for degradation of phenolic compounds. Cyanide is very toxic and inhibits respiration by binding with cytochrome oxidase, and the presence of cyanide can be monitored by an oxygen electrode exploiting immobilized bacteria. Further, biosensors have been developed for the detection of pesticides like OPs. Similarly, herbicides and triazines that inhibit the process of photosynthesis can be detected by amperometric sensors and optical transducers. Amperometric biosensors can compromise a viable, low-cost solution for field monitoring and environmental analysis. The exact balance of immobilization, transduction, and biorecognition strategies can yield better quality affinity electrochemical biosensors for environmental applications. In short, low detection limit and specificity for a specific analyte are key issues using biosensors in environmental analysis.

5.6 CONCLUSION

Even though bioremediation is not intrinsically a green technology, it can be conducted in a greener way. Recognizing and identifying the potential for adverse human health effects in bioremediation operations can lead to strategies to mitigate these effects. Enzymes are versatile catalysts and powerful tools to clean up and preserve the environment. Enzymes like laccases, LiP, and MnP produced from various fungi have been proven to be effective for the remediation of TNT, PAHs, PCBs, OPs, phenols, dyestuffs, wastewater, and so on. They are exploited in industries like agro-food, chemical, biochemical, and pharmaceutical industries. Enzymes envisage an unpolluted environment through effective waste management. They act either as molecular detectors or biosensors for environmental monitoring. The production of biodiesel, bioethanol, and biohydrogen from renewable agricultural products or wastes using biocatalysts has promptly emerging contribution to a cleaner environment. The bioremediation technology is greener, clean, and safe technology for the cleanup of contaminated site. To ensure the use of bioremediation as a green technology, the monitoring for toxicity reduction should be expanded to include all bioremediation efforts.

KEYWORDS

- **bioremediation**
- **biocatalyst**
- **clean energy**
- **biodiesel**
- **biohydrogen**

REFERENCES

1. Vidali, M. Bioremediation. An Overview. *Pure Appl. Chem.* **2001,** *73* (7), 1163–1172.
2. Shinde, S. Bioremediation. An Overview. *Recent Res. Sci. Technol.* **2013,** *5* (5), 67–72.
3. Gianfreda, L.; Rao, M. A.; Scelza, R.; De La Luz Mora, M. Role of Enzymes in Environment Cleanup/Remediation. In *Agro-Industrial Wastes as Feedstock for Enzyme Production*; Academic Press: San Diego, CA, 2016; pp. 133–155.
4. Karigar, C. S.; Rao, S. S. Role of Microbial Enzymes in the Bioremediation of Pollutants: A Review. *Enzyme Res.* **2011,** *2011*, 1–11.

5. Duran, N.; Esposito, E. Potential Applications of Oxidative Enzymes and Phenoloxidase-like Compounds in Wastewater and Soil Treatment: A Review. *Appl. Catal., B.* **2000,** *28* (2), 83–99.
6. Park, J. W.; Park, B. K.; Kim, J. E. Remediation of Soil Contaminated with 2, 4-Dichlorophenol by Treatment of Minced Shepherd's Purse Roots. *Arch. Environ. Contam. Toxicol.* **2006,** *50* (2), 191–195.
7. Newman, L. A.; Doty, S. L.; Gery, K. L.; Heilman, P. E.; Muiznieks, I.; Shang, T. Q.; Gordon, M. P. Phytoremediation of Organic Contaminants: A Review of Phytoremediation Research at the University of Washington. *J. Soil Contam.* **1998,** *7* (4), 531–542.
8. Leung, M. Bioremediation: Techniques for Cleaning Up a Mess. *J. Biotechnol.* **2004,** *2*, 18–22.
9. Fetzner, S. Oxygenases without Requirement for Cofactors or Metal Ions. *Appl. Microbiol. Biotechnol.* **2002,** *60* (3), 243–257.
10. Nebe, J.; Baldwin, B. R.; Kassab, R. L.; Nies, L.; Nakatsu, C. H. Quantification of Aromatic Oxygenase Genes to Evaluate Enhanced Bioremediation by Oxygen Releasing Materials at a Gasoline-contaminated Site. *Environ. Sci. Technol.* **2009,** *43* (6), 2029–2034.
11. Fetzner, S.; Lingens, F. Bacterial Dehalogenases: Biochemistry, Genetics, and Biotechnological Applications. *Microbiol. Mol. Biol. Rev.* **1994,** *58* (4), 641–685.
12. Jones, J. P.; O'Hare, E. J.; Wong, L. L. Oxidation of Polychlorinated Benzenes by Genetically Engineered CYP101 (Cytochrome P450cam). *Eur. J. Biochem.* **2001,** *268* (5), 1460–1467.
13. Cirino, P. C.; Arnold, F. H. Protein Engineering of Oxygenases for Biocatalysis. *Curr. Opin. Chem. Biol.* **2002,** *6* (2), 130–135.
14. Arora, P. K.; Srivastava, A.; Singh, V. P. Application of Monooxygenases in Dehalogenation, Desulphurization, Denitrification and Hydroxylation of Aromatic Compounds. *J. Biorem. Biodegrad.* **2010,** *1*, 112.
15. Dua, M.; Singh, A.; Sethunathan, N.; Johri, A. Biotechnology and Bioremediation: Successes and Limitations. *Appl. Microbiol. Biotechnol.* **2002,** *59* (2–3), 143–152.
16. Quejr, L.; Ho, R. Y. Dioxygen Activation by Enzymes with Mononuclear Non-heme Iron Active Sites. *Chem. Rev.* **1996,** *96* (7), 2607–2624.
17. Couto, S. R.; Herrera, J. L. T. Industrial and Biotechnological Applications of Laccases: A Review. *Biotechnol. Adv.* **2006,** *24* (5), 500–513.
18. Mai, C.; Schormann, W.; Milstein, O.; Hüttermann, A. Enhanced Stability of Laccase in the Presence of Phenolic Compounds. *Appl. Microbiol. Biotechnol.* **2000,** *54* (4), 510–514.
19. Ullah, M. A.; Bedford, C. T.; Evans, C. S. Reactions of Pentachlorophenol with Laccase from *Coriolus versicolor*. *Appl. Microbiol. Biotechnol.* **2000,** *53* (2), 230–234.
20. Demiralp, B.; Büyük, İ.; Aras, S.; Cansaranduman, D. Industrial and Biotechnological Applications of Laccase Enzyme. *Turk. Bull. Hyg. Exp. Biol.* **2015,** *72* (4), 351–368.
21. Kim, J. S.; Park, J. W.; Lee, S. E.; Kim, J. E. Formation of Bound Residues of 8-Hydroxybentazon by Oxidoreductive Catalysts in Soil. *J. Agric. Food Chem.* **2002,** *50* (12), 3507–3511.
22. Rodriguez, E.; Pickard, M. A.; Vazquez-Duhalt, R. Industrial Dye Decolorization by Laccases from Ligninolytic Fungi. *Curr. Microbiol.* **1999,** *38* (1), 27–32.

23. Champagne, P. P.; Ramsay, J. A. Dye Decolorization and Detoxification by Laccase Immobilized on Porous Glass Beads. *Bioresour. Technol.* **2010,** *101* (7), 2230–2235.
24. Chakroun, H.; Mechichi, T.; Martinez, M. J.; Dhouib, A.; Sayadi, S. Purification and Characterization of a Novel Laccase from the Ascomycete *Trichoderma atroviride*: Application on Bioremediation of Phenolic Compounds. *Process Biochem.* **2010,** *45* (4), 507–513.
25. Xu, F. Catalysis of Novel Enzymatic Iodide Oxidation by Fungal Laccase. *Appl. Biochem. Biotechnol.* **1996,** *59* (3), 221–230.
26. Koua, D.; Cerutti, L.; Falquet, L.; Sigrist, C. J.; Theiler, G.; Hulo, N.; Dunand, C. Peroxibase: A Database with New Tools for Peroxidase Family Classification. *Nucleic Acids Res.* **2008,** *37*(Database), D261-D266.
27. Hiner, A. N.; Ruiz, J. H.; López, J. N. R.; Cánovas, F. G.; Brisset, N. C.; Smith, A. T.; Acosta, M. Reactions of the Class II Peroxidases, Lignin Peroxidase and *Arthromyces ramosus* Peroxidase, with Hydrogen Peroxide: Catalase-like Activity, Compound III Formation, and Enzyme Inactivation. *J. Biol. Chem.* **2002,** *277* (30), 26879–26885.
28. Sumithran, S.; Sono, M.; Raner, G. M.; Dawson, J. H. Single Turnover Studies of Oxidative Halophenol Dehalogenation by Horseradish Peroxidase Reveal a Mechanism Involving Two Consecutive One Electron Steps: Toward a Functional Halophenol Bioremediation Catalyst. *J. Inorg. Biochem.* **2012,** *117*, 316–321.
29. Alneyadi, A. H.; Ashraf, S. S. Differential Enzymatic Degradation of Thiazole Pollutants by Two Different Peroxidases—A Comparative Study. *Chem. Eng. J.* **2016,** *303*, 529–538.
30. Xu, P.; Liu, L.; Zeng, G.; Huang, D.; Lai, C.; Zhao, M., Zhang, C. Heavy Metal-induced Glutathione Accumulation and Its Role in Heavy Metal Detoxification in *Phanerochaete chrysosporium*. *Appl. Microbiol. Biotechnol.* **2014,** *98* (14), 6409–6418.
31. Ten Have, R.; Teunissen, P. J. Oxidative Mechanisms Involved in Lignin Degradation by White-rot Fungi. *Chem. Rev.* **2001,** *101* (11), 3397–3414.
32. Tang, L.; Zeng, G. M.; Shen, G. L.; Zhang, Y.; Huang, G. H.; Li, J. B. Simultaneous Amperometric Determination of Lignin Peroxidase and Manganese Peroxidase Activities in Compost Bioremediation Using Artificial Neural Networks. *AnalyticaChimicaActa.* **2006,** *579* (1), 109–116.
33. Abdel-Hamid, A. M.; Solbiati, J. O.; Cann, I. K. Insights into Lignin Degradation and Its Potential Industrial Applications. In *Advances in Applied Microbiology*, Vol. 82; Academic Press: Cambridge, MA, 2013; pp. 1–28.
34. Piontek, K.; Smith, A. T.; Blodig, W. Lignin Peroxidase Structure and Function. *Biochem. Soc. Trans.* **2001,** *29* (2), 111–116.
35. Ruiz-Duenas, F. J.; Morales, M.; Pérez-Boada, M.; Choinowski, T.; Martínez, M. J.; Piontek, K.; Martínez, Á. T. Manganese Oxidation Site in Pleurotus Eryngii Versatile Peroxidase: A Site-directed Mutagenesis, Kinetic, and Crystallographic Study. *Biochemistry.* **2007,** *46* (1), 66–77.
36. Wong, D. W. Structure and Action Mechanism of Ligninolytic Enzymes. *Appl. Biochem. Biotechnol.* **2009,** *157* (2), 174–209.
37. Tsukihara, T.; Honda, Y.; Sakai, R.; Watanabe, T.; Watanabe, T. Exclusive Overproduction of Recombinant Versatile Peroxidase MnP2 by Genetically Modified White Rot Fungus, Pleurotus Ostreatus. *J. Biotechnol.* **2006,** *126* (4), 431–439.

38. Vasileva-Tonkova, E.; Galabova, D. Hydrolytic Enzymes and Surfactants of Bacterial Isolates from Lubricant-contaminated Wastewater. *Zeitschriftfürnaturforschung C.* **2003,** *58*(1–2): 87–92.
39. Sánchez-Porro, C.; Martin, S.; Mellado, E.; Ventosa, A. Diversity of Moderately Halophilic Bacteria Producing Extracellular Hydrolytic Enzymes. *J. Appl. Microbiol.* **2003,** *94* (2), 295–300.
40. Schmidt, O. In *Wood Tree Fungi*; Springer-Verlag: Berlin Heidelberg, 2006; p. 334.
41. Sharma, D.; Sharma, B.; Shukla, A. K. Biotechnological Approach of Microbial Lipase: A Review. *Biotechnology*. **2011,** *10* (1): 23–40.
42. Hermansyah, H.; Wijanarko, A.; Gozan, M.; Surya, R. A. T.; Utami, M. K.; Shibasaki-Kitakawa, N.; Yonemoto, T. Consecutive Reaction Model for Triglyceride Hydrolysis Using Lipase. *Jurnalteknologi*. **2007,** *2*, 151–157.
43. Adriano-Anaya, M.; Salvador-Figueroa, M.; Ocampo, J. A.; García-Romera, I. Plant Cell-wall Degrading Hydrolytic Enzymes of *Gluconacetobacter diazotrophicus*. *Symbiosis*. **2005,** *40* (3), 151–156.
44. Yang, C.; Xia, Y.; Qu, H.; Li, A. D.; Liu, R.; Wang, Y.; Zhang, T. Discovery of New Cellulases from the Metagenome by a Metagenomics-guided Strategy. *Biotechnol. Biofuels*. **2016,** *9* (1), 138.
45. Sun, Y.; Cheng, J. Hydrolysis of Lignocellulosic Materials for Ethanol Production: A Review. *Bioresour. Technol.* **2002,** *83* (1), 1–11.
46. Imran, M.; Anwar, Z.; Irshad, M.; Asad, M. J.; Ashfaq, H. Cellulase Production from Species of Fungi and Bacteria from Agricultural Wastes and Its Utilization in Industry: A Review. *Adv. Enzyme Res.* **2016,** *4* (2), 44–55.
47. Annamalai, N.; Rajeswari, M. V.; Balasubramanian, T. Thermostable and Alkaline Cellulases from Marine Sources. In *New and Future Developments in Microbial Biotechnology and Bioengineering*; Elsevier: Los Angeles, CA, 2016; pp. 91–98.
48. Khan, M. N.; Luna, I. Z.; Islam, M. M.; Sharmeen, S.; Salem, K. S.; Rashid, T. U.; Rahman, M. M. Cellulase in Waste Management Applications. In *New and Future Developments in Microbial Biotechnology and Bioengineering*; Elsevier: Los Angeles, CA, 2016; pp. 237–256.
49. Romeh, A. A.; Hendawi, M. Y. Bioremediation of Certain Organophosphorus Pesticides by Two Biofertilizers, Paenibacillus (Bacillus) Polymyxa (Prazmowski) and *Azospirillum lipoferum* (Beijerinck). *J. Agric. Sci. Technol.* **2014,** *16*, 265–276.
50. Alves, N. J.; Moore, M.; Johnson, B. J.; Dean, S. N.; Turner, K. B.; Medintz, I. L.; Walper, S. A. Environmental Decontamination of a Chemical Warfare Simulant Utilizing a Membrane Vesicle-encapsulated Phosphotriesterase. *ACS Appl. Mater. Interfaces*. **2018,** *10* (18), 15712–15719.
51. Koudelakova, T.; Bidmanova, S.; Dvorak, P.; Pavelka, A.; Chaloupkova, R.; Prokop, Z.; Damborsky, J. Haloalkane Dehalogenases: Biotechnological Applications. *Biotechnol. J.* **2013,** *8* (1), 32–45.
52. Beena, A. K.; Geevarghese, P. I. A Solvent Tolerant Thermostable Protease from a Psychrotrophic Isolate Obtained from Pasteurized Milk. *Dev. Microbiol. Mol. Biol.* **2010,** *1*, 113–119.
53. Singh, C. J. Optimization of an Extracellular Protease of *Chrysosporium keratinophilum* and Its Potential in Bioremediation of Keratinic Wastes. *Mycopathologia*. **2003,** *156* (3), 151–156.

54. Barrett, A. J. An Introduction to the Endopeptidases. In *Biochemistry of Pulmonary Emphysema*; Springer: London, 1992; pp. 27–34.
55. Rao, M. B.; Tanksale, A. M.; Ghatge, M. S.; Deshpande, V. V. Molecular and Biotechnological Aspects of Microbial Proteases. *Microbiol. Mol. Biol. Rev.* **1998,** *62* (3), 597–635.
56. Ahuja, S. K.; Ferreira, G. M.; Moreira, A. R. Utilization of Enzymes for Environmental Applications. *Crit. Rev. Biotechnol.* **2004,** *24* (2–3): 125–154.
57. Kumar, S.; Mathur, A.; Singh, V.; Nandy, S.; Khare, S. K.; Negi, S. Bioremediation of Waste Cooking Oil Using a Novel Lipase Produced by *Penicillium chrysogenum* SNP5 Grown in Solid Medium Containing Waste Grease. *Bioresour. Technol.* **2012,** *120*, 300–304.
58. Moeser, A. J.; Van Kempen, T. A. T. G. Dietary Fibre Level and Enzyme Inclusion Affect Nutrient Digestibility and Excreta Characteristics in Grower Pigs. *J. Sci. Food Agric.* **2002,** *82* (14), 1606–1613.
59. Godfrey, T.; West, S.; Poldermans, B. Industrial Enzymology (2nd Edn). *Trends Food Sci. Technol.* **1997,** *8* (5), 178–178.
60. Krawczyk, T. Surfactants & Detergents: The Continuing Evolution of Yesteryear's Powdered Detergent Is Continuing through the Further Modification of Liquid and Concentrated Products. *Inform-Champaign.* **1996,** *7*, 6–29.
61. Magnin, L.; Delpech, P.; Lantto, R. Potential of Enzymatic Deinking. *Process Biotechnol.-Amsterdam.* **2002,** 323–332.
62. Yachmenev, V. G.; Bertoniere, N. R.; Blanchard, E. J. Intensification of the Bio-processing of Cotton Textiles by Combined Enzyme/Ultrasound Treatment. *J. Chem. Technol. Biotechnol.* **2002,** *77* (5), 559–567.
63. Harris, R. E.; Mckay, I. D. Stimulation Fluids-New Applications for Enzymes in Oil and Gas Production-New Enzyme-based Oilfield Processes Offer Potential Performance, Cost and Environmental Benefits. *Pet. Eng. Int.* **1999,** *72* (4), 65–69.
64. Ruggaber, T. P.; Talley, J. W. Enhancing Bioremediation with Enzymatic Processes: A Review. *Pract. Period. Hazard.; Toxic, Radioact. Waste Manage.* **2006,** *10* (2), 73–85.
65. Pritchard, P.; Lin, J.; Mueller, J.; Shields, M. Bioremediation Research in EPA: An Overview of Needs, Directions and Potentials. In *Biotechnology in Industrial Waste Treatment and Bioremediation*; Lewis Publishers: Boca Raton, FL, 1996; pp. 3–26.
66. Heitkamp, M. A.; Freeman, J. P.; Miller, D. W.; Cerniglia, C. E. Pyrene Degradation by a Mycobacterium Sp.: Identification of Ring Oxidation and Ring Fission Products. *Appl. Environ. Microbiol.* **1988,** *54* (10), 2556–2565.
67. Wammer, K. H.; Peters, C. A. Polycyclic Aromatic Hydrocarbon Biodegradation Rates: A Structure-based Study. *Environ. Sci. Technol.* **2005,** *39* (8), 2571–2578.
68. Harayama, S. Polycyclic Aromatic Hydrocarbon Bioremediation Design. *Curr. Opin. Biotechnol.* **1997,** *8* (3), 268–273.
69. Esteve-Núñez, A.; Caballero, A.; Ramos, J. L. Biological Degradation of 2, 4, 6-Trinitrotoluene. *Microbiol. Mol. Biol. Rev.* **2001,** *65* (3), 335–352.
70. Alshabib, M.; Onaizi, S. A. A Review on Phenolic Wastewater Remediation Using Homogeneous and Heterogeneous Enzymatic Processes: Current Status and Potential Challenges. *Sep. Purif. Technol.* **2019,** *219*, 186–207.
71. Ohtsubo, Y.; Kudo, T.; Tsuda, M.; Nagata, Y. Strategies for Bioremediation of Polychlorinated Biphenyls. *Appl. Microbiol. Biotechnol.* **2004,** *65* (3), 250–258.

72. Chacko, J. T.; Subramaniam, K. Enzymatic Degradation of Azo Dyes—A Review. *Int. J. Environ. Sci.* **2011,** *1* (6), 1250–1260.
73. Manco, G.; Porzio, E.; Suzumoto, Y. Enzymatic Detoxification: A Sustainable Means of Degrading Toxic Organophosphate Pesticides and Chemical Warfare Nerve Agents. *J. Chem. Technol. Biotechnol.* **2018,** *93* (8), 2064–2082.
74. Mugdha, A.; Usha, M. Enzymatic Treatment of Wastewater Containing Dyestuffs Using Different Delivery Systems. *Sci. Rev. Chem. Commun.* **2012,** *2* (1), 31–40.
75. Sissell, K. Cleaning Up in the Enzymes Business. *Chem. Week.* **2001,** *163* (9), 45–46.
76. Srirangan, K.; Akawi, L.; Moo-Young, M.; Chou, C. P. Towards Sustainable Production of Clean Energy Carriers from Biomass Resources. *Appl. Energy.* **2012,** *100*, 172–186.
77. Zheng, X. J.; Yu, H. Q. Biological Hydrogen Production by Enriched Anaerobic Cultures in the Presence of Copper and Zinc. *J. Environ. Sci. Health, Part A.* **2004,** *39* (1), 89–101.
78. Lionetto, M. G.; Caricato, R.; Giordano, M. E.; Erroi, E.; Schettino, T. Carbonic Anhydrase as Pollution Biomarker: An Ancient Enzyme with a New Use. *Int. J. Environ. Res. Public Health.* **2012,** *9* (11): 3965–3977.
79. Alkorta, I.; Aizpurua, A.; Riga, P.; Albizu, I.; Amézaga, I.; Garbisu, C. Soil Enzyme Activities as Biological Indicators of Soil Health. *Reviews on Environmental Health.* **2003,** *18* (1), 65–73.
80. Koedrith, P.; Thasiphu, T.; Weon, J. I.; Boonprasert, R.; Tuitemwong, K.; Tuitemwong, P. Recent Trends in Rapid Environmental Monitoring of Pathogens and Toxicants: Potential of Nanoparticle-based Biosensor and Applications. *Sci. World J.* **2015,** *2015*, 1–12.
81. Dennison, M. J.; Turner, A. P. Biosensors for Environmental Monitoring. *Biotechnol. Adv.* **1995,** *13* (1), 1–12.
82. Nigam, V. K.; Shukla, P. Enzyme Based Biosensors for Detection of Environmental Pollutants—A Review. *J. Microbiol. Biotechnol.* **2015,** *25* (11), 1773–1781.
83. Verma, N.; Singh, M. A Disposable Microbial Based Biosensor for Quality Control in Milk. *Biosens. Bioelectron.* **2003,** *18* (10), 1219–1224.

CHAPTER 6

GREEN CATALYSIS, GREEN CHEMISTRY, AND ORGANIC SYNTHESES FOR SUSTAINABLE DEVELOPMENT

DIVYA MATHEW[1*], BENNY THOMAS[2], and K. S. DEVAKY[3]

[1]*FDP Substitute, Department of Chemistry, St. Berchmans College, Changanassery, Kerala, India*

[2]*Assistant Professor, Department of Chemistry, St. Berchmans College, Changanassery, Kerala, India*

[3]*Professor, School of Chemical Sciences, Mahatma Gandhi University, Kottayam, Kerala, India*

[*]Corresponding author. E-mail: divyabennythomas@gmail.com

ABSTRACT

Green catalysts offer an environmentally attractive catalyst–solvent system under mild reaction conditions for chemo-, regio-, and stereoselective design of products and processes. The fundamental pillars of green chemistry thus reduce the practice and generation of hazardous materials. The design and application of new catalytic systems achieve the dual goals of environmental protection and economic benefit. Green catalysis offers benefits like lower energy requirements, increased selectivity, decreased practice of processing agents, and platform for less toxic materials. Green catalysis and synthesis addresses environmentally benign process by providing the ease of separation of product and catalyst, thereby eliminating the necessity for separation through distillation or extraction.

6.1 INTRODUCTION

The welfare of modern society is incredible without the myriad products of industrial organic synthesis.[1] The history of organic synthesis is mostly traced back to Wohler's synthesis of urea from ammonium isocyanate in 1828, which laid to rest the vital force theory that preserves that a compound produced by a living organism could not be produced synthetically. The discovery had monumental significance, since, theoretically, all organic compounds are agreeable to synthesize in the laboratory. The next landmark in the progress of organic synthesis was the preparation of aniline purple, the first synthetic dye by Perkin in 1856. In the mid-1990s, the concept of *benign by design* that addressed the environmental issues of both chemical products and the processes was developed. Preferably exploiting renewable raw materials, green chemistry evades the practice of hazardous reagents and solvents in the production and application of chemical products, and eradicates waste materials effectively.

6.2 CATALYSIS AS A FOUNDATIONAL PILLAR OF GREEN CHEMISTRY

Catalysis relies on the design of chemical processes using fewer amounts of hazardous raw materials affording the generation of hazardous substances.[2] The design and application of new catalytic systems simultaneously achieve the dual goals of environmental protection and cost-effectiveness. Green chemistry seeks more benign ways to produce materials from feedstock to solvents. The design of a catalyst endeavors to optimize the stability of the catalyst, turnover number, solubility, and ease of separation of the spent catalyst from the product. The application of catalysts is ubiquitous in the chemical industry in areas ranging from pharmaceuticals to polymers to petroleum processing. Catalytic methods can provide safer substitutes to reagents frequently used in oxidation reactions, such as chromium, a suspected carcinogen. More than 90% of all industrial processes are based on catalysis. The extensive utilization of catalytic processes by industry reflects the profitable and environmental benefits achieved through catalysis. An additional benefit from the use of catalysis is that it enables the implementation of numerous green chemistry goals ranging from the use of alternative solvents such as supercritical CO_2 or water, to utilization of bio-based renewable feedstock such as glucose, to energy minimization. Of course, catalysts will endure to offer even greater improvements in efficiency, selectivity, energy reduction, and rate enhancement. Catalysis is undoubtedly one of the foundational pillars of green chemistry, and it will continue to be one of the main vehicles that take the chemical enterprise into a future of sustainability.[3]

6.3 GREEN SOLVENTS IN ORGANIC SYNTHESIS

Solvents used in various organic syntheses are rationally hazardous to the environment. Volatile organic solvents are discharged into the environment by evaporation in extensive amounts, since they are used in much higher proportions than the reagents themselves. Problems related to the utilization of solvents can be alleviated by carrying out the chemical reactions without solvents or by the use of nonvolatile harmless solvents to a greater extent. An ideal "green" solvent should have high boiling point. Furthermore, it must be nontoxic, dissolve numerous organic compounds, cheap, natural, readily available, and recyclable.[4] As per green chemistry, an ideal solvent facilitates the mass transfer but does not get dissolved. Unmistakably such necessities strongly limit the selection of substance as a green solvent. The extensive efforts of research groups throughout the world have led to the establishment of noteworthy alternatives to the common organic solvents. This includes supercritical liquids (SCLs), ionic liquids, low-melting polymers, perfluorinated or fluorous solvents, and water.[5]

6.3.1 *WATER*

Water is the basis and bearer of life on earth. Numerous biochemical, organic, and inorganic reactions that affect the living systems have inevitably occurred in an aqueous medium.[5] The basis of modern organic chemistry is that "organic reactions in organic solvents" or "like dissolves like." Chemists have focused their attention on carrying out organic reactions in the aqueous medium because of some potential advantages of low cost, safety, synthetic efficiency, simple handling, and environmental benefits. Water is the cheapest solvent available on earth, and using aqueous solvents in chemical processes becomes more economical. Many organic solvents are flammable, potentially explosive, mutagenic, or carcinogenic. In the point of view of these adverse effects, water is a safer solvent. The protection and deprotection of functional groups in organic synthesis is time- and energy-consuming. But water displays high synthetic efficiency as water-soluble substrates can be used directly, especially in carbohydrate and protein chemistry. In large industrial processes, isolation of the organic products can be executed by simple phase separation in the aqueous medium. Moreover, because of the highest heat capacity of water, it is easier to control the reaction temperature. The use of water may alleviate the challenge of pollution created by organic solvents owing to its ready recyclability. Furthermore, it is benign when discharged into the environment without any harmful residues. On the basis of the above features, water is undoubtedly the greenest solvent.[6] The insolubility of many organic compounds in water is one of the main drawbacks of using water. But water is highly reactive with organometallic compounds. The use of cosolvents or surfactants may increase the solubility of nonpolar reagents in water by disrupting the dense hydrogen bonding network of pure water. Wittig olefination reactions with stabilized ylides can be performed in an aqueous medium without using a phase-transfer catalyst. Recently, water-soluble phosphonium salts are synthesized, and their Wittig reactions with substituted benzaldehydes can be carried out in aqueous sodium hydroxide solution.

6.3.2 IONIC LIQUIDS

Ionic liquids are one of the most widely explored alternatives to organic solvents. They display negligible vapor pressure, good chemical and thermal stability, inflammability and high ionic conductivity, wide electrochemical potential, and catalytic properties compared to conventional organic solvents.[4,5] Additionally, ionic liquids consist of ions in contrast to conventional solvents consisting of single molecules. They are liquid at room temperature or they have low melting temperatures, usually below 100 °C. Enormous varieties of ionic liquids can be envisaged by simple combination of different cations and anions on reasonable variation of physical properties like hydrophobicity, viscosity, density, and solvating ability (Scheme 1).[7] The ionic liquids can not only be used as a green alternative to organic solvents but also can act as reagents, catalysts, or media for catalyst immobilization. Highly expensive methods of preparation, difficulty in purification process, and scale-up are some of the limitations of using ionic liquids. The impurities already present or the addition of a catalyst and substrate may increase the viscosity of the highly viscous ionic liquids making them gel-like and hence difficult to process. Ionic liquids like chloroaluminates are highly sensitive to oxygen and water, and hence, all substrates must be dried and degassed before use. Catalysts immobilized in ionic liquids may show leaching out problems in the product phase. Despite these problems, ionic liquids are attracting substitutes to volatile organic solvents in many different reactions, including oligomerization and polymerization, hydrogenation, hydroformylation and oxidation, C–C coupling, and metathesis. Ionic liquids containing BF_4^- or PF_6^- anions have been very widely used because of the formation of separate phases with many organic materials, biphasic catalysis, non-nucleophilicity, and the presence of an inert environment increasing their lifetime.

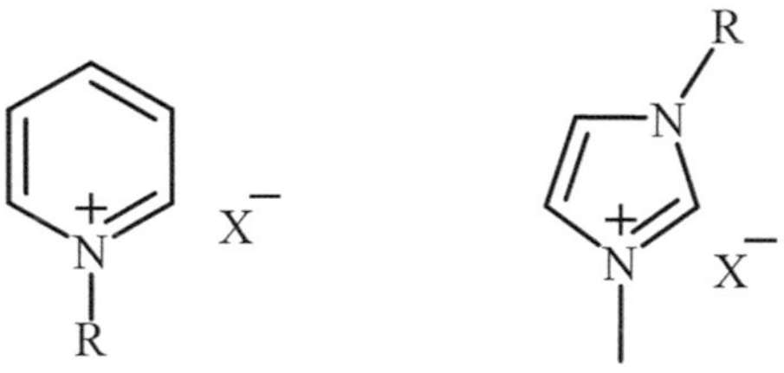
R
N
X
R

R = alkyl & $\bar{X}$ = PF_4, Br, OTs

SCHEME 1 Chemical structure of some widely used ionic liquids.

Ionic liquids may behave as an acidic, basic, or organocatalyst, depending upon the functional group attached to the cation and or anion.[8] Different types of Bronsted acidic ionic liquids (BAILs) have been synthesized and used as recyclable reaction media and acid promoters (Scheme 2).[9]

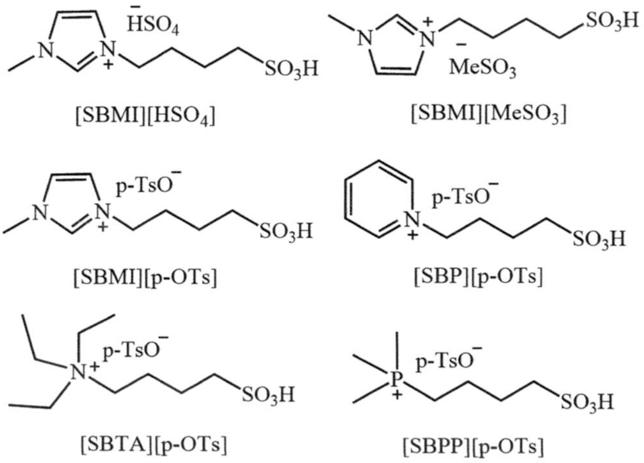
HSO4
SO3H
[SBMI][HSO4]
SO3H
MeSO3
[SBMI][MeSO3]
p-TsO
SO3H
[SBMI][p-OTs]
p-TsO
SO3H
[SBP][p-OTs]
p-TsO
SO3H
[SBTA][p-OTs]
p-TsO
SO3H
[SBPP][p-OTs]

SCHEME 2 Structure of BAILs.

The potential of BAILs has been exploited in many organic transformations like Beckmann rearrangement, aldol condensation, Michael reaction, and so on. Sulfonyl-containing ionic liquid can also be used as a green, solvent-free, metal-free, mild, and efficient recyclable catalyst. Acidic ionic liquids are reported to be ideal solvent for Friedel–Crafts acylation reactions. Acidic chloroaluminate ionic liquids can generate acylium ions, as the Lewis acid catalysts like $AlCl_3$ or $FeCl_3$ do with acyl chloride. Basic ionic liquids are found to be effective to catalyze a number of reactions like aza-Michael addition reaction, Michael addition of active methylene compounds, condensation reaction of aldehydes and ketones with hydroxylamine, and synthesis of quinolines and pyrroles. Hydrogen bonding interactions are one of the promising strategies in organocatalysis.

6.3.3 POLY(ETHYLENE GLYCOL) (PEG)

Poly(ethylene glycol) is a cheap, thermally stable, and recyclable material obtained by the polymerization of ethylene oxide. It is nontoxic and biocompatible. PEG is a green alternative to volatile organic solvents. Moreover, it is a convenient medium for organic reactions because of its low vapor pressures, inflammability, and ease of separation from the reaction medium. Furthermore, PEG can be used as an effective medium for phase-transfer catalysis.[10] Even though PEGs are not widely accepted, they are commercial products much cheaper than ionic liquids. But their properties cannot be changed easily. The PEG-promoted reactions have attracted organic chemists due to the higher solvating power of these compounds and their ability to act as a phase-transfer catalyst, negligible vapor pressure, easy recyclability, reusability, ease of work-up, eco-friendly nature, and low cost. The necessity of organic solvent for the extraction of the reaction products is one of the greatest disadvantages with PEGs. In short, PEG-promoted reactions have fascinated organic chemists due to the solvating power of these compounds, their ability to act as a phase-transfer catalyst, negligible vapor pressure, easy recyclability, reusability, ease of work-up, eco-friendly nature, and low cost.

6.3.4 PERFLUORINATED SOLVENTS

Horváth and Rabai introduced the term “fluorous” for the first time by analogy with “aqueous” or “aqueous medium.” Gladysz and Curran defined the fluorous compounds as highly fluorinated substances with sp^3-hybridized carbon atoms. Perfluoroalkanes, perfluoroalkyl ethers, and perfluoroalkyl amines are some of the widely used perfluorous solvents.[11,12,13] They are chemically stable and harmless to the environment due to their nontoxicity, inflammability, thermal stability, and recyclability. These solvents can dissolve oxygen, and hence, they are very beneficial in medical technology. Chemically, fluorous solvents are fluorine substituted organic compounds. Fluorous liquids have quite unusual properties including high density, high stability of the C–F bond, and extremely low solubility in water and organic solvents. The low solubility of the perfluorinated solvents in water and organic solvents can be enlightened in terms of their low surface tension, weak intermolecular interactions, high densities, and low dielectric constants. Furthermore, they are miscible with the organic solvents at higher temperatures. But they are inappropriate to perform most chemical reactions due to extreme nonpolarity. Hence, they are used in combined form with conventional organic solvents to give biphasic mixtures. In such a biphasic mixture, the soluble part may be in the fluorous phase while the starting materials are dissolved in the immiscible solvent phase (organic, water, or nonorganic). On heating, these two distinct layers become homogenized and further get separated again upon cooling. On cooling, the reaction products will remain in the organic phase while the unreacted substances and the catalyst will remain in the perfluorous phase. This separation paves a way for an easy separation of the reaction products and recycling of the spent catalyst without the use of an organic solvent for extraction. Such a system combines the advantages of a monophasic system with the ease of product separation in the biphasic system. Sonogashira coupling in a liquid/liquid fluorous biphasic system is reported with higher efficiency. The highly expensive nature and the requirement of toxic gaseous fluorine or hydrogen fluoride (HF) for their production are the main disadvantages associated with fluorous solvents.

6.3.5 SCLs

An SCL is defined as any substance at a temperature and pressure above its critical point, where distinct liquid and gas phases do not exist. SCLs have properties in between those of its liquid and gaseous phases. These properties can be specifically changed by varying the temperature and pressure.[14] The most widely accepted SCL is supercritical carbon dioxide ($scCO_2$). The critical point of $scCO_2$ is at 73 atm and 31.1 °C. Because of the extreme conditions essential to realize the critical point, other supercritical solvents are not as beneficial; for instance, the critical point of water is at 218 atm and 374 °C. The selectivity of a chemical reaction can be dramatically improved when conducted in SCLs when compared to the use of traditional organic solvents. $scCO_2$ has been successfully employed in cationic polymerizations. One example is the polymerization of isobutyl vinyl ether (IBVE) using an adduct of acetic acid and IBVE as the initiator, ethylaluminum dichloride as a Lewis acid, and ethyl acetate as a Lewis base deactivator. The reaction proceeded via a heterogeneous precipitation process in $scCO_2$ to form poly(IBVE) in 91% yield with a molecular weight distribution of 1.8. Stoichiometric and catalytic Diels–Alder reactions in $scCO_2$ have been studied extensively and the first report appeared in 1987. The reaction of maleic anhydride and isoprene can be successfully conducted in $scCO_2$ with higher rate.

The major advantages of $scCO_2$ are (1) inflammable and less toxic than most organic solvents, (2) relatively inert toward reactive substances, (3) natural gas found in the atmosphere, (4) no regulations concerning its use, (5) easily removed by decreasing the pressure from the reaction products, (6) high gas-dissolving ability, (7) low solvating ability, (8) high diffusion rate and (9) good mass transfer properties.

However, there are some disadvantages associated with $scCO_2$. Some of them are (1) reactivity toward strong nucleophiles, (2) requirement of specialized and expensive equipment to achieve the critical conditions, (3) low dielectric constant and low dissolving ability, and (4) hydrocarbon solvent nature and difficulty to dissolve catalysts and/or reagents.

6.3.6 BIOSOLVENTS

Biosolvents have been developed as an alternative to volatile organic compounds, which are usually harmful to the environment and to human health. The most important chemical classes of biosolvents are esters of naturally occurring acids and fatty acids, bioethanol, terpenic compounds, isosorbide, glycerol, and glycerol derivatives. They offer the advantage of being produced from renewable sources such as vegetable, animal, or mineral raw materials by chemical and physical processes without the consumption of fossil resources. They are already widely used in cosmetics, cleaning agents, paint, inks, and agricultural chemicals. Bio-based solvents have successfully been employed in multicomponent reactions (MCRs) with seemingly synergistic effects. For example, glycerol exercises a promoting effect on the oxo-Diels–Alder reaction of the intermediate methylendimedone with styrene due to its polar and protic properties.

A bio-based solvent has to meet certain criteria for their application such as (1) optimal technical specifications like dissolution capability, volatility, and flash point; (2) environmental safety; (3) eco-compatible production; and (4) availability and cost of the renewable raw materials.

6.3.7 ORGANIC CARBONATES

Organic carbonates represent esters of carbonic acids and are a class of compounds with a broad field of application due to their unusual properties. They are easily available in large amounts, inexpensive; possess low toxicity; and are completely biodegradable. Organic carbonates are widely accepted for extraction purposes, pharmaceutical, and medical applications. Moreover, they are extensively exploited in batteries too. Two subclasses of carbonates are differentiated: open chain and cyclic organic carbonates.[15] Cyclic carbonates display a wider temperature range in the liquid state, and therefore, they are more suitable as solvents. Exclusively, such cyclic carbonates fulfill the necessities of green solvents such as low flammability, volatility, and low toxicity.[16] Propylene carbonate is an aprotic, highly dipolar solvent with low viscosity and a very large liquid state range. Since propylene carbonate has a high molecular dipole moment of 4.9 D, it is susceptible to microwave irradiation. It can be considered as a very interesting solvent for microwave-assisted organic synthesis; however, unfortunately, it has been hardly investigated yet. Furthermore, if the pure enantiomeric carbonate is utilized, its stereogenic center can be exploited for the enantioselectivity of a stereoselective reaction.

6.4 GREEN CATALYSTS IN ORGANIC SYNTHESIS

"Green" in chemistry is used to promote the practice of eco-friendly procedures, reusable and recyclable reagents with minimum waste production. Catalyst is simply a substance that increases the rate of a chemical reaction without itself undergoing any permanent chemical change. Furthermore, catalysis is a key technology to achieve the objectives of sustainable chemistry. The practice of catalysts paves way for minimizing the necessity for large quantities of reagents for chemical transformations and results in ultimate reduction in the discharge of wastes. The eco-friendly green catalysts can be regenerated and reused multiple times to minimize waste production during the process.

6.4.1 GREEN LEWIS ACID CATALYSTS

Lewis acid catalysis is of great interest in organic synthesis.[17] But the Lewis acids have a higher affinity toward water rather than toward the substrates, and hence, the presence of even a minute amount of water may stop the reaction. Consequently, Lewis acids have restricted application in organic synthesis. However, undeniably, water is a safe, harmless, and environmentally benign solvent.[18] From a viewpoint of green chemistry, nevertheless, it is desirable to use water instead of organic solvents as a reaction solvent. Rare earth metal triflates can be used as water-stable Lewis acids in water-containing solvents, for instance, $Sc(OTf)_3$ and $Yb(OTf)_3$. Furthermore, Lewis acid–surfactant-combined catalyst has also been developed without using any organic solvents. The first example of Lewis acid-catalyzed reaction is the hydroxymethylation of silyl enol ethers by commercial formaldehyde in aqueous solution. Moreover, $Yb(OTf)_3$ is found to be the most effective catalyst among various $Ln(OTf)_3$ tested for the aldol reaction because the catalytic amount of $Yb(OTf)_3$ can drive the reaction to completion. Furthermore, $Yb(OTf)_3$ can activate aldehydes other than formaldehyde with silyl enol ethers in aqueous solvents. Some important types of Lewis acid catalysts are discussed here.

6.4.1.1 LEWIS ACID–SURFACTANT-COMBINED CATALYSTS (LASCS)

Lewis acid-catalyzed aldol reactions in aqueous solvents are smoothly catalyzed by several metal salts. An organic solvent like THF is also helpful in the smooth progress of aldol-type reactions. A new reaction system of metal triflates is developed in combination with a small amount of a surfactant such as sodium dodecyl sulfates to avoid the practice of harmful organic solvents.

6.4.1.2 LEWIS ACID CATALYSIS IN $SCCO_2$

Lewis acids in $scCO_2$ are appropriate substitutes for toxic organic solvents to accomplish benign chemical reactions. One of the most widely accepted

Lewis acid catalysts in $scCO_2$ is scandium tris(heptadecafluorooctanesulfo nate) ($Sc(O_3SC_8F_{17})_3$). Diels–Alder reactions of carbonyl dienophiles with dienes and aza Diels–Alder reactions of imines with a diene are found to be successfully carried out using $Sc(O_3SC_8F_{17})_3$ as catalyst in $scCO_2$.

6.4.1.3 INDIUM TRIBROMIDE

Indium tribromide is a novel type of water-tolerant green Lewis acid catalyst for organic synthesis. It displays high chemo-, regio-, and stereoselectivity. The advantages of $InBr_3$ compared to conventional Lewis acids include (1) its water stability, (2) easy recovery and recyclability, (3) operational simplicity, and (4) strong tolerance to oxygen- and nitrogen-containing substrates and functional groups. This catalyst promotes various transformation reactions including alkynylation of aldehydes and N,O-acetals, Michael addition, synthesis of 1,3-dioxane derivatives, conversion of oxiranes to thiiranes, sulfonation of indoles, alkynylation of acid chlorides, and synthesis of 2-halo-1,3-dienes.

6.4.1.4 NAFION–Fe

The applicability of Strecker reaction relies on the synthesis of various α-aminonitriles, the precursors of α-amino acids. The substrates for Strecker reaction are ketones, aliphatic/aromatic amines, and trimethylsilyl cyanide. This reaction can be carried out in moderate to high yields and high purity using a new "green" Lewis acid catalyst, Nafion–Fe under conventional thermal, as well as microwave conditions. Iron Nafionate is Fe(III) salt of Nafion–H, a solid polymeric perfluoroalkanesulfonic acid. Simple reaction setup, easy work-up procedure, mild reaction conditions, shorter reaction times, and high purity of the products are the notable features of the methodology using Nafion–Fe. It is a reusable, environmentally benign, highly effective, and easily accessible catalyst. It has a significant contribution in green chemistry satisfying the emergent demand for environmentally benign and clean synthetic processes.[19] Rare earth metal triflates and LASCs can function as easily recoverable and effectually reusable Lewis acids in aqueous media. Lewis acid catalysis can also be successfully carried out in $scCO_2$ to contribute to the development of "greener" reactions.

6.4.2 GREEN NANOCATALYSTS

Catalysis meets all the requirements of green chemistry. Nanocatalysis is regarded as an emerging field of science due to its high activity, selectivity, and productivity. The shape, exceptionally large surface area-to-volume ratio, and nanoscale size impart unique properties to nanocatalysts.[20] Synthesis of stabilized nanoparticles (NPs) sized between 1 and 100 nm is the main task of the nanochemistry. NPs may be synthesized by top-down technologies and bottom-up technologies. Selectivity and reactivity of NPs can be enhanced by controlling the crystal structure, morphology, and surface composition. The activity of NPs is a concern of surface structure, exposure of different crystallographic facets, and increased no. of edges, corners, and faces. The efficient control of morphology of the NPs is a fascinating task all over the world in recent years. Various elements and materials like aluminum, iron, titanium dioxide, clays, and silica have been used as nanoscaled catalysts. Structure and shape-based properties and large surface area of any materials at its nanoscale size can affect the catalytic activity of a material. The shape, size, and fine-tuning of nanocatalysts have accomplished greater selectivity. It enables the industrial chemical reactions to become more resource-efficient by less energy consumption and minimal waste production. Thus, it counters the environmental impact caused by our reliance on the chemical process. NPs have widespread application as the most important industrial catalyst ranging from chemical manufacturing to energy conversion and storage. The heterogeneity and their individual differences in size and shape are responsible for the variable and particle-specific catalytic activity of NPs.

The catalytic activity of nanoscale copper can be effectively utilized for the production of hydrogen from formaldehyde at room temperature. The reductive dechlorination of lindane can be catalyzed by nanoscale Fe–Pd bimetallic particles in aqueous medium. Impregnation of palladium on the surface of iron NPs increases the catalytic activity of Fe–Pd bimetallic system to enhance the dechlorination of lindane. Dechlorination occurs via adsorption of chlorinated compounds on the particle surface. Palladium on the surface acts as a collector of hydrogen gas that is produced by the reduction of water molecule in the presence of iron NPs. Nanoscale palladium reacts with hydrogen gas to form either metal hydride or hydrogen radicals. Both are highly reactive toward C–Cl bonds and finally replace the entire chlorine atoms from lindane to form cyclohexane. Surface functionalized nanomagnetite supported NPs in catalysis, green chemistry, and pharmaceutically significant reactions. These NPs act as a bridge between heterogeneous and homogeneous catalysis. Magnetite-supported metal nanocatalysts have been successfully exploited in a variety of organic syntheses.

6.4.3 NATURE'S CATALYSTS "ENZYMES"

Enzyme-catalyzed organic synthesis is a green alternative to conventional methods of chemical catalysis. Enzymes offer advantages like high selectivity, mild reaction conditions, recyclability, and biocompatibility.[21] Lipases are the most widely employed enzymes in organic reactions, for example, esterification, aminolysis, and Michael addition. Transesterification reactions are reversible, but when enolate esters are used, they liberate unstable enols as by-products, which instantly tautomerize to give the corresponding aldehydes or ketones. Steric hindrance offered by isopropenyl esters, vinyl esters are favored and result in better reaction rates. Enzymes can also catalyze Michael addition of nucleophiles to α,β-unsaturated carbonyl compounds. Organomercurial lyase and mercuric ion reductase have evolved efficient strategies to carry out organometallic and bioinorganic chemistry on mercury species based on the affinity of cysteinyl thiols for the ligation of RHgX and Hg(I1), and make the environment "green." Moreover, enzymes can be effectively used for polymerization reactions like polycondensation, oxidative polymerization, and ring-opening polymerization.

6.4.3.1 POLYCONDENSATION

In polycondensation reaction, as a result of polymerization, small molecules like water, ammonia, or alcohol get eliminated. Enzymes are reported as effective catalyst in polycondensation reaction, for instance, lipases. Polyesters can be effectively synthesized via lipase-catalyzed polycondensation of hydroxy acids or their esters.

6.4.3.2 OXIDATIVE POLYMERIZATION

Peroxidase and laccase can be employed for the oxidative polymerization of phenols and aniline derivatives yielding polyaromatic compounds. Horseradish peroxidase with iron protoporphyrin as the active site is the most widely used catalyst in oxidative polymerizations.

6.4.3.3 RING-OPENING POLYMERIZATION

A common methodology in which enzymes are employed is the synthesis of polyesters by lipase-catalyzed ring-opening polymerization of lactones.

Thus, the enzymatic polymerization is desirable and contributes to "green polymer chemistry" for maintaining a sustainable society.

6.4.4 FRUIT JUICES AS GREENER CATALYSTS

The design of facile and green synthetic approaches is crucial in organic synthesis. Green chemistry offers a motivational and inspirational tool for organic chemists to develop benign pathways for the synthesis of bioactive molecules. Among various green chemistry aspects, the selection of catalysts under the framework of mild reaction conditions with optimum yield is the most important part of the reaction procedure. Fruit juice plays a significant role as homogenous biocatalyst in many of the chemical reactions in an easy and smooth way following all the parameters of green chemistry.[22] Fruit juices are natural source for various acids, act as natural acid catalyst in organic synthesis and catalyzed reactions at room temperature and under solvent-free conditions, and provide an alternative for harmful catalysts for organic synthesis. Fruit juice allows mild and highly selective transformation in a facile and environmentally friendly manner. Moreover, they are easily available and comparatively inexpensive, and their juice can be extracted easily. Their acidic properties, enzymatic activity, benign environmental character, and commercial availability are responsible for the growing interest of fruit juice in organic synthesis. Their catalytic activity includes the formation of C–C, C–N bonds and breaking of C–O, C–N bonds in different synthetically important organic compounds which have been studied. Fruit juice of lemon, pineapple, tamarind, and coconut is extensively used in organic synthesis. Most people are familiar with the traditional uses for lemon juice such as culinary, medicinal, and industrial purposes. Lemon juice is reported to catalyze Knoevenagel condensation reaction to synthesize arylidene-malononitriles, which shows antibacterial and antifungal activity. Further, it is useful for the three-component synthesis of dihydropyrimidinones, triazoles, synthesis of Schiff bases, and bis-, tris-, and tetraindoles. Pineapple juice and tamarind juice are used for the synthesis of dihydropyrimidinones and bis-, tris-, and tetraindoles, respectively. Anilides and aldamines are also synthesized by *Acacia concinna* fruit juice and *Sapindus trifoliatus* fruit juice, respectively.

Coconut juice is used as a biocatalyst for reduction of carbonyl compounds and hydrolysis of esters, amides, and anilides.[23]

6.4.5 BASIC AMINO ACIDS AS GREEN CATALYSTS

Amino acids comprise the elements of amino N-terminal and carboxylic C-terminals. Even though the majority of the amino acids are amphoteric, the side chains can make them weak bases or acids. Arginine, lysine, and histidine comprise the basic amino acid triad.[24] Arginine is uncertainly an essential amino acid, while lysine and histidine are essential amino acids. They are widely found in a variety of foods such as milk and wheat, and can also be produced from fermentation of glucose and fructose. Being part of our food system, the basic amino acids are not toxic, unlike the reported organic amines and other chemical catalysts. The basic amino acid, arginine, can act as a nontoxic and efficient green catalyst for the isomerization of glucose to fructose in water. Arginine is the most effective isomerization catalyst among three basic amino acids, which can achieve similar fructose yields with better selectivity to other toxic isomerization catalysts. The presence of carboxylic moiety is presumably responsible for the better fructose selectivity. Mechanistically, arginine isomerized glucose through an enediol intermediate pathway similar to those of sodium hydroxide and other strong bases.[25]

6.4.6 PHOTOCATALYSTS AS GREENER CATALYSTS

Light can be considered as an ideal component for eco-friendly, "green" chemical synthesis. It is an abundant, inexpensive, nonpolluting, and boundlessly renewable source of clean energy.[26] As the society becomes progressively aware of the adverse impacts of industry on the environment, the design of methods to efficiently utilize the solar energy has emerged as one of the central scientific challenges of the 21st century. The connection between solar energy and environmental sustainability, however, is a much older idea that dates to the turn of the last century. The observation that light alone could affect exclusive chemical changes in organic compounds led early 20th century photochemists to recognize that the sun might represent an inexhaustible source of clean synthesis. Yet, the necessity for high-energy ultraviolet radiation restricts the environmental benefits of photochemical synthesis on industrially relevant scales. Transition metal photocatalysts offer a promising substitute for the development of practical and scalable industrial processes with great environmental benefits. Despite the widespread recognition of the potential economic and ecological benefits of photochemistry, however, mainly due to the inability of most organic molecules to absorb visible wavelengths of light, the chemical industry is slow to adopt photochemical reactions for the large-scale production of fine chemicals. The photochemistry of $Ru(bpy)_3^{2+}$ has been especially well studied for the synthesis of complex organic molecules.[27]

6.4.7 MAGNETIC NPs (MNPs) AS POWERFUL GREEN CATALYSTS

Green chemistry encouraged the scientific community to design efficient ways for the separation and subsequent recycling of the spent homogenous catalysts from the reaction mixture. MNPs are efficient supports for catalysis. They comprise mainly two components, a magnetic material, frequently iron, nickel, and cobalt, and a chemical component bearing catalytic functionality. MNPs are highly benign reactants for the correlation of homogeneous inorganic and organic catalysts.[28] The physicochemical properties of MNPs display strong dependence on their chemical structure and method of synthesis. NPs with size ranging from 1 to 100 nm may show superparamagnetism. For instance, ferrite NPs used in catalytic reactions possess sizes ranging from 2 to 100 nm. They can be synthesized by coprecipitation methods, simple calcination at high temperatures, hydrothermal routes, or self-combustion techniques. The easy removal of ferrite nanocatalysts from reaction systems is due to their magnetic properties. Ferrite NPs with distinct shapes varying from spherical to nanorods and nanotubes are available. Further, shape of NPs, doping ratios, presence of impurities, and specific surface area may extensively impact the catalytic activity of ferrites. Ferrite NPs are quite effectual in the synthesis of series of glycinamides. Moreover, oxidation reactions, alkylation, and dehydrogenation reactions can be effectively carried out using ferrite NPs. The catalytic activity of Fe_3O_4 NPs is well established under solvent-free conditions in a one-pot three-component condensation reaction, for example, the condensation of aromatic aldehyde, urea or thiourea, and β-dicarbonyl. This reaction affords the corresponding dihydropyrimidinones or thiones in high to excellent yields. Compared with the classical Biginelli reactions, Fe_3O_4 NP-catalyzed reaction gives better yield in short reaction time with ease of subsequent recycling and catalyst reusability.

6.5 GREENER ORGANIC SYNTHESES

Efficiency and environmental sustainability are the two principal issues in science and industry. Green chemistry comprises a new approach to the synthesis, processing, and application of chemical substances. Thus, it excludes the hazards materials that cause environmental pollution. To be green, organic syntheses must meet, if not all, at least some of the following requirements: (1) minimal or no discharge of waste, (2) atom efficiency, (3) reduction in the practice and generation of toxic chemicals, (4) production of biodegradable compounds, (5) avoiding solvents, (6) reduction in energy requirements, (7) utilization of renewable materials, and (8) exploiting catalysts rather than stoichiometric reagents.

6.5.1 POLYMERS IN GREEN SYNTHESIS

Solid-supported catalysts are the keystone for transition metal-based catalysis. The transition metal resources are scarcer and more expensive.[29] Higher reactivity and selectivity along with the ease of recycling and recovery of the spent catalyst are the major advantages compared to nonsupported catalysts. The reuse and recycling of transition metal catalysts is currently an important focus area in organic and inorganic chemistry. It would be highly desirable if one could not only reuse and remove supported metal catalysts but also tune the desired catalytic activity and selectivity. Clearly, cooperativity between the support and the catalytic moiety or at least a detailed understanding of the support/catalyst interface and easy tunability of the support are key elements that could allow achievement of this goal. The rigidity of polymer backbone, the nature of the crosslinker used, the density of reactive functionality at catalyst site, and the mode of catalyst attachment make the solid supports superior. For a monometallic reaction, one can envision enhanced activity with a polymeric system that not only ensures easy access to reagents but also facilitates isolation of catalytic sites. For bimetallic reactions, a flexible structure that allows catalytic sites to be in close proximity to each other is desirable. Even though the transition metals and sophisticated ligands used are expensive, the capability to remove toxic metals from the waste stream and the potential to control costs via catalyst recovery and recycle lead to a trend in society and industry toward "green" chemistry. Metallosalen complexes are effectual in facilitating a wide variety of asymmetric synthetic transformations. Polymer-supported salen catalysts can be broadly classified based on their synthetic design. The Co-salen and Al-salen complexes are vital examples of polymer design in improving cooperative catalysis. Soluble and highly flexible supports not only increase the access of reagents but also increase catalyst proximity thereby facilitating site–site interactions. Furthermore, high catalyst density or augmented local concentration of catalysts improves salen–salen interactions. From the point of catalyst attachment, catalysts linked to the polymer backbone in a pendant fashion are desirable. Longer and more flexible linkers facilitate the bimetallic pathway. In addition, the presence of dendrimeric branching units in the linker can also position catalysts in an orientation that favors their bimolecular interactions.

6.5.2 GREENER ONE-POT SYNTHESIS

Waste reduction through clever, green, clean, and smart chemical technologies and catalysis are at the heart of green chemistry. One-pot multistep reactions are very attractive since they significantly lower the cost of the synthetic routes, by reducing the number of purification and separation steps, as well as reducing the amount of waste and solvents.[30] However, the application of this type of synthesis is limited since the catalysts utilized in these reactions have to be compatible with each other. Different pairs of opposing reagents or catalysts like acid/base, oxidant/reducing catalyst, and enzyme/metal complex catalyst are proved to be effective for the one-pot synthesis. MCRs are usually associated with a number of benefits such as procedural simplicity, shorter reaction times, high atom economy, energy saving, lower costs, and avoidance of time-consuming expensive purification processes. In addition, MCRs have appeared to be much more environmentally friendly in comparison with conventional multistep synthesis. Further, MCRs are capable to offer rapid and easy access to libraries of heterocyclic "drug-like" scaffolds with high structural diversity finding applications in the field of drug discovery and development. The main advantages of MCRs include mild reaction conditions, operational simplicity, excellent yields, simple separation procedures, high atom economy and inexpensive, and environmentally benign catalyst. Moreover, reusability of the reaction media without significant loss of activity is an added advantage. Thus, MCRs have emerged as fascinating and well-accepted strategies to the scientific community at large for the construction of novel and complex molecular structures. The multicomponent coupling reactions are emerging as a useful source for accessing small drug-like molecules with several levels of structural diversity. They are welcome too in terms of economic and practical considerations. Multicomponent reactions involve three or more compounds reacting in a single event, but consecutively to form new products, which contains the essential parts of all the starting materials.

The formation of carbon–carbon bond is one of the most fundamental operations in organic chemistry. Formation of two carbon–carbon bonds in one reaction vessel is the basis of some of the most often used organic reactions like cycloaddition, annulations, and so on. The formation of three to six bonds effectively in one reaction vessel is definitely a severe challenge due to the multiplicity of reaction pathways available to reactive polyfunctional molecules and to numerous monofunctional molecules in the same reaction vessel. However, the success would provide rapid and efficient means for transforming simpler molecules into structurally much more complex, nonpolymeric, useful compounds. Over the years, various procedures have been developed for constructing three to six bonds in one-pot annulation reactions. The occurrence of imidazole moiety in natural products and pharmacologically active compounds has established a diverse array of synthetic approaches to these heterocycles. There are quite a lot of methods for the synthesis of highly substituted imidazole derivatives. The most used methods in the last decade are condensation of diones, aldehydes, primary amines and ammonia, N-alkylation of trisubstituted imidazoles, condensation of benzoin or benzoin acetate with aldehydes, and so on. The one-pot synthetic strategy offers the possibility of cost-effective and environmentally friendlier ways for large-scale synthesis of pyran annulated heterocyclic molecules of pharmaceutical interest.[31]

6.5.3 GREEN PHASE-TRANSFER CATALYSIS

Organic synthesis is the foremost mode to harvest chemical products of practical applications. Organic transformations to desired products usually necessitate sequence of chemical operations utilizing additional reagents, catalysts, solvents, and so on. These transformations are not quantitative and selective processes. Hence, during the course of synthesis, many waste materials are produced besides the desired products. The regeneration, destruction, and disposal of these waste materials are an energy-consuming process and create a heavy burden on the environment. It is therefore of crucial to develop synthetic methodologies that minimize these problems.

Phase-transfer catalysis is one of the most general and efficient methodologies that fulfill the requirements of green chemistry.[32] It consists of heterogeneous two-phase systems with a reservoir phase and an organic phase. The reservoir phase consists of reacting anions or base for generation of organic anions, whereas the organic phase locates organic reactants and catalysts. The reacting anions are continuously introduced into the organic phase in the form of lipophilic ion pairs with lipophilic cations supplied by the catalyst. Most often tetraalkylammonium cations serve this purpose. PTC reactions are mechanistically more complicated. The inorganic phase contains aqueous or solid NaOH or KOH or solid K_2CO_3, whereas the organic phase contains the anion precursor, an electrophilic reactant, and eventually a solvent.

Major advantages of PTC in industrial applications are (1) exclusion of organic solvents, (2) exclusion of hazardous, inconvenient, and expensive reactants; (3) high reactivity and selectivity of the active species; (4) better yields and high purity of products; (5) simplicity of the procedure; (6) low investment cost; (7) low energy consumption; (8) opportunity to mimic countercurrent process; and (9) minimization of industrial wastes.

PTC can be considered as the most efficient and general green technology due to the previously mentioned specific features. Alkylation of phenylacetonitrile exemplifies the application of this methodology. In the absence of catalyst, no reaction occurs among phenylacetonitrile, alkyl halide, and 50% aq. NaOH. Upon introduction of a tetraalkylammonium halide in a catalytic amount, the exothermic reaction produces phenylalkylacetonitrile.[33]

6.5.4 GREENER MICROWAVE SYNTHESIS

Green chemistry utilizes a set of principles that reduces or eliminates the use or generation of hazardous substances in the design, manufacture, and applications of chemical products. One of the key areas of green chemistry is the elimination of solvents in chemical processes or the replacement of hazardous solvents with relatively benign solvents. Microwave-assisted chemistry has blossomed into a useful technique for a variety of applications in organic synthesis and transformations. Although MW-assisted reactions in organic solvents have developed rapidly, the focus is now shifted to environmentally friendlier methods using greener solvents and supported reagents. There are many examples of the successful application of MW-assisted chemistry to organic synthesis. These include the use of benign reaction media, the use of solvent-free conditions, and the use of solid-supported reusable catalysts.[34]

Two subfields of microwave chemistry—open-vessel microwave chemistry (OVMC) and closed-vessel microwave chemistry—are developed independently based on the common objective of safe and clean application of microwaves in organic synthesis by capturing the benefits and minimizing the risks.

Well-recognized general advantages of microwave heating with either open or closed vessels, with or without solvents, are (1) intensely shorter reaction times than conventional heating, (2) remote introduction of energy without contact between the source and the sample, (3) immediate switch on and off of the energy input to the sample, (4) thermal inertia is lower than with conventional conductive heating, (5) transfer of energy throughout the entire mass of the product not at the surfaces, (6) higher heating rates than conventional methods, and (7) application in sequential or parallel synthesis.

Green MW-assisted syntheses have the potential to increase efficiency and safety and to minimize waste generation through cleaner processing. These advantages are significant in green and sustainable chemistry.

At the outset, OVMC employs domestic microwave ovens, commercially available vessels typically glass beakers, watch glasses, and conical flasks, and it is easy to implement. Solvent-free conditions can be employed in MW synthesis using neat reactants, reactants adsorbed onto solid supports, or reactants in the presence of phase-transfer catalysts as in the case of anionic reactions. Besides the apparent potential benefits in solvent usage, domestic microwave ovens conduct the reactions more conveniently and rapidly without temperature measurements.[35] Some of the major benefits claimed for microwave-heated reactions under solvent-free conditions are (1) circumvention of large volumes of solvent and hence no need for redistillation; (2) simple work-up by extraction, distillation, or sublimation; (3) use of recyclable solid supports instead of polluting mineral acids and oxidants; (4) absence of solvent facilitates scale-up; (5) enhanced safety by reducing risks of overpressure and explosions; (6) quite cleaner, faster, and better yield than conventional synthesis; and (7) multiscale production.

Neutral amphoteric alumina can be used as a catalyst in MW synthesis. Alumina behaves as an amphoteric support due to the acidity connected to OH surface groups resulting from hydration of Al_2O_3 and basicity associated with the accessible lone pair electrons on oxygen atoms. The combination of alumina supported reactions and microwave irradiation can be successfully applied to deprotect ester functionality of alcoholic groups in multistep organic synthesis. Deacylation of alcoholic and phenolic acetates under solvent-free conditions on neutral alumina surfaces is another example.

Alumina impregnated with KF acts as a strong base, and it is able to abstract a proton from rather weakly acidic carbon acids up to pKa of 35. Reactions with basic alumina in conjunction with microwave irradiation can lead to efficient procedures for base-catalyzed reactions. Synthesis of 2-aroylbenzofurans from α-tosyloxy ketones and salicylaldehydes is an example.[36]

Acidic supports like K10 and KSF montmorillonites are effective in MW-assisted organic synthesis. Clays are made up of layered silicates, and the imbalance in the exchange of interlayer cations is attributable to their acidities. Commercially available inexpensive K10 and KSF montmorillonites behave as strong acids and can be used as substitutes for HCl or H_2SO_4. For example, in the acetalization of galactono-lactone with long-chain aldehydes, K10 or KSF clays are solvent-free green alternative for conventional conditions using solvent and sulfuric acid.

6.5.5 GREENER CLAY CATALYSTS

Catalysts play a central role in achieving the goals of green chemistry in synthetic chemistry. They could be simple or complex, synthetic, or natural chemicals. They create a favorable path for a reaction to occur under the mildest possible conditions. For example, clays and zeolites have greater acceptance as catalyst in synthetic chemistry.[37] Clays are solid acidic catalysts, and it can function as both Bronsted and Lewis acid catalysts in their natural and ion-exchanged form. Modified smectite or swelling clays are effective catalysts for a diverse range of organic transformations. They become sticky and plastic in the presence of moisture, but they become hard and cohesive under dry conditions. Chemically, they are hydrous aluminosilicates with crystalline structure. They contain various other cations too. On the basis of their chemical composition and crystal structure, they are of four types, namely, illite, smectite, vermiculite, and kaolinite. The smectite clay is most useful as a catalyst in organic synthesis and is sold under the name montmorillonite. Montmorillonite clay is the main constituent of bentonites and Fuller's earth.

In organic transformations, the clay catalysts can function as either Bronsted or Lewis acids or both. The Lewis acid nature of clay is due to the presence of Al^{3+} and Fe^{3+} at the crystal edges. They exhibit an enhanced rate of ion exchange of the interlayer cations like Na^+ and Ca^{2+} by Al^{3+} ions in $AlCl_3$ solution. The dissociation of the intercalated water molecules coordinated to the cations is responsible for the Bronsted acidity of clays. As the water content increases, Bronsted acidity is found to be decreasing. But the exchange of interlayer cations with highly polarizing ions like Cr^{3+} increases the Bronsted acidity. Further, the surface area and pore volume in the clay structure have a significant contribution to the efficiency of the catalyst. On treating the clay with dilute acid, total acidity can be further increased by the process of "proton-exchange." Bronsted acidic clay catalyst can be used as a green substitute to highly corrosive mineral acids in organic reactions.

The montmorillonite lattice is composed of gibbsite and silicates. Here, an octahedrally coordinated gibbsite $[Al_2(OH)_6]$ is sandwiched between two sheets of tetrahedrally coordinated silicate $[SiO_4]^{4-}$. The three-sheet layer repeats itself in the lattice. The interlayer space is responsible for the physicochemical properties of the clay. Because of charge imbalances aroused by the exchange of trivalent Al^{3+} ions with tetravalent Si^{4+} ions in the tetrahedral sheets and of bivalent Mg^{2+} ions with trivalent Al^{3+} ions in the octahedral sheets, montmorillonite stems show a high degree of M^+ cation exchange efficiency. Montmorillonite clay catalyst is reported as an effective substitute for Bronsted acids or Lewis acids in a variety of organic reactions under milder conditions with greater selectively and better yields in shorter reaction times. Moreover, the easy separation of the catalyst by mere filtration from the reaction mixture makes the work-up and purification procedures simpler. Further, the spent catalyst can be regenerated and reused for a number of repeated cycles of reactions without much loss in efficiency. Thus, in short, the entire synthetic activity of clays is both economical and environmentally benign.

Furthermore, the organic reactions occurring on the surface and interstitial space of the clays can be effectively catalyzed by them. Structural and functional modifications of clays can be done by incorporating different metal cations, molecules, or complexes. These modifications in properties make them more useful catalysts with higher selectivity in product structure and yield.

6.5.5.1 FORMYLATION OF PHENOLS

The introduction of the –CHO group on to the aromatic ring is termed as formylation of phenols. It can be effectively carried out with formaldehyde in the presence of clay catalyst controlled by R_3N. Moreover, clays are the platform for oxidation reactions at some stage.

6.5.5.2 OXIDATION–REDUCTION REACTIONS

Clays can serve as excellent oxidizing reagents for alcohols, thio-compounds, and many other compounds. K-I0 clay supported iron(III) nitrate (Clayfen) and copper(II) nitrate reagents (claycop) are two well-known examples.

6.5.5.3 PYROLYTIC ELIMINATION

Pyrolytic reactions are temperature sensitive cleavage reactions. On pyrolysis, esters are converted to olefins by the elimination of acid functionalities. Pyrolytic reactions necessitate higher temperatures of about 500 °C. Generally, pyrolytic reactions occur in the absence of catalysts. However, aluminum-exchanged montmorillonite can give better results in xylene medium at −150 °C in many organic transformations.

6.5.5.4 REARRANGEMENT REACTIONS

Clay materials act as effective catalysts for rearrangement or isomerization reactions under very mild conditions. For instance, clay-catalyzed isomerization of n-alkanes to branched-chain alkanes is of key interest in the petrochemical industry. Clays having either the Bronsted acidity or the Lewis acidity can bring about the desired result in rearrangement/isomerization reactions, for example, pinacol–pinacolone rearrangement.

6.5.5.5 PINACOL–PINACOLONE REARRANGEMENT

Pinacol–pinacolone rearrangement reactions are acid-catalyzed conversion of tertiary 1,2-glycols. This reaction proceeds through alkyl/aryl migration to an adjacent carbon. Montmorillonite clay with Bronsted acidic properties is well suited for catalyzing such reactions with excellent results.

6.6 CONCLUSION

Efficiency and environmental sustainability are the two principal challenges in science and industry. Our future challenges in resource and environmental sustainability demand more proficient and benign scientific technologies for chemical processes and products. Waste reduction through clean chemical technologies and catalysis are at the heart of green chemistry. Though complete greenness may be difficult to reach, it is a goal chemists must aim at, through the improvement of several aspects and parameters of a given reaction, from the synthesis and availability of its reactants and reagents to the separation and purification of the product. Green chemistry overcomes such challenges by designing new synthetic and minimizing by-products and pursuing greener solvents. Enzyme catalysis, phase-transfer catalysis, microwave-assisted synthesis, Lewis acid catalysis, polymer-supported synthesis, and one-pot synthesis are some greener strategies in organic synthesis.

KEYWORDS

- **green solvents**
- **green catalysts**
- **enzymes**
- **nanocatalysts**
- **supercritical liquids**
- **green synthesis**

REFERENCES

1. Sheldon, R. Introduction to Green Chemistry, Organic Synthesis and Pharmaceuticals. In *Green Chemistry in the Pharmaceutical Industry*; Wiley-VCH Verlag GmbH: Weinheim, Germany, 2010; pp 1–20.
2. Anastas, P. T.; Kirchhoff, M. M.; Williamson, T. C. Catalysis as a Foundational Pillar of Green Chemistry. *Appl. Catal., A* **2001,** *221* (1–2), 3–13.
3. Anastas, P. T.; Warner, J. C. Principles of Green Chemistry. In *Green Chemistry: Theory and Practice*; Oxford University Press: New York, 1998; pp 29–56.
4. Desimone, J. M. Practical Approaches to Green Solvents. *Science* **2002,** *297* (5582), 799–803.
5. Li, C. J.; Chan, T. H. *Comprehensive Organic Reactions in Aqueous Media*. John Wiley & Sons: Hoboken, NJ, 2007.
6. Castro-Puyana, M.; Marina, M. L.; Plaza, M. Water as Green Extraction Solvent: Principles and Reasons for Its Use. *Curr. Opin. Green Sustainable Chem.* **2017,** *5*, 31–36.
7. Sureshkumar, M.; Lee, C. K. Biocatalytic Reactions in Hydrophobic Ionic Liquids. *J. Mol. Catal. B: Enzym.* **2009,** *60* (1–2), 1–12.

8. Ratti, R. Ionic Liquids: Synthesis and Applications in Catalysis. *Adv. Chem.* **2014,** *2014*, 729842.
9. Shanab, K.; Neudorfer, C.; Schirmer, E.; Spreitzer, H. Green Solvents in Organic Synthesis: An Overview. *Curr. Org. Chem.* **2013,** *17* (11), 1179–1187.
10. Kianpour, E.; Azizian, S. Polyethylene Glycol as a Green Solvent for Effective Extractive Desulfurization of Liquid Fuel at Ambient Conditions. *Fuel* **2014,** *137*, 36–40.
11. Suryakiran, N.; Ramesh, D.; Venkateswarlu, Y. Synthesis of 3-Amino 1 H-pyrazoles Catalyzed by P-toluene Sulphonic Acid Using Polyethylene Glycol-400 as an Efficient and Recyclable Reaction Medium. *Green Chem. Lett. Rev.* **2007,** *1* (1), 73–78.
12. Gladysz, J. A.; Curran, D. P. Fluorous Chemistry: From Biphasic Catalysis to a Parallel Chemical Universe and Beyond. *Tetrahedron* **2002,** *58* (20), 3823–3825.
13. Horváth, I. T. Fluorous Biphase Chemistry. *Acc. Chem. Res.* **1998,** *31* (10), 641–650.
14. Wells, S. L.; Desimone, J. CO_2 Technology Platform: An Important Tool for Environmental Problem Solving. *Angew. Chem. Int. Ed.* **2001,** *40* (3), 518527.
15. Bandres, M.; De Caro, P.; Thiebaud-Roux, S.; Borredon, M. E. Green Syntheses of Biobased Solvents. *Comptesrenduschimie* **2011,** *14* (7–8), 636–646.
16. Vollmer, C.; Thomann, R.; Janiak, C. Organic Carbonates as Stabilizing Solvents for Transition-metal Nanoparticles. *Dalton Trans.* **2012,** *41* (32), 9722–9727.
17. Kobayashi, S.; Manabe, K. Green Lewis Acid Catalysis in Organic Synthesis. *Pure Appl. Chem.* **2000,** *72* (7), 1373–1380.
18. Corma, A.; Garcia, H. Lewis Acids: From Conventional Homogeneous to Green Homogeneous and Heterogeneous Catalysis. *Chem. Rev.* **2003,** *103* (11), 4307–4366.
19. Zhang, Z. H. Indium Tribromide: A Water-tolerant Green Lewis Acid. *Synlett* **2005,** *2005* (4), 711–712.
20. Kalidindi, S. B.; Jagirdar, B. R. Nanocatalysis and Prospects of Green Chemistry. *Chemsuschem* **2012,** *5* (1), 65–75.
21. Puskas, J. E.; Sen, M. Y.; Seo, K. S. Green Polymer Chemistry Using Nature's Catalysts, Enzymes. *J. Polym. Sci., Part A: Polym. Chem.* **2009,** *47* (12), 2959–2976.
22. Gulati, S.; Singh, R. Green and Environmentally Benign Organic Synthesis by Using Fruit Juice as Biocatalyst: A Review. *Int. Res. J. Pure Appl. Chem.* **2018,** *16*, 1–15.
23. Pal, R. Fruit Juice: A Natural, Green and Biocatalyst System in Organic Synthesis. *Open J. Org. Chem.* **2013,** *1*, 47–56.
24. Watanabe, Y.; Sawada, K.; Hayashi, M. A Green Method for the Self-aldol Condensation of Aldehydes Using Lysine. *Green Chem.* **2010,** *12* (3), 384–386.
25. Yang, Q.; Sherbahn, M.; Runge, T. Basic Amino Acids as Green Catalysts for Isomerization of Glucose to Fructose in Water. *ACS Sustainable Chem. Eng.* **2016,** *4* (6), 3526–3534.
26. Michelin, C.; Hoffmann, N. Photocatalysis Applied to Organic Synthesis—A Green Chemistry Approach. *Curr. Opin. Green Sustainable Chem.* **2018,** *10*, 40–45.
27. Yoon, T. P.; Ischay, M. A.; Du, J. Visible Light Photocatalysis as a Greener Approach to Photochemical Synthesis. *Nat. Chem.* **2010,** *2* (7), 527.
28. Abu-Dief, A. M.; Abdel-Fatah, S. M. Development and Functionalization of Magnetic Nanoparticles as Powerful and Green Catalysts for Organic Synthesis. *Beni-Suef Univ. J. Basic Appl. Sci.* **2018,** *7* (1), 55–67.
29. Madhavan, N.; Jones, C. W.; Weck, M. Rational Approach to Polymer-supported Catalysts: Synergy between Catalytic Reaction Mechanism and Polymer Design. *Acc. Chem. Res.* **2008,** *41* (9), 1153–1165.

30. Abu-Reziq, R.; Wang, D.; Post, M.; Alper, H. Separable Catalysts in One-pot Syntheses for Greener Chemistry. *Chem. Mater.* **2008,** *20* (7), 2544–2550.
31. Brahmachari, G.; Laskar, S.; Banerjee, B. Eco-friendly, One-pot Multicomponent Synthesis of Pyran Annulated Heterocyclic Scaffolds at Room Temperature Using Ammonium or Sodium Formate as Non-Toxic Catalyst. *J. Heterocycl Chem.* **2014,** *51* (S1), E303–E308.
32. Abdel-Malek, H. A.; Ewies, E. F. Phase-transfer Catalysis in Organic Syntheses. *Global J. Curr. Res.* **2014,** *3* (1), 1–21.
33. Makosza, M. Phase-transfer Catalysis. A General Green Methodology in Organic Synthesis. *Pure Appl. Chem.* **2000,** *72* (7), 1399–1403.
34. Polshettiwar, V.; Varma, R. S. Microwave-assisted Organic Synthesis and Transformations Using Benign Reaction Media. *Acc. Chem. Res.* **2008,** *41* (5), 629–639.
35. Strauss, C. R.; Varma, R. S. Microwaves in Green and Sustainable Chemistry. In *Microwave Methods in Organic Synthesis*; Springer: Berlin, Heidelberg, 2006; pp 199–231.
36. Varma, R. S.; Varma, M.; Chatterjee, A. K. Microwave-assisted Deacetylation on Alumina: A Simple Deprotection Method. *J. Chem. Soc., Perkin Trans. 1* **1993,** (9), 999–1000.
37. Nagendrappa, G. Organic Synthesis Using Clay Catalysts. *Resonance* **2002,** *7* (1), 64–77.

CHAPTER 7

CHARACTERIZATION OF SOME CARBOHYDRATES IN AQUEOUS SOLUTIONS AS SEEN BY MUTUAL DIFFUSION

M. MELIA RODRIGO[1], ANA C. F. RIBEIRO[2], LUIS M. P. VERISSIMO[2], ANA M. T. D. P. V. CABRAL[3], ARTUR J. M. VALENTE[2], and MIGUEL A. ESTESO[1*]

[1]*U.D. Química Física, Universidad de Alcalá 28871. Alcalá de Henares, Madrid, Spain*

[2]*Department of Chemistry, University of Coimbra, 3004-535 Coimbra, Portugal*

[3]*Faculty of Pharmacy, University of Coimbra, 3000-295 Coimbra, Portugal*

[*]*Corresponding author. E-mail: miguel.esteso@uah.es*

ABSTRACT

In this paper, we report a set of physical chemical parameters, including activity coefficients, activation energies, and hydrodynamic radii estimated from diffusion coefficients, for some carbohydrates—glucose, fructose, sucrose, lactose and cyclodextrins, and its derivatives. The values are discussed on the basis of intermolecular interactions and thus allowing a better understanding of the structure and thermodynamic behavior of these systems.

7.1 INTRODUCTION

The ever-increasing development of technology and science demands accurate data concerning the fundamental thermodynamic and transport properties of aqueous solutions containing carbohydrates.[1–7] Several fields, such as food technology, chemical, biological, and biochemical phenomena, which involve these solutions, have been moving toward a more scientific treatment. For example, the lactose (disaccharide derived from the condensation of galactose and glucose) (Fig. 7.1) is not only a technological important compound, but also plays an important role in biological, pharmaceutical, medical, food, and biomedical applications.[8,9]

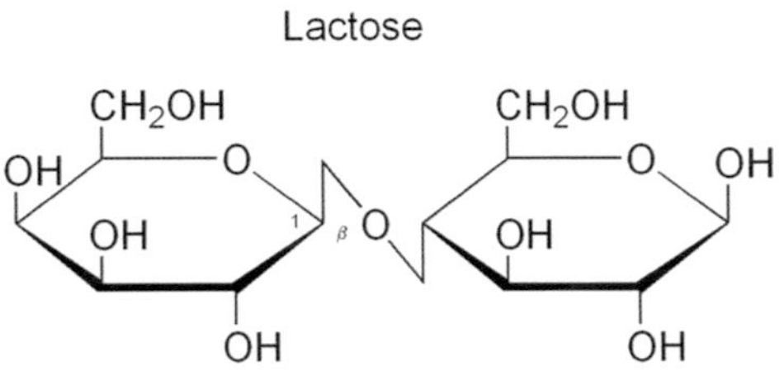

Lactose

FIGURE 7.1 Structure of lactose.

Other systems containing different carbohydrates, such some monosaccharides (e.g., fructose), disaccharides (e.g., sucrose), or cyclodextrins (Fig. 7.2) and its derivatives (Fig. 7.3), have also relevant importance at both the technological and fundamental areas. The last ones are one of the most commonly used groups of carriers to encapsulate drug molecules of limited water solubility, increasing their stability and improving bioavailability.[10] They are cyclic oligosaccharides, consisting of a variable number of D-(+)-glucose molecules, linked together by α-(1,4)-type bonds. Native cyclodextrins (Fig. 7.2), composed by 6, 7, or 8 glucose units, are referred to as α-, β-, γ-cyclodextrins, respectively.[10]

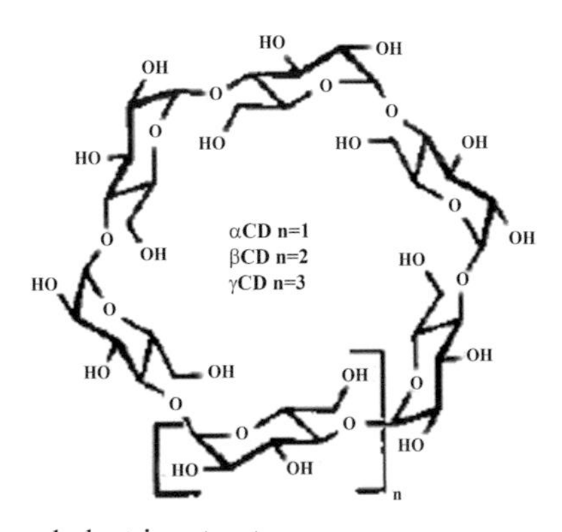

FIGURE 7.2 α-, β-, γ-cyclodextrins structures.

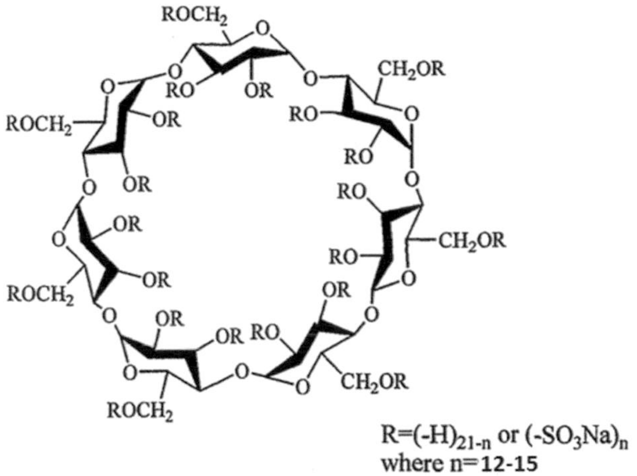

$R=(-H)_{21-n}$ or $(-SO_3Na)_n$
where n=12-15

FIGURE 7.3 Structure of β-cyclodextrin sulfated sodium salt.

As a result, there has been a growing demand by scientists and technologists of accurate values of thermodynamic and transport data concerning carbohydrates in solutions.

Despite considerable works that have been carried out up to now, the transport behavior of these systems is still poorly understood. Because this information is essential for a better understanding of the physical chemistry conditions underlining the diffusion phenomena in different systems (e.g., in human biological systems) and for the design of controlled-release systems, we have proposed to perform, in the past years, a comprehensive study of the diffusion of some carbohydrates, that is, fructose, glucose, sucrose, lactose, native cyclodextrins, and some derivatives.[11–13] These studies have provided a comprehensive information—both kinetic and thermodynamic as well as structural—for the design and operation of different systems, for example, for those controlled release and delivery of drugs. The present work intends to summarize that information, supplying the scientific and technological communities with data on these important parameters in solution transport processes.

Thus, we pursue the view that the understanding of transport properties in different systems as well as thermodynamic properties will help us to understand diffusion and interactions mechanisms of those processes and, consequently, will contribute to a better design of systems with pharmaceutical, medical, and industrial applications. In other words, to know how the properties related to both mobility (diffusion, conductivity, viscosity, and transport number) and thermodynamics (optimum carrier concentration, drug solubility as a function of carrier concentration, binding constants, pH, temperature, and activity coefficients) of different aqueous systems.

7.2 A SIMPLE OVERVIEW OF EXPERIMENTAL TECHNIQUES TO MEASURE DIFFUSION COEFFICIENTS

Many techniques are used to study diffusion in solutions. It is very common to find misunderstandings concerning the meaning of a parameter, frequently just denoted by *D* and merely called diffusion coefficient, in the scientific literature, communications, meetings, or simple discussions among researchers. In fact, it is necessary to distinguish self-diffusion (intradiffusion, tracer diffusion, single ion diffusion, ionic diffusion) and mutual diffusion (interdiffusion, concentration diffusion, salt diffusion).[1–5] In our case, the mutual diffusion is analyzed.

NMR and capillary-tube, the most popular methods, can only be used to measure intradiffusion coefficients.[5]

Experimental methods, such as diaphragm-cell (inaccuracy 0.5%–1%), conductimetric (inaccuracy 0.2%), Gouy and Rayleigh interferometry (inaccuracy <0.1%), and Taylor dispersion (inaccuracy 1%–2%), can be employed to determine mutual diffusion coefficients. While the first and second methods consume days in experimental time, the last ones imply just hours.[5] The conductimetric technique follows the diffusion process by measuring the ratio of electrical resistances of the electrolyte solution in two vertically opposed capillaries as time proceeds.[5,14] Despite this method has previously given us reasonably precise and accurate results, it is limited to studies of mutual diffusion in electrolyte solutions, and like diaphragm-cell experiments, the run times are inconveniently long (ca. 12 h). The Gouy method also has high precision, but when applied to microemulsions they are prone to gravitational instabilities and convection. Thus, the Taylor dispersion has become increasingly popular for measuring diffusion in solutions, because of its experimental short time and its major application to the different systems (electrolytes or nonelectrolytes). In addition, with this method it is possible to measure multicomponent mutual diffusion coefficients.

7.3 THE TAYLOR DISPERSION TECHNIQUE

In present work, we present some data of mutual diffusion for carbohydrates in aqueous solutions obtained from Taylor technique, which has been described in detail elsewhere.[15–18] Basically, this method is based on the dispersion of small amounts of solution injected into laminar carrier streams of solvent or solution of different composition, flowing through a long capillary tube. The length of the Teflon dispersion tube used in the present study was measured directly by stretching the tube in a large hall and using two high quality theodolites and appropriate mirrors to accurately focus on the tube ends. This technique gave a tube length of 3.2799 (±0.0001) $\times 10^4$ mm, in agreement with less-precise control measurements using a good-quality measuring tape. The radius of the tube, 0.5570 (± 0.00003) mm, was calculated from the tube volume obtained by accurately weighing (resolution 0.1 mg) the tube when empty and when filled with distilled water of known density.

At the start of each run, a 6-port Teflon injection valve (Rheodyne, model 5020) is used to introduce 0.063 mL of solution into the laminar carrier stream of slightly different composition. A flow rate of 0.17 mL min^{-1} is maintained by a metering pump (Gilson model Minipuls 3) to give retention times of about 1.1×10^4 s. The dispersion tube and the injection valve are kept at 298.15 K and 303.15 K (± 0.01 K) in an air thermostat.

Dispersion of the injected samples is monitored using a differential refractometer (Waters model 2410) at the outlet of the dispersion tube. Detector voltages, $V(t)$, are measured at accurately 5 s intervals with a digital voltmeter (Agilent 34401 A) with an IEEE interface. Binary diffusion coefficients are evaluated by fitting the dispersion equation

$$V(t) = V_0 + V_1 t + V_{max} (t_R/t)^{1/2} \exp[-12D(t - t_R)^2/r^2 t] \tag{7.1}$$

to the detector voltages. The additional fitting parameters are the mean sample retention

The Taylor dispersion technique (accuracy 1%), a rapid and convenient flow technique, to measure the mutual diffusion in solutions of different compounds such as carriers (emphasizing cyclodextrins and block copolymer micelles) and different drugs, such as ethambutol dihydrochloride, L-Dopa, and Carbidopa.

7.4 HYDRODYNAMIC RADII

Regarding the interpretation and analysis of the diffusion behavior of nonelectrolytes (e.g., carbohydrates) in aqueous solutions, particular relevance has been attributed to the equations resulting from a hydrodynamic analysis of the phenomenon of diffusion. According to this treatment, the diffusion coefficient of a particle in a liquid depends basically on two factors: the size of the diffusing entity and the resistance that the liquid offers to the diffusion (generically, its viscosity). For the case of the flow of a spherical particle in a liquid, these factors are considered in the Stokes–Einstein relation, described in eq (7.2),[2,5]

$$D_i^0 = \frac{k_B T}{6\pi\eta_0 R_H} \quad (7.2)$$

where T is the absolute temperature, η_0, represents the viscosity of solvent and R_H and D_i^0, the hydrodynamic radius of an equivalent spherical particle, and its limiting diffusion coefficient at infinitesimal concentration, also known as tracer diffusion coefficient. Table 7.1 shows the values for mutual diffusion coefficient at infinitesimal concentration, D^0, and the hydrodynamic radius, R_H, for 10 carbohydrates.

This hydrodynamic relationship is in principle of very limited validity for diffusing molecules in solution, since the liquid is not a continuous medium, but is supposed to consist of discrete molecules of the same magnitude as those of the diffusing molecules. In addition, the molecules are generally not spherical, and even the most symmetrical molecules can only be considered spheres in approach. The approximation given by Stokes' frictional law to diffusive molecules improves with increasing solute molecule size compared to the solvent. Thus, this ratio often provides a good approximation, as is the case of some carbohydrates such as cyclodextrins in solution.[11] In conclusion, in spite of its limitations, it can be used to analyze the diffusion of molecules, especially when an approximate estimate of the molecular sizes, already demonstrated in previous studies (e.g., calculation of the Stokes' radius of the β-cyclodextrin in aqueous solution).[11]

TABLE 7.1 Estimation of the Mutual Diffusion Coefficients at Infinitesimal Concentration, D^0, of Some Carbohydrates and the Respective Hydrodynamics Radius, R_H [a].

Carbohydrate	D^0 (10^{-9}m^2 s^{-1})[a]	R_H (10^{-9}m)
Glucose	0.679	0.357
Fructose	0.686	0.361
Sucrose	0.523	0.469
Lactose	0.566	0.433
α-CD	0.353	0.700
β-CD	0.326	0.752

TABLE 7.1 *(Continued)*

Carbohydrate	D^0 ($10^{-9}m^2 s^{-1}$)[a]	R_H (10^{-9}m)
γ-CD	0.358	0.685
HP-α-CD	0.344	0.710
HP-β-CD	0.322	0.760
β-cyclodextrin sulfated	0.713	0.344

[a] See Refs. 11, 12, 13, 19, and 20.

7.5 ACTIVITY COEFFICIENTS

Since diffusion is an irreversible process, we can say that the force responsible for such phenomenon is the chemical potential gradient of the diffusing substance that is quantified in ideal solutions by the concentration gradient at constant temperature, it is not in real solutions.[1–4] Hartley, Onsager, and Fuoss[6] have derived an expression for the coefficient of mutual diffusion whose value depends on the concentration (eq 7.3). That is

$$D_{AB} = D_{AB}^{0} B_B^C \tag{7.3}$$

where

$$B_B^C = \left\{ 1 + \left(\frac{\partial \ln \gamma_B}{\partial \ln c_B} \right)_{T,p} \right\} \tag{7.4}$$

B^c_B, γ_B and D_{AB}^0 represent the thermodynamic factor, the solute activity coefficient and the limit value of the diffusion coefficient for infinitesimal concentration, respectively. This relationship has proved to be valid for dilute solutions of various electrolytes and nonelectrolytes.[5,6] However, this equation neglects some effects of great importance for concentrated solutions, such as varying the viscosity of the solution with the concentration and diffusion flux of the solvent opposite that of the solute. In this sense, Gordon[1–6] modified this equation to be applied at higher concentrations, introducing a third factor, the viscosity of the solution, (eq 7.5),

$$D_{AB} = D^{0}_{AB} \frac{\eta_{H_2O}}{\eta_S} B^{c}_{B} \tag{7.5}$$

Recent studies have demonstrated not only the successful applicability of these equations to dilute nonelectrolyte systems, but also their importance in revealing data on the effects of concentration and temperature on the behavior of activity coefficients.[11,12] Their estimates are indeed important, since they are macroscopic parameters that reflect the various molecular processes (solute–solute and solute–solvent interactions).

By combining Hartley's equation (eq 7.5), and eq. (7.6),

$$\ln \gamma = B(c/\text{mol dm}^{-3}) \tag{7.6}$$

(only valid in general for dilute solutions ($c < 0.1$ mol dm^{-3}), the activity coefficients for different types of carbohydrates (monosaccharides, disaccharides, and one polysaccharides) have been estimated. Table 7.2 gives the estimations of the activity coefficients, γ for one particular monosaccharide, disaccharide and polysaccharide (that is, glucose, sucrose, and β-CD). The analysis of these parameters allows us to assess how the systems are deviating from ideal behavior. From the results shown in Table 7.2, it is verified that both the increase of concentration of the carbohydrate as well as the increase of the molecular weight of different saccharides contribute to the decrease of the activity coefficients. We may interpret these changes on the basis of increasing solute–water interactions with increasing concentration, and that this reflects all contributions from water molecules, including those beyond the first layer, which are weakly bound and unlikely to diffuse as a unit.

TABLE 7.2 Activity Coefficients, γ, for Aqueous Solutions Containing One Monosaccharide (Glucose), One Disaccharide (Sucrose), and One Polysaccharide (β-CD), Estimated from Equation (7.6) at 298.15 K.

Carbohydrate	**c (mol dm^{-3})**	γ
Glucose	0.001	0.9997
	0.002	0.9994
	0.005	0.9986
	0.010	0.9972
	0.100	0.9724

TABLE 7.2 *(Continued)*

Carbohydrate	c (mol dm^{-3})	γ
Sucrose	0.001	0.9998
	0.002	0.9996
	0.005	0.9991
	0.010	0.9981
	0.100	0.9812
β-CD	0.001	0.9990
	0.002	0.9980
	0.005	0.9965
	0.010	0.9915

7.6 ACTIVATION ENERGIES

The interpretation and analysis of the diffusion behavior of nonelectrolytes in aqueous solutions also allows the determination of activation energies through an Arrhenius-like relationship between the kinetic parameter, E_D and the temperature, T, (Eyring eq (7.7)[2–5]),

$$E_D = -R \frac{d \ln (D^0 / T)}{d(1/T)} \tag{7.7}$$

where D_0 is the limiting diffusion coefficient at infinitesimal concentration and R is the ideal gas constant.

The logarithm of the limiting D^0/T values plotted against $1/T$ is linear and through of its slope the activation energies are estimated. Table 7.3 shows some values obtained for some carbohydrates. In general, for carbohydrates of similar size and structure, these E_a values are also similar in magnitude.

TABLE 7.3 Estimations of Activation Energies.[a]

Carbohydrate/water	E_a/kJ.mol^{-1}
Sucrose/water	20.13[b]
Glucose/water	17.75[b]
Fructose/water	15.64[b]
α-CD/water	17.11[c]
β-CD/water	18.72[c]
HP-α-CD/water	19.40[d]
HP-β-CD/water	19.40[d]

[a]Eyring equation (eq 7.7).[5]
[b]See Ref. 12.
[c]See Refs. 11, 13.

7.7 CONCLUSIONS

We have presented the limiting diffusion coefficients values for some carbohydrates, and, from them, we have estimated the respective hydrodynamic radii, activity coefficients and activation energies. These data permitted us a better understanding of the thermodynamic and transport behavior of these aqueous carbohydrates systems.

ACKNOWLEDGMENTS

The authors in Coimbra are grateful for funding from "The Coimbra Chemistry Centre," which is supported by the Fundação para a Ciência e a Tecnologia (FCT), Portuguese Agency for Scientific Research, through the programs UID/QUI/UI0313/2019 and COMPETE. M.M.R.L. is thankful to the University of Alcalá (Spain) for the financial assistance (Mobility Grants for Researchers-2018).

KEYWORDS

- **carbohydrates**
- **diffusion**
- **solutions**
- **hydrodynamic radius**
- **activity coefficients**

REFERENCES

1. Harned, H. S.; Owen, B. B. *The Physical Chemistry of Electrolytic Solutions*, 3rd ed; Reinhold Pub. Corp.: New York, 1964.
2. Erdey-Grúz, T. *Transport Phenomena in Aqueous Solutions*; Adam Hilger: London, 1974.
3. Bockris, J. O'M.; Reddy, A. K. N. *Modern Electrochemistry. An Introduction to an Interdisciplinary Area*, Vol. 1, 6th printing; Plenum Press: New York, 1977.
4. Cussler, E. L. *Diffusion: Mass Transfer in Fluid Systems*; Cambridge University Press: Cambridge, 1984.
5. Tyrrell, H. J. V.; Harris, K. R. *Diffusion in Liquids: A Theoretical and Experimental Study*; Butterworths: London, 1984.
6. Robinson, R. A.; Stokes, R. H. *Electrolyte Solutions*, 2nd revised ed; Dover Publications, Inc.: New York, 2002.
7. Lobo, V. M. M. *Handbook of Electrolyte Solutions*; Elsevier: Amsterdam, 1990.
8. Sano, Y.; Yamamoto, S. Mutual Diffusion Coefficient of Aqueous Sugar Solutions. *J. Chem. Eng. Jpn.* **1993,** *26*, 633–636.
9. Banipal, P. K.; Banipal, T. S.; Lark, B. S.; Ahluwalia, J. C. Partial Molar Heat Capacities and Volumes of Some Mono-, Di- and Trisaccharides in Water at 298.15, 308.15 and 318.15 K. *J. Chem. Soc. Faraday Trans.* **1997,** *93*, 81–87.
10. Uekama, K.; Hirayama, F.; Irie, T. Cyclodextrin Drug Carrier Systems. *Chem. Rev.* **1998,** *98*, 2045–2076.
11. Ribeiro A. C. F.; Leaist, D. G.; Esteso M. A.; Lobo V. M. M.; Valente, A. J. M.; Santos C. I. A. V.; Ana, M. T. D. P. V.; Cabral and Francisco, J. B. Veiga. Binary Diffusion Coefficients for Aqueous Solutions of β-Cyclodextrin at Temperatures from 298.15 K and 312.15 K. *J. Chem. Eng. Data* **2006,** *51*, 1368–1371.
12. Ribeiro, A. C. F.; Ortona, O.; Simões, S. M. N.; Santos, C. I. A. V.; Prazeres, P. M. R. A.; Valente, A. J. M.; Lobo V. M. M.; Burrows, H. D. Binary Mutual Diffusion Coefficients of Aqueous Solutions of Sucrose, Lactose, Glucose and Fructose in the Temperature Range 298.15 K to 328.15 K. *J. Chem. Eng. Data* **2006,** *51*, 1836–1840.
13. Ribeiro, A. C. F.; Valente, A. J. M.; Santos, C. I. A. V.; Prazeres, P. M. R. A.; Lobo, V. M. M.; Burrows, H. D.; Esteso, M. A.; Cabral, A. M. T. D. P. V.; Veiga, F. J. B. Binary Mutual Diffusion Coefficients of Aqueous Solutions of α-cyclodextrin, 2-hydroxypropyl-α-cyclodextrin and 2-hydroxypropyl-β-cyclodextrin at Temperatures from 298.15K to 312.15 K. *J. Chem. Eng. Data* 2007, *52*, 586–590.
14. Lobo, V. M. M.; Valente, A. J. M.; Ribeiro, A. C. F. Differential Mutual Diffusion Coefficients of Electrolytes Measured by the Open-Ended Conductimetric Capillary Cell: A Review. In *Focus on Chemistry and Biochemistry*; Zaikov, G. E., Lobo, V. M. M., Guarrotxena, N. Eds; Nova Science Publishers: New York, 2003; pp. 15–38.
15. Aris, R. On the Dispersion of a Solute in a Fluid Flowing through a Tube. *Proc. R. Soc. L.* **1956,** *235*, 67–77. doi:10.1098/rspa.1956.0065.
16. Barthel, J.; Gores, H. J.; Lohr, C. M.; Seidl, J. J. Taylor Dispersion Measurements at Low Electrolyte Concentrations. I. Tetraalkylammonium Perchlorate Aqueous Solutions. *J. Solut. Chem.* **1996,** *25*, 921–935.
17. Loh, W. Taylor Dispersion Technique for Investigation of Diffusion in Liquids and Its Applications. *Quim. Nova* **1997,** *20*, 541–545.

18. Callendar, R.; Leaist, D. G. Diffusion Coefficients for Binary, Ternary, and Polydisperse Solutions from Peak-Width Analysis of Taylor Dispersion Profiles. *J. Solut. Chem.* **2006,** *35*, 353–379. doi: 10.1007/s10953-005-9000-2.
19. Ribeiro, A. C. F.; Valente, A. J. M.; Santos, C. I. A. V.; Prazeres, P. M. R. A.; Esteso, M. A.; Lobo, V. M. M.; Burrows, H. D.; Ascenso, O. S.; Cabral, A. C. F.; Veiga, F. J. B. Some Transport Properties of Gamma-cyclodextrin Aqueous Solutions at 298.15 K and 310.15 K. *J. Chem. Eng. Data* **2008,** *53*, 755–759.
20. Barros, M. C. F.; Veríssimo, L. M. P.; Ribeiro, A. C. F.; Esteso, M. A. Diffusion in Ternary Aqueous {L-dopa + $(NaSO_3)_n$-β-cyclodextrin} Solutions Using the Pseudo-binary Approximation. *J. Chem. Thermodyn.* **2018,** *123*, 7–21.

CHAPTER 8

ADVANCEMENTS IN BIOREMEDIATION AND BIOTECHNOLOGY—A CRITICAL OVERVIEW

SUKANCHAN PALIT

43, Judges Bagan, Post-Office - Haridevpur, Kolkata-700082, India
E-mail: sukanchan68@gmail.com; sukanchan92@gmail.com; sukanchanp@rediffmail.com

ABSTRACT

In present-day human civilization science and technology are moving at a rapid pace. Biological sciences and biotechnology are in the avenues of newer scientific ingenuity and vast scientific and engineering profundity. Bioremediation is a wonder of science as regards environmental protection and environmental engineering science. The main vision of this paper is to elucidate the vast technological applications of bioremediation and nanotechnology in human progress. Civilization and science today stand in the middle of vision, scientific forbearance, and alacrity. Biotechnology and biological sciences are today merged with the visionary areas of environmental protection. This is the main objective of this treatise. Other areas of research pursuit are the domains of recent advances in the field of bioremediation, biotechnology, and biological sciences. Scientific and engineering verve, and the world of validation of engineering sciences will surely lead a visionary way in the true unraveling of nanotechnology and biotechnology today. Nanobiotechnology is a frontier of science and another area of research endeavor.

8.1 INTRODUCTION

The domain of biological sciences and nanotechnology are in the avenues of newer scientific thoughts and new scientific regeneration. Mankind today is in the grip of immense scientific introspection and vast scientific determination. Bioremediation is an avenue of immense scientific comprehension today. In this treatise, the author deeply elucidates the need of environmental protection techniques in the furtherance of global science and engineering scenario. Environmental engineering, chemical engineering, and biological sciences are in the crucial juncture of scientific comprehension and vast engineering vision. Conventional and nonconventional environmental engineering tools are changing the broad scientific firmament today. The provision of basic human needs such as sustainability, water, energy, food, education, and shelter are the areas of immense scientific introspection as well. This paper deeply interprets the areas of membrane science and advanced oxidation processes in the true emancipation of environmental engineering and chemical process engineering.

8.2 THE AIM AND OBJECTIVE OF THIS STUDY

The success of science and technology today lies in the hands of the civil society, technologists, engineers, and governments around the world. The main aim and objective of this treatise is to depict profoundly the areas of conventional and nonconventional environmental engineering tools such as bioremediation, membrane science, and advanced oxidation processes. Novel separation processes such as nanofiltration, ultrafiltration, reverse osmosis, and other membrane separation techniques are creating scientific wonders and vast scientific ingenuity in present-day civilization. The world of science and technology are moving through difficult phases due to severe water shortage and global warming in many regions of the world. The reenvisioning and the sharpening of environmental engineering tools is the need of the hour. The deliberations as regards the efficiency of conventional and nonconventional environmental engineering tools are to be reorganized and revamped with the march of science and mankind. The author deeply reiterates the role of environmental engineering in the progress of science and technology. Mankind today is in deep need of basic amenities such as water, electricity, food, shelter, and education. Sustainability, whether it is energy or environmental, is also the need of human endeavor today. Engineering and technology are immensely retrogressive in present-day human civilization. Thus comes the need of biological treatment of environmental protection such as bioremediation. The author also delineates the success and the efficacy of bioremediation and biotechnology in removing hazardous wastes in industrial wastewater and drinking water. These challenges, the targets and the ingenuity of environmental engineering tools, are reorganized in this treatise.

8.3 WHAT DO YOU MEAN BY BIOREMEDIATION?

Bioremediation is the use of microorganism metabolism to remove hazardous pollutants. Bioremediation can occur on its own (natural attenuation or intrinsic bioremediation) or can be spurred via the addition of fertilizers to increase the bioavailability within the medium. Biotechnology and biological sciences are the marvels of science and engineering. At present, the needs of civilization and scientific progress are the areas of energy and environmental sustainability. Here comes the importance of biotechnology, environmental biotechnology, and nanobiotechnology. Bioremediation can be used at the site of contamination (in situ) or on contamination removed from the original site (ex situ). In the vast case of contaminated soils, sediments, and sludges it can involve land tilling in order to make the nutrients and oxygen more available to the microorganisms. Some of the diverse areas of bioremediation technologies are phytoremediation, bioventing, bioleaching, land farming, bioreactor, composting, bioaugmentation, rhizofiltration, and biostimulation. Bioremediation is of immense scientific importance as it destroys or renders harmless various contaminants using natural biological activity. Engineering importance and the world of scientific enquiry and deep profundity will all be the veritable forerunners in the field of biotechnology and bioremediation. Bioremediation uses low-cost, low-technology tools, which generally have a high public acceptance and often be carried out in-site. Environmental sustainability is integrated with environmental engineering techniques such as bioremediation. Although the methodologies employed are not scientifically complex, considerable experience and vast scientific expertise may be required to design and implement a successful bioremediation program. The conventional tools used for bioremediation have been to dig up contaminated soil and remove it to a landfill, or to cap and contain the contaminated areas of the site. Some technologies that have been used are high-temperature incineration and various types of chemical decomposition.

Currently, technological advancements in the field of bioremediation are the challenges and the vision of civilization. In this chapter, the author rigorously points toward the scientific success, the scientific provenance, and the deep ingenuity in the field of biotechnology and biological sciences.[22,23]

8.4 THE VAST SCIENTIFIC DOCTRINE OF BIOTECHNOLOGY

Biotechnology is a marvel of science today. Biotechnology integrated with nanotechnology is also a major pillar toward scientific emancipation. The vast scientific doctrine in the field of biotechnology needs to be reorganized with the passage of scientific history and time. Biotechnology, biological sciences, and biological engineering are in the process of new scientific and engineering rejuvenation. The challenges and the vision of the doctrine of biotechnology are truly far-reaching and needs to be revamped as civilization moves forward. At present, biotechnology and bioremediation are the true needs of groundwater remediation and groundwater decontamination.

Bioremediation is defined as the process whereby organic wastes are biologically treated and degraded under controlled conditions to an innocuous state, or to levels below concentration limits established by global regulatory authorities. Technological stance, scientific enquiry, and the vast scientific ingenuity are the pillars of biotechnological research globally. Biotechnology is the broad area of biology involving living systems and organisms to develop and make a product or any technological application that uses biological systems, living organisms, or derivatives to make or modify products or processes for specific use. Research frontiers in the field of biotechnology are of tremendous importance as civilization moves forward. Bio-engineering and biomedical engineering are the challenges and the vision of research paradigm in today's world. Depending on the tools and applications, biotechnology often overlaps with the fields of molecular biology, bio-engineering, biomedical engineering, biomanufacturing, molecular engineering, and so forth. For thousands of years, civilization has used and developed biotechnology in agriculture, food production, and medical science. Today biotechnology has expanded to include new and diverse areas of science and technology such as genomics, recombinant gene techniques, applied immunology, and development of pharmaceutical therapies. Scientific retribution and deep scientific provenance form the backbone of biotechnological research and development initiatives globally. The vast and wide concept of "biotechnology" encompasses a wide range of processes for modifying living processes according to human purposes. The challenges, the vision, and the targets of biotechnology applications in modern science are absolutely inspiring and far-reaching. In this well-researched treatise the author reiterates the scientific success, the vision, and the purpose of biotechnology and its applications in human scientific progress. This treatise explains in minute details the interface of environmental sustainability with biological sciences and biotechnology. Technological travails in modern civilization are immense as regards biological treatment of industrial wastewater. This paper opens up new thoughts and newer scientific avenues in the field of biotechnology also.[22,23] The recent advances in biotechnology and medicine are stem cell research, human genome project, targeted cancer therapies, human papilloma virus vaccine, face transplants, nerve regeneration, and the research of brain signals to audible speech. Human scientific stance, the scientific verve, and determination will all surely be the path-breakers toward a new epoch in biotechnology and bioengineering. Bioengineering is the next-generation technology and research and development initiatives are vastly inspiring and groundbreaking. This chapter opens up new a vision in biology.

8.5 SUSTAINABLE DEVELOPMENT AND THE VAST VISION FOR THE FUTURE

Sustainable development whether it is energy or environmental is the imminent need of the hour. Science and technology today stand in the middle of vision, scientific integrity, and vast scientific enquiry. Social and economic sustainability are the other areas of scientific endeavor today. Provision of basic human needs such as water, electricity, food, education, and shelter are in a state of immense disaster. The vision for the future needs to be revamped as regards application of nanotechnology and biotechnology to diverse areas of scientific and academic rigor. The areas of environmental remediation and chemical process engineering are in the avenues of newer scientific hope, grit, and ingenuity. In this treatise, the author deeply enunciates the scientific intricacies and the vast scientific profundity in the field of biotechnology, bioremediation, conventional and nonconventional environmental engineering techniques. Environmental and energy sustainability are the zenith of scientific endeavor globally today. The vast vision of the progress of human civilization lies in the hands of scientists, engineers, and the civil society. This treatise widely opens up new thoughts and newer scientific imagination in the field of bioremediation. Treatment of drinking water and industrial wastewater by bioremediation is a novel application as regards its efficiency and the effectiveness of the process. The author vastly pronounces the scientific success, the vision, and the profundity of environmental engineering tools in its quest toward true realization of science and engineering.[22,23]

8.6 THE VAST WORLD OF ENVIRONMENTAL REMEDIATION AND THE MARCH OF SCIENCE

The vast world of environmental remediation today is in the midst of vision, might, and scientific forbearance. The march of science needs to be reenvisioned and reorganized as civilization marches forward. Environmental and energy sustainability are the pallbearers toward a new visionary era in the field of environmental protection and energy engineering. Chemical process engineering and environmental engineering in the similar vein are in the avenues of newer scientific refurbishment. Environmental biotechnology and nanobiotechnology are the needs of human scientific progress. Developing as well as industrialized nations around the world are today in the middle of a monstrous catastrophe, that is, arsenic and heavy metal groundwater contamination. Here comes the importance of biotechnology, biological sciences, and the vast world of bioremediation. Severe scarcity of drinking water is destroying the scientific fabric in many countries around the globe. Thus, the need of an effective environmental engineering technique. The environmental engineering tools that will be dealt with are novel separation processes and advanced oxidation processes. Civilization's immense scientific vision and stance, the scientific integrity, and the scientific grit and determination will all lead a visionary way in the true realization of energy and environmental sustainability. The scientific imagination in the field of bioremediation needs to be reenvisioned and reorganized as water science and technology enters into a new arena of vision and scientific acuity.[22,23]

8.7 RECENT ADVANCES IN THE FIELD OF BIOREMEDIATION

Bioremediation is the major field of scientific pursuit today. Bioremediation is applied in environmental protection with immense vision, insight, and deep scientific ingenuity. The author deeply comprehends the scientific intricacies and the scientific doctrine of the application of bioremediation in environmental engineering. Technological and engineering provenance and the world of scientific challenges and profundity are the forerunners toward a new epoch in the field of bioremediation. Science of bioremediation is witnessing immense challenges and this area is elucidated in details in this section.

Sharma[1] discussed with cogent insight features, strategies, and applications in the field of bioremediation. The problems associated with contaminated sites are now increasing in prominence in many countries around the world.[1] Contaminated lands generally result from past industrial activities when awareness of the health and environmental effects connected with the production, use, and disposal of hazardous substances were less understood than today.[1] Bioremediation is the technology that uses microorganism metabolism to remove pollutants and it uses relatively low-cost, low-technology techniques that can be carried out on site. The author deliberated in minute details principle of bioremediation, factors of bioremediation, environmental factors, bioremediation strategies, in-situ bioremediation, biosparging, bioventing, bioaugmentation, biopiling, ex-situ bioremediation, bioreactors, special features, and limitations of bioremediation.[1] The vast domain of phytoremediation are the other avenues of this research pursuit. There are five types of phytoremediation techniques, namely, phytoextraction, phytotransformation, phytodegradation, rhizofiltration, and phytostabilization.[1] Science and engineering of bioremediation are today surpassing vast and versatile scientific frontiers.[1] The scientific importance and the deep scientific ingenuity of bioremediation are today in the path of new rejuvenation. This paper rigorously points toward the success of bioremediation in diverse areas of biological sciences.[1]

Kensa[2] discussed with lucidity and cogent insight bioremediation. Any processes that uses microorganisms, fungi, green plants, or their enzymes to return the natural environment to its original conditions is termed as bioremediation.[2] This paper depicts profoundly what is bioremediation, principles of bioremediation, factors of bioremediation, strategies, types, genetic engineering approaches, monitoring of bioremediation, and the vast world of advantages and disadvantages of bioremediation.[2] Vast quantities of organic and inorganic compounds are released to the environment each year as a result of human activities.[2] Currently, the challenges and the deep scientific vision of biological sciences and ecological engineering are of immense importance. Basic types of bioremediation tools are elaborated in minute details in this paper. Genetic engineering approaches in bioremediation are the other hallmarks of this well-researched treatise.[2] In today's scientific world, bioremediation is a far less expensive than other technologies that are used to remove hazardous substances from industrial wastewater. This is a text that opens new questions and new enquiries in the field of bioremediation, biological sciences, and biological engineering.[2]

Sardrood et al.[3] discussed in minute details an introduction to bioremediation. Technological vision and verve stands as major pillars in the quest to bioremediation scientific enquiry. The vision and the truth of science of bioremediation are today in the middle of scientific introspection. The vast rise of human population has led to the enhancement of exploitation of natural resources and the demands of the population for food, energy, and other basic human needs.[3] In the last century, the Union Carbide (Dow) Bhopal disaster had shocked the civilization and destroyed the vast scientific firmament of might and vision.[3] The contaminants known to be biologically degraded by microorganisms so far known and applied in bioremediation have been categorized into five groups:

a) Halogenated aromatic hydrocarbons
b) Munitions wastes
c) Organic solvents
d) Pesticides
e) Polyaromatic hydrocarbons[3]

The authors in this paper discussed in minute details the role of environmental biotechnology in pollution management, types of bioremediation, in situ bioremediation, ex situ bioremediation, bioremediation techniques, land farming/land treatment/prepared bed bioreactors, organisms involved in bioremediation processes, mycoremediation, and the holistic domain of integrated water resource management.[3] Technological profundity and verve, the challenges of science, and the march of mankind will veritably lead a long way in the unraveling of the scientific truth behind biological sciences. With the burgeoning population of the world, the science and engineering of bioremediation is going to be a necessity of today's modern civilization.[3] The diversity of bioremediants, the challenges of the available techniques, the variation of the substrates used by bioremediants in different types of aqueous and terrestrial habitats all seem as good signs of this well-established branch of science and technology. The march of biological sciences and biological engineering are vast and versatile. The authors rigorously pinpoint the scientific success, the intricacies, and the vision behind applications of bioremediation in environmental protection.[3]

The vision, the targets, and the scientific retribution of biological sciences and bioremediation science are immense, far-reaching, and versatile. The status of environment is extremely grave as mankind trudges forward toward a new epoch of scientific might and vision. In this paper, the authors deeply elucidate the success of biological sciences and biological engineering in the furtherance of science, engineering, and mankind.

8.8 RECENT SCIENTIFIC RESEARCH PURSUIT IN THE FIELD OF BIOTECHNOLOGY

Biotechnology and biological sciences are in the critical juncture of vision, scientific forbearance, and vast scientific and engineering insight. In this section the author elucidates with cogent insight and vision some of the recent advances in the field of biotechnology. Technological integrity and scientific validation are the pillars of scientific endeavor today. Biotechnology is today integrated with the domain of nanotechnology, which is termed as nano-biotechnology. Environmental engineering and biotechnology in the similar manner are merged together to form a newer branch of scientific pursuit, that is, environmental biotechnology.

Khan et al.[4] discussed and deliberated in minute details recent advances in medicinal plant biotechnology. Medicinal plants are the most important source of life-saving drugs for the majority of human civilization.[4] Technological motivation and scientific validation stands as major imperatives toward a new world of biotechnology and biological sciences. Plant secondary metabolites are vastly important as drugs, fragrances, pigments, food additives, and pesticides.[4] The biotechnological tools are important to select, multiply, and improve medicinal plants. Today genetic transformation is a powerful tool toward furtherance of science and engineering.[4] The authors discussed in details plant tissue culture, combinatorial biosynthesis, genetic transformation technology, bioanalytics, and metabolomics. This review deeply highlights the vast and intricate world of genetic transformation. Genetic transformation may provide increased and efficient system for in vitro production of secondary metabolites and is a visionary area of science and technology. The roots of science of biotechnology lies in the domain of biological sciences and biological engineering today. It is estimated that 80% of people worldwide rely chiefly on conventional, largely herbal medicines to meet their primary health care and basic medical science needs.[4] The global demand for herbal medicine is not only large but is ever-growing and far-reaching. Various innovations and technologies have been adopted for enhancing bioactive molecules in medicinal plants.[4] Here comes the need of forays in biotechnology. Genetic engineering and genetic transformation technology are the pillars of intense scientific endeavor today. Genetic transformation technology has been proved to be an effective tool for the production of plants with the desired traits in crops and agricultural yields. It highly promises to overcome the substantial agronomic and environmental issues that have not been solved using traditional plant-breeding programs.[4] Technological and engineering vision and provenance, the challenges of plant biotechnology and the futuristic vision of biological sciences, will open new doors of innovation and scientific instinct in the field of biotechnology and biology. The authors discussed in deep details the scientific intricacies of genetic engineering and plant biotechnology. The improved in vitro plant cell culture systems have the immense potential for commercial exploitation of secondary metabolites. Design of bioreactors in plant biotechnology is the other hallmark of this paper. This review also highlights the vast possibilities for the use of bioconversion for the production and development of plant natural products by different pathways. The splendor, the ingenuity, and the truth of the science of plant biotechnology is elaborated in deep details in this paper.[4]

Jube et al.[5] deeply discussed with immense scientific far-sightedness recent advances in food biotechnology research. Modern biotechnology involves molecular and sub-molecular techniques that use whole or parts of living organisms to produce or improve commercial products and processes. It is a visionary area of molecular biology, which started with the creation of the first recombinant gene 30 years ago. Science and technology surpassed vast and versatile frontiers after that era. These food biotechnology tools change the eating habits of humans and the beverages they drink. They have also enhanced other aspects of our lives through the detection of early diagnosis of many serious ailments such as arteriosclerosis, cancer, diabetes, Parkinson's and Alzheimer's.[5] The science of medicine is thus in the process of new scientific regeneration. Biotechnology is the scientific field that offers the greatest potential to stop human hunger and provide basic human needs.[5] Biotechnology has today merged with environmental protection, water, and wastewater treatment. The authors of this paper discussed in minute details bioengineered plants, essential vitamins, essential minerals, essential amino acids, lysine, essential phytochemicals, isoflavonoids, and the world of bioengineered animals.[5] The application of genetic engineering in the food industry are discussed in details in this paper. Animals and microorganisms also have been extensively researched to provide better food products.[5] Here comes the vast importance of food biotechnology. Despite the vast benefits provided by modern biotechnology, this relatively new domain is not fully merged with the human society. There are still many issues of safety, reliability, and efficacy in the research endeavor in food biotechnology. These are discussed in detail in this paper.

Khattak et al.[6] discussed and deliberated in deep detail the recent advances in genetic engineering in a review. Classical fields of genetic engineering and some of its advancements are discussed in this review. The rapid and promising field of genetic engineering have given new impetus to biotechnology. The challenges and the vision lies in the hands of biologists and biotechnologists.[6] The authors in this paper discussed in detail recombinant DNA technology, advances in genetic engineering and the utmost importance of genetic engineering.[6] Genetic engineering being a field of biotechnology deals with genes. This field has also helped create thousands of organisms and processes that can be tailor-made for use in medicine, research, emancipation of science and full-scale manufacturing. Although the risk of disaster caused by the misuse of genetic engineering is extremely high, the potential of the benefits are path-breaking and visionary.[6] Today, genetic engineering is linked with chemical process technology and environmental remediation. A new era in the field of biotechnology will usher in as civilization treads forward.[6]

Biotechnology and biological engineering are the necessities of scientific research pursuit today. The success and the vision of biotechnology are changing the scientific firmament. In this chapter, the author deeply addresses the immense success and the scientific integrity in the field of bioengineering and biological sciences. Newer thoughts and new scientific imagination will transform the global scientific paradigm in biotechnology.[7–12]

8.9 THE CHALLENGES AND THE VISION OF HEAVY METAL GROUNDWATER REMEDIATION

Heavy metal and arsenic remediation of groundwater and de-poisoning of drinking water are the immediate need of the hour. The challenges and the vision of groundwater remediation are veritably immense and thought-provoking as science and engineering moves toward a newer scientific era. Developing as well as industrialized nations around the world are in the grip of a disaster of massive proportions, that is, the arsenic groundwater poisoning and the lack of pure drinking water. Here comes the need of integrated water resource management, human-factor engineering and technology management. Developing as well as developed nations around the world are in the grip of lack of proper implementation of environmental sustainability. Provision of clean drinking water is an absolute need to the human society. In the similar vein, environmental and energy sustainability are the needs of the scientific progress and the march of human civilization. The innovations, the scientific instinct and the ingenuity are the pillars of science and engineering today. Heavy metal groundwater poisoning will really change the very firmament of science and technology today. This challenge and this vision is elaborated in minute details in this chapter.[13–20]

8.10 BIOREMEDIATION AND GROUNDWATER REMEDIATION

Bioremediation and groundwater remediation are the scientific vision of tomorrow's scientific endeavor in the field of environmental engineering and chemical process engineering.[21–23] The domains of biotechnology and nanotechnology are today in the avenues of newer scientific profundity and deep introspection. Water treatment, drinking water treatment, and industrial wastewater treatment are the burdens and scientific intricacies of modern civilization today. Technological fervor, scientific candor, and the vast scientific ingenuity are the forerunners toward a new epoch in the field of nanotechnology, biotechnology, and environmental engineering. Groundwater remediation is the need of human mankind in today. Developing and disadvantaged nations around the world are today in the grip of a monstrous issue—provision of pure drinking water. Hence the needs of integrated water resource management and new innovations in biotechnology, bioremediation, and biological sciences. The success of science and engineering of bioremediation lies in the hands of biologists, environmental engineers, chemical engineers, and nanotechnologists. Water science and technology should be integrated with biotechnology and biological sciences. In this paper, the author rigorously points toward the success of science and technology in mitigation of global issues such as water shortage, global warming, and groundwater remediation. Science and engineering today are highly challenged as mankind treads forward toward a new epoch. This paper uncovers the scientific ingenuity and the vast scientific integrity, determination and grit in the field of biological sciences and environmental engineering sciences. Groundwater remediation and drinking water treatment are the immediate need of the hour. The success of science and civilization lies in the efficacy and efficiency of environmental protection tools. Today groundwater remediation and bioremediation are integrated with each other in application, scientific vision, and vast scientific profundity. Technologically, the poor and disadvantaged nations around the world are in the grip of one disaster over another. Scarcity of drinking water is a veritable burden of human civilization. Bioremediation and biotechnology will surely open up new doors of scientific thoughts, scientific enquiry and deep scientific ingenuity in environmental remediation.[20–23]

8.11 FUTURE RECOMMENDATIONS OF THIS STUDY AND THE FUTURE FLOW OF SCIENTIFIC THOUGHTS

The domains of nanotechnology and biotechnology are witnessing immense scientific introspection. The future of science and technology are highly challenged and path-breaking. In this treatise, the author with immense scientific insight depicts the scientific success, the vast scientific intricacies, and the ingenuity in the field of nanobiotechnology and its applications in human society. Today the global scenario in the field of drinking water treatment and industrial wastewater treatment is extremely grave. The challenges and the vision of groundwater remediation are immense, scientific path-breaking, and are replete with scientific acuity. Technological and scientific validation in the field of environmental remediation needs to be reorganized and reenvisioned with the passage of deep scientific history and time. Civilization today stands in the doors of scientific and technological validation. Future recommendations and future flow of thoughts should be directed toward greater scientific realization of environmental protection and mitigation of global climate change. Integrated water resource management and integrated wastewater management are the feasible solutions of environmental remediation today. The world of environmental engineering today stands in midst of deep trouble as arsenic and heavy metal groundwater and drinking water contamination devastates the scientific and academic rigor. The technology and engineering of nano-science and nano-engineering needs to be revamped and merged with diverse areas of science and engineering such as environmental engineering and chemical process engineering. Human scientific vision is highly challenged as developing and poor countries around the world are in the critical juncture of water scarcity and less sustainable development. The focus of future research should be toward greater scientific and engineering realization and motivation toward energy and environmental sustainability. These areas need to be widely and vehemently addressed as science and engineering surges ahead toward a new era of scientific might, vision, and scientific insight. Water science and technology is an area of immense concern and the civil society and the scientific domain needs to garner resources and act profoundly for future scientific endeavor. Future research pursuit should involve newer innovations and newer scientific rejuvenation in the field of environmental engineering, biotechnology, biological sciences, and chemical process engineering. Future recommendations and future flow of thoughts should be targeted toward a newer scientific imagination in the field of environmental protection and global water challenges. In this respect, human factor engineering and technology management in the field of environmental remediation should be other areas of research pursuit.[21–23]

8.12 CONCLUSION AND SCIENTIFIC PERSPECTIVES

The world of science and technology is today in the middle of deep scientific vision, introspection, and might. The scientific perspectives in the field of environmental protection need to be reframed and revamped with the progress of scientific rigor. Today scientific and academic rigor in the field of environmental engineering science globally are of highest order. In this paper, the author deeply elucidates the application of biological sciences and biotechnology in environmental remediation and other allied areas. The thrust areas of science and technology are the diverse branches of biotechnology and bioremediation. This well-researched treatise unfolds the scientific intricacies and the vision of the science of bioremediation. Today nanotechnology is integrated with diverse areas of science and engineering. Nanotechnology merged with biotechnology is the area of nanobiotechnology. The authors deeply discussed in this well-researched treatise the need of environmental remediation in the furtherance of science and engineering globally. Arsenic and heavy metal groundwater poisoning are challenging the scientific scenario in developing and developed nations around the globe. This challenge and the vision are enumerated in minute details in this chapter. The world of scientific validation and the futuristic vision of environmental engineering will surely open new doors of innovation and scientific instinct in environmental protection in decades to come.

KEYWORDS

- **bioremediation**
- **biotechnology**
- **biology**
- **vision**
- **environment**
- **engineering**

REFERENCES

1. Sharma, S. Bioremediation: Features, Strategies and Applications. *Asian J. Pharm. Life Sci.* **2012,** *2* (2), 202–213.
2. Kensa, V. M. Bioremediation—An Overview. *J. Ind. Pollut. Control* **2011,** *27* (2), 161–168.
3. Sardrood, B. P.; Goltapeh, E. M.; Varma, A. An Introduction to Bioremediation. In *Fungi as Bioremediators, Soil Biology 32*; Eds.; Goltapeh, E. M. et al.; Springer Verlag: Berlin Heidelberg, Germany, 2013.
4. Khan, M. Y.; Aliabbas, S.; Kumar, V.; Rajkumar, S.2009 Recent Advances in Medicinal Plant Biotechnology. *Indian J. Biotechnol.* **2009,** *8*, 9–22.
5. Jube, S.; Borthakur, D. Recent Advances in Food Biotechnology Research. In *Food Biochemistry and Food Processing*; Hui, Y. H., Nip, W-K, Nollet, LML, Paliyath, G., Sahlstrom, S., Simpson, B. K., Eds.; Blackwell Publishing: Oxford, United Kingdom, 2006; pp. 35–70.
6. Khattak, J. Z. K.; Rauf, S.; Anwar, Z.; Wahedi, H. M.; Jamil, T. Recent Advances in Genetic Engineering—A Review. *Curr. Res. J. Biol. Sci.* **2012,** *4* (1), 82–89.
7. Palit, S. Microfiltration, Groundwater Remediation and Environmental Engineering Science—A Scientific Perspective and a Far-reaching Review. *Nat., Environ. Pollut. Technol.* **2015,** *14* (4), 817–825.
8. Palit, S.; Hussain, C. M. Biopolymers, Nanocomposites, and Environmental Protection: A Far-reaching Review.- In *Bio-based Materials for Food Packaging*; Shakeel, A., Ed.; Springer Nature Singapore. Pvt. Ltd.: Singapore, 2018; pp. 217–236.
9. Palit, S.; Hussain, C. M. Nanocomposites in Packaging: A Groundbreaking Review and a Vision for the Future.-In *Bio-based Materials for Food Packaging*; Shakeel A., Ed.; Springer Nature Singapore. Pvt. Ltd.: Singapore, 2018; pp. 287–303.
10. Palit, S. Advanced Environmental Engineering Separation Processes, Environmental Analysis and Application of Nanotechnology—A Far-reaching Review. In *Advanced Environmental Analysis- Application of Nanomaterials*; C. M. Hussain, Kharisov, B., Eds.; The Royal Society of Chemistry: Cambridge, United Kingdom, 2017; Vol. 1, pp. 377–416.

11. Hussain, C. M.; Kharisov, B. *Advanced Environmental Analysis—Application of Nanomaterials, Volume-1*; The Royal Society of Chemistry: Cambridge, United Kingdom, 2017.
12. Hussain, C. M. Magnetic Nanomaterials for Environmental Analysis. In *Advanced Environmental Analysis—Application of Nanomaterials*; Hussain, C. M., Kharisov, B., Eds.; The Royal Society of Chemistry: Cambridge, United Kingdom, 2017; Vol.1, pp. 3–13.
13. Hussain, C. M. *Handbook of Nanomaterials for Industrial Applications*; Elsevier: Amsterdam, Netherlands, 2018.
14. Palit, S.; Hussain, C. M. Environmental Management and Sustainable Development: A Vision for the Future. In *Handbook of Environmental Materials Management*; Hussain, C. M., Ed.; Springer Nature: Switzerland A.G., 2018; pp. 1–17.
15. Palit, S.; Hussain, C. M. Nanomembranes for Environment. In *Handbook of Environmental Materials Management*; Hussain, C. M., Ed.; Springer Nature: Switzerland A.G., 2018; pp. 1–24.
16. Palit, S.; Hussain, C. M. Remediation of Industrial and Automobile Exhausts for Environmental Management. In *Handbook of Environmental Materials Management*; Hussain, C. M., Ed.; Springer Nature: Switzerland A.G., 2018; pp. 1–17.
17. Palit, S.; Hussain, C. M. Sustainable Biomedical Waste Management. In *Handbook of Environmental Materials Management*; Hussain, C. M. , Ed.; Springer Nature: Switzerland A.G., 2018; pp. 1–23. 18.Palit, S. Industrial vs. Food Enzymes: Application and Future Prospects. In *Enzymes in Food Technology: Improvements and Innovations*; Kuddus, M., Ed.; Springer Nature Singapore Pvt. Ltd.: Singapore, 2018; pp. 319–345.
18. Palit, S.; Hussain, C. M. Green Sustainability, Nanotechnology and Advanced Materials—a Critical Overview and a Vision for the Future. In *Green and Sustainable Advanced Materials Applications*; Shakeel, A., Hussain, C. M., Eds.; Wiley Scrivener Publishing: Massachusetts, USA, 2018; Vol. 2, pp. 1–18.
19. Palit, S. Recent Advances in Corrosion Science: A Critical Overview and a Deep Comprehension. In *Direct Synthesis of Metal Complexes*; Kharisov, B. I., Eds.; Elsevier: Amsterdam, Netherlands, 2018; pp. 379–410.
20. Palit, S. Nanomaterials for Industrial Wastewater Treatment and Water Purification. In *Handbook of Ecomaterials*; Springer International Publishing: Switzerland A.G., 2017; pp. 1–41.
21. www.wikipedia.com (accessed on April 25, 2019).
22. www.google.com (accessed on April 25, 2019).

CHAPTER 9

POROUS MATERIALS, METAL-ORGANIC FRAMEWORKS AND CONDUCTION

FRANCISCO TORRENS[1*] and GLORIA CASTELLANO[2]

[1]*Institut Universitari de Ciència Molecular, Universitat de València, Edifici d'Instituts de Paterna, P.O. Box 22085, E-46071 València, Spain*

[2]*Departamento de Ciencias Experimentales y Matemáticas, Facultad de Veterinaria y Ciencias Experimentales, Universidad Católica de Valencia San Vicente Mártir, Guillem de Castro-94, E-46001 València, Spain*

[*]*Corresponding author. E-mail: torrens@uv.es*

ABSTRACT

Porous materials attract considerable attention owing to their wide variety of scientific and technological applications, e.g., catalysis, shape- and size-selective absorption and adsorption, gas storage, and electrode materials. Considering the large variety of materials that could be classified as porous, this work focuses on nanostructured micro- and mesoporous materials. Via the approach, it is offered a more focused and practical analysis of key porous materials, which are considered relatively homogeneous from an electrochemical viewpoint. The key porous materials include (1) porous silicates and aluminosilicates; (2) porous metal oxides and related compounds; (3) porous polyoxometalates; (4) metal-organic frameworks; (5) porous carbons, nanotubes and fullerenes; (6) porous polymers and certain hybrid materials.

9.1 INTRODUCTION

Setting the scene: Porous materials (PMs), concept, classifications, mixed PMs, metal-organic frameworks (MOFs), PM oxides, related materials, PM Cs, organic-inorganic hybrid materials (HMs), nanocomposites (NCs), and conducting enhancement in functional polymeric composites.

The PMs continue to attract considerable attention owing to their wide variety of scientific and technological applications, e.g., catalysis, shape- and size-selective absorption and adsorption, gas storage and electrode materials. Considering the large variety of materials that could be classified as PMs, the present work focuses on nanostructured micro- and mesoporous materials. Via the approach, it was offered a more focused and practical analysis of key PMs, which are considered relatively homogeneous from an electrochemical viewpoint. The key PMs included (1) PM silicates and aluminosilicates; (2) PM metal oxides (MOs) and related compounds; (3) PM polyoxometalates; (4) MOFs; (5) PM Cs, nanotubes (NTs) and fullerenes; (6) PM polymers and certain HMs.

In earlier publications, it was informed the effects of type, size, and elliptical deformation on molecular polarizabilities of model single-wall carbon NTs (SWNTs) from atomic increments,[1–4] SWNTs periodic properties and table based on the chiral vector,[5,6] calculations on SWNTs solvents, cosolvents, cyclopyranoses[7–10] and organic-solvent dispersions,[11,12] packing effect on cluster nature of SWNTs solvation features,[13] information entropy analysis,[14] cluster origin of SWNTs transfer phenomena,[15] asymptotic analysis of coagulation–fragmentation equations of SWNT clusters,[16] properties of fullerite and symmetric C-forms, similarity laws,[17] fullerite crystal thermodynamic characteristics, law of corresponding states,[18] cluster nature of nanohorns (SWNHs) solvent features,[19] SWNTs (co-)solvent selection, *best* solvents, acids, superacids, host–guest inclusion complexes,[20] C/BC_2N/BN fullerenes/SWNTs/nanocones (SWNCs)/SWNHs/buds (SWNBs)/GRs cluster solvation models in organic solvents,[21–27] elementary polarizability of Sc/*fullerene*/GR aggregates, di/GR–cation interactions,[28] and the world of materials from layered ones to conductive two-dimensional (2D) MOFs.[29–33] The purpose of this report is to review PMs, concepts, classifications, mixed PMs, MOFs, PM oxides, related materials, PM Cs, organic-inorganic HMs, NCs and conducting enhancement in functional polymeric composites.

9.2 POROUS MATERIALS, CONCEPT, AND CLASSIFICATIONS

The PMs attracted considerable attention since the 1960s, owing to their wide variety of scientific and technological applications.[34] The term pore means a limited space or cavity in a continuous material. The PMs comprise inorganic compounds, e.g., aluminosilicates, to biomembranes and tissues. Pores are classified into three categories: micropores (less than 2nm), mesopores (2–50nm) and macropores (larger than 50nm). The PMs include clay minerals, silicates, aluminosilicates, organosilicas, metals, Si, MOs, Cs and C-NTs (CNTs), polymers and coordination polymers, or MOFs (Fig. 9.1), metal and MO nanoparticles (NPs), thin films (TFs), membranes, monoliths, etc.

FIGURE 9.1 Artistic representation of a metal-organic framework.
Source: Mónica Giménez.

Fundamental and applied research dealing with novel PMs is addressed to improve template-synthesis strategies, PM chemical modification via molecular chemistry, construction of metal, and MO nanostructures (NSs) with controlled interior nanospace, reticular MOF design with pore sizes ranging from the micropore to the mesopore scales, etc. The PMs are useful for sensing, catalysis, shape- and size-selective-reagents absorption and adsorption, gas storage, electrode materials, etc. Owing to the considerable variety of materials that could be classified as PM, several classifications were proposed. According to PM pores distribution, one can distinguish between regular and irregular PMs, whereas according to pores size distribution, one can separate between uniformly and nonuniformly sized PMs. From a structural viewpoint, PMs were viewed as building-blocks (BBs) result, following a construction order that could extend from the centimeter to the nanometre levels. The PMs range from highly ordered crystalline materials (e.g., aluminosilicates, MOFs) to amorphous sol–gel (SG) compounds, polymers, and fibres. This work focuses on materials that present PM structures but ion-insertion solids showing no micro- or mesoporous structures, e.g., metal polycyanometalates. In order to present a systematic approach from the electrochemical viewpoint, PMs are divided into (1) PM silicates and aluminosilicates; (2) PM MOs and related compounds [e.g., pillared oxides, laminar hydroxides, polyoxometalates (POMs)]; (3) MOFs; (4) PM Cs, NTs, and fullerenes; (5) PM organic polymers and HMs. Although the list does not exhaust the entire PM range, it covers those that could be described in terms of extended PM structures, whose electrochemistry was studied. A growing interest existed in the preparation of metal and MO NSs with controlled interior nanospace, whereas a variety of nanoscopic porogens (e.g., dendrimers, cross-linked and core–corona NPs, HM copolymers, cage supramolecules) are under research. A number of NS systems are treated in this work. The most relevant PM characteristic is the disposal of a high effective-surface/volume ratio, usually expressed in terms of their specific surface area (SSA, area per mass unit), which can be determined from $N_{2(g)}$ adsorption/desorption data. Different methods are available for determining SSA (Brunauer–Emmett–Teller, Langmuir, Kaganer), micropore volume (*t*-plot, α_s, Dubinin–Astakhov), and mesopore diameter (Barrett–Joyner–Halenda). Table 9.1 summarizes SSA values for selected PMs.

TABLE 9.1 Typical Values for Specific Surface Area of Selected Porous Materials.

Material	Specific surface area [$m^2 \cdot g^{-1}$]
Zeolite X	700
SBA-15	650
MCM-41	850
Activated carbon	2000
Nanocubes MOF-5	3500

Source: Ref. [34].

9.3 MIXED POROUS MATERIALS

Chemistry of PMs involves a variety of systems, which are termed as mixed systems, resulting from different-structural-moieties combination, ensuing in significant pristine-PM-properties modifications, in which one includes quite different PMs, i.e. (1) Composites formed by a binder addition to PMs and eventually other components, developing mixtures for definite applications, which system type is frequently used for preparing composite electrodes. (2) Functionalized PMs prepared by functional-groups attachment to a PM matrix. (3) The PMs with encapsulated species where molecular guests are entrapped in PM-host cavities. (4) Doped PMs where a material structural component becomes partially substituted by a dopant species, or when external species ingress in the original material as an interstitial ion. The term doping is applied to, e.g., yttria (Y_2O_3)-doped zirconias (ZrO_2) used for potentiometric O_2 determination, but also to describe Li^+ incorporation in polymers and NS Cs. (5) Intercalation PMs in which different NS components are attached to PM matrix (e.g., metal and MO NPs generated into zeolites and mesoporous silicates, organic polymers intercalated between laminar hydroxides). From several applications, it is convenient to describe many systems above as resulting from the parent-PM modification by a second component, in which sense one separates net modification, building, and functionalization processes. Net modification exists when the parent-PM final structure is modified via its combination with the second component, resulting in the formation of a new link system. Net building occurs when PM is formed assembling both-components units. Functionalization involves selected-molecular-groups attachment to the host PM without modification of its structure.

9.4 METAL-ORGANIC FRAMEWORKS

The MOFs are crystalline systems that could be described as infinite nets, resulting from the bonding of metal ions, which act as coordination centers, with polyfunctional organic molecules. First prepared by Tomic[35] and Yaghi et al.,[36] MOFs could be regarded as coordination polymer sponges by virtue of their large surface area, high porosity, and permeability to guest molecules. In contrast with other microporous materials, no volume inaccessible to guest molecules or ions exists. Organic linkers contain carboxylate $RCOO^-$, cyanide $C{\equiv}N^-$, or pyridine C_6H_5N coordinating groups.

Figure 9.2 shows a MOF with hexagonal pore apertures formed with triphenylene ligands. The singular MOF porosity allows for a significant redox conductivity that, in contrast with zeolites and other microporous aluminosilicates, could involve all PM units, e.g., Cu^{2+}- and Zn^{2+}-based MOFs with terephthalic acid $C_6H_4(COOH)_2$, in which metal centers and organic units are potentially electroactive in contact with suitable electrolytes. Relevant MOF properties prompted their use for catalysis, gas storage and separation, fuel cells, Li-based batteries, and electrocatalysis. Conversely, MOFs were electrochemically synthesized. The MOFs are indirectly related to other electrochemical applications, acting as a template for C-PM synthesis to be applied as a double-layer electrochemical capacitor.

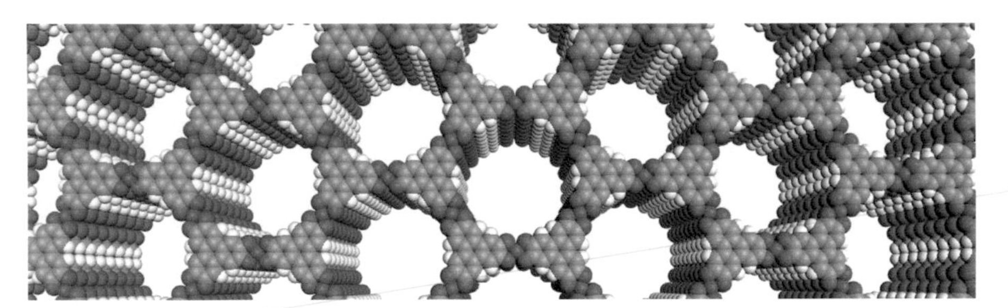

FIGURE 9.2 MOF with hexagonal pore apertures formed with triphenylene ligands. *Source*: K. A. Mirica.

9.5 POROUS OXIDES AND RELATED MATERIALS

Most oxide materials adopt microscopic structures formed by oxometal units tailored to form cage and tunnel cavities. The electrochemistry of MOs in contact with aqueous electrolytes is dominated by surface redox reactions, involving tightly bound hydrated oxometal groups. Porous oxides, notwithstanding, undergo electron transfer processes coupled with ion insertion issue from/to the electrolyte, with the possibility of adopting interstitial positions, or binding to metal centers via hydroxo- and aquo-groups formation. In contact with aqueous alkaline media, hydroxylation processes dominate MO electrochemistry. Notwithstanding, in contact with acidic media, proton and cation insertion processes occur, eventually leading to complicated responses where reductive or oxidative dissolution processes frequently take place. As far as such processes involve PM-structure disintegration, electrochemically-assisted dissolution processes are taken tangentially. Related materials comprise layered double hydroxides (LDHs), POM-group-based compounds, etc. The LDHs form lamellar structures entrapping charge-balancing anions. Polyoxovanadates, molybdates, and tungstates are POM-compound representatives, where complicated geometries are obtained from MO_6 octahedral (O_h) units containing charge-balancing ions occupying cavities. A fourth group of materials of electrochemical interest is that obtained doping the above and related materials. Typically, electroactive ions substitute parent metal centers in original-material lattice.

9.6 POROUS CARBONS

The term porous Cs is used to designate the materials prepared by organic felts or fibers (cellulose, polyacrylonitrile) carbonization under inert atmosphere or reducing conditions. Complex pyrocarbon structures are formed depending on temperature. Ordered mesoporous Cs (OMCs) posses a uniform, periodic mesopores distribution with a high surface area. They can be prepared nanocasting ordered mesoporous silicas (SiO_2, e.g., MCM-48, SBA-15) and mesocellular SiO_2 foams. During the OMC replication process, SiO_2-template internal PM structure is inversely replicated while the primary-particle external morphology is preserved, particle-size control being of interest. Chemical and thermal treatments are used for C-activation, which chemical activation, usually via KOH, introduces O-functionalities in C-surface.

9.7 ORGANIC-INORGANIC HYBRID MATERIALS AND NANOCOMPOSITES

The label of HMs is assigned to a variety of systems, presenting in common an organic-inorganic component combination blended on the molecular scale. After Kickelbick (2007), one distinguishes between organic-inorganic HMs sensu stricto, when inorganic units are formed in situ by molecular precursors, and NCs when discrete structural units are used in the respective size regime, which are usually formed embedding NPs, nanowires (NWs), nanorods (NRs), CNTs, etc., in an organic polymer. Preparation strategies involve either the combination of BBs, which retain to a great extent their molecular integrity throughout material-formation process or BB-precursors transformation into a novel material. Most HM development derives from SG-technology use. The HMs involve relatively weak interactions [e.g., van der Waals (VDW) forces, H-bonding (class I), covalent bonding (class-II HMs)] between BBs. A wide variety of NC/HMs were prepared and electrochemically characterized. A fraction of the materials are presented in this work. In the case of HMs that could be described as a result of the attachment of an organic moiety to an inorganic net, one distinguishes between net modification, building, and functionalization. In the first, the organic units are anchored to the inorganic support, which results in modifying its surface properties. In a functionalized system, a reactive organic group is bonded to the inorganic support, whereas if combining both organic and inorganic moieties forms a mixed net, both are termed as net builders, in which context the term NC is used when the size of one structural unit falls in 1–100nm. The NPs, NRs, metal–oxo clusters, fullerenes, CNTs, spherosilicates, and oligomeric silsesquioxanes are included in the group. Applications of HMs and NCs comprise catalysis, batteries, optical-electronic (optoelectronic) devices, corrosion inhibition, etc. In most cases, multifunctional materials can be prepared from different molecular precursors.

9.8 CONDUCTING ENHANCEMENT IN FUNCTIONAL POLYMERIC COMPOSITES

Study of H^+ conductivity in materials science gained attention because of their potential applications in chemical sensors, electrochemical devices, and energy generation, in which regard, efforts were diverged to high-performing (HP)-polymeric-membranes development as H^+ exchange membranes (PEMs) for fuel cell (FC) (PEMFC) applications. High conductivity at different humidity and temperature, and enhanced chemical and mechanical stability under operative conditions are considered main PEM goals. A common approach for improved-conductive-PEMs fabrication involves the combination of conductive polymers with fillers, ionic liquids (ILs) and nanofibres (NFs). In particular, MOFs experienced rapid growth as fillers for conductive composite membranes, because of their intrinsic stability and structural versatility. Compañ group filled different polymer matrices [sulphonated poly(ether ether ketone) (SPEEK), polybenzimidazole (PBI)] with an MOF subclass, zeolitic imidazole frameworks (ZIFs), and evaluated their use as PEMFCs.[37] The NC functional materials showed an improved conductivity at moderate temperatures and enhanced chemical and mechanical stability.

9.9 FINAL REMARKS

From the present results and discussion, the following final remarks can be drawn.

1. Porous materials continue to attract considerable attention owing to their wide variety of scientific and technological applications, e.g., catalysis, shape- and size-selective absorption and adsorption, gas storage, and electrode materials.
2. Considering the large variety of materials that could be classified as porous, this work focuses on nanostructured micro- and meso-porous materials. Via the approach, it was offered a more focused and practical analysis of key porous materials, which are considered relatively homogeneous from an electrochemical viewpoint.
3. The key porous materials included (1) porous silicates and aluminosilicates; (2) porous metal oxides and related compounds; (3) porous polyoxometalates; (4) metal-organic frameworks; (5) porous carbons, nanotubes, and fullerenes; (6) porous polymers and certain hybrid materials.

ACKNOWLEDGMENTS

The authors thank the support from Generalitat Valenciana (Project No. PROMETEO/2016/094) and Universidad Católica de Valencia San Vicente Mártir (Project No. UCV.PRO.17-18.AIV.03).

KEYWORDS

- **concept**
- **classification**
- **mixed porous material**
- **porous oxide**
- **porous carbon**
- **organic-inorganic hybrid material**
- **nanocomposite**

REFERENCES

1. Torrens, F. Effect of Elliptical Deformation on Molecular Polarizabilities of Model Carbon Nanotubes from Atomic Increments. *J. Nanosci. Nanotech.* **2003,** *3*, 313–318.
2. Torrens, F. Effect of Size and Deformation on Polarizabilities of Carbon Nanotubes from Atomic Increments. *Future Generation Comput. Syst.* **2004,** *20*, 763–772.
3. Torrens, F. Effect of Type, Size and Deformation on Polarizability of Carbon Nanotubes from Atomic Increments. *Nanotechnology* **2004,** *15*, S259–S264.
4. Torrens, F. Corrigendum: Effect of type, size, and deformation on polarizability of carbon nanotubes from atomic increments. *Nanotechnology* **2006,** *17*, 1541–1541.
5. Torrens, F. Periodic Table of Carbon Nanotubes Based on the Chiral Vector. *Internet Electron. J. Mol. Des.* **2004,** *3*, 514–527.
6. Torrens, F. Periodic Properties of Carbon Nanotubes Based on the Chiral Vector. *Internet Electron. J. Mol. Des.* **2005,** *4*, 59–81.
7. Torrens, F. Calculations on *Cyclo*pyranoses as Co-solvents of Single-wall Carbon Nanotubes. *Mol. Simul.* **2005,** *31*, 107–114.
8. Torrens, F. Calculations on Solvents and Co-solvents of Single-wall Carbon Nanotubes: *Cyclo*pyranoses. *J. Mol. Struct. (THEOCHEM)* **2005,** *757*, 183–191.
9. Torrens, F. Calculations on Solvents and Co-solvents of Single-wall Carbon Nanotubes: *Cyclo*pyranoses. *Nanotechnology* **2005,** *16*, S181–S189.
10. Torrens, F. Some Calculations on Single-wall Carbon Nanotubes. *Probl. Nonlin. Anal. Eng. Syst.* **2005,** *11* (2), 1–16.
11. Torrens, F. Calculations of Organic-solvent Dispersions of Single-wall Carbon Nanotubes. *Int. J. Quantum Chem.* **2006,** *106*, 712–718.
12. Torrens, F.; Castellano, G. Cluster Origin of the Solubility of Single-wall Carbon Nanotubes. *Comput. Lett.* **2005,** *1*, 331–336.
13. Torrens, F.; Castellano, G. Cluster Nature of the Solvation Features of Single-wall Carbon Nanotubes. *Curr. Res. Nanotech.* **2007,** *1*, 1–29.
14. Torrens, F.; Castellano, G. Effect of Packing on the Cluster Nature of C Nanotubes: An Information Entropy Analysis. *Microelectron. J.* **2007,** *38*, 1109–1122.
15. Torrens, F.; Castellano, G. Cluster Origin of the Transfer Phenomena of Single-wall Carbon Nanotubes. *J. Comput. Theor. Nanosci.* **2007,** *4*, 588–603.
16. Torrens, F.; Castellano, G. Asymptotic Analysis of Coagulation–fragmentation Equations of Carbon Nanotube Clusters. *Nanoscale Res. Lett.* **2007,** *2*, 337–349.
17. Torrens, F.; Castellano, G. Properties of Fullerite and other Symmetric Carbon Forms: Similarity Laws. *Symmetry Cult. Sci.* **2008,** *19*, 341–370.
18. Torrens, F.; Castellano, G. Fullerite Crystal Thermodynamic Characteristics and the Law of Corresponding States. *J. Nanosci. Nanotechn.* **2010,** *10*, 1208–1222.
19. Torrens, F.; Castellano, G. Cluster Nature of the Solvent Features of Single-wall Carbon Nanohorns. *Int. J. Quantum Chem.* **2010,** *110*, 563–570.
20. Torrens, F.; Castellano, G. (Co-)solvent Selection for single-wall carbon nanotubes: *Best* Solvents, Acids, Superacids and Guest–Host Inclusion Complexes. *Nanoscale* **2011,** *3*, 2494–2510.
21. Torrens, F.; Castellano, G. *Bundlet* Model for Single-wall Carbon Nanotubes, Nanocones and Nanohorns. *Int. J. Chemoinf. Chem. Eng.* **2012,** *2* (1), 48–98.
22. Torrens, F.; Castellano, G. Solvent features of Cluster Single-wall C, BC_2N and BN Nanotubes, Cones and Horns. *Microelectron. Eng.* **2013,** *108*, 127–133.

23. Torrens, F.; Castellano, G. Corrigendum to: Solvent Features of Cluster Single-wall C, BC_2N and BN Nanotubes, Cones and Horns. *Microelectron. Eng.* **2013,** *112*, 168–168.
24. Torrens, F.; Castellano, G. *Bundlet* Model of Single-wall Carbon, BC_2N and BN Nanotubes, Cones and Horns in Organic Solvents. *J. Nanomater. Mol. Nanotech.* **2013,** *2*, 1000107–1–9.
25. Torrens, F.; Castellano, G. C-nanostructures Cluster Models in Organic Solvents: Fullerenes, Tubes, Buds and Graphenes. *J. Chem. Chem. Eng.* **2013**, *7*, 1026–1035.
26. Torrens, F.; Castellano, G. Cluster Solvation Models of Carbon Nanostructures: Extension to Fullerenes, Tubes and Buds. *J. Mol. Model.* **2014,** *20*, 2263–1–9.
27. Torrens, F.; Castellano, G. Cluster Model Expanded to C-nanostructures: Fullerenes, Tubes, Graphenes and Their Buds. *Austin J. Nanomed. Nanotech.* **2014,** *2* (2), 7–1–7.
28. Torrens, F.; Castellano, G. Elementary Polarizability of Sc/*fullerene*/*graphene* Aggregates and di/Graphene–Cation Interactions. *J. Nanomater. Mol. Nanotech.* **2013,** *S1*, 001–1–8.
29. Torrens, F.; Castellano, G. Conductive Layered Metal-organic Frameworks: A Chemistry Problem. *Int. J. Phys. Study Res.* **2018,** *1* (2), 42–42.
30. Torrens, F.; Castellano, G. Conductive Two-dimensional Nanomaterials: Metal-organic Frameworks. *J. Appl. Phys. Nanotech.* **2019,** *2* (2), 52–52.
31. Torrens Zaragozá, F. From Layered Materials to Bidimensional Metal-organic Frameworks. *Nereis* **2019,** *2019* (11), 65–80.
32. Torrens, F.; Castellano, G. Conductive Layered Metal-organic Frameworks: A Chemistry Problem. *Madridge J. Phys.*, in press.
33. Torrens, F.; Castellano, G. World of Conductive Two-dimensional Metal-organic Frameworks. In *Green Materials and Environmental Chemistry: New Production Technologies, Unique Properties, and Applicatons*; Yaser, A. Z., Khullar, P., Haghi, A. K., Eds.; Apple Academic–CRC: Waretown, NJ, in press.
34. Doménech-Carbó, A. *Electrochemistry of Porous Materials*; CRC: Boca Raton, FL, 2010.
35. Tomic, E. A. J. Glass Transition Temperatures of Poly(methyl methacrylate) Plasticized with Low Concentrations of Monomer and Diethyl Phthalate. *Appl. Polym. Sci.* **1965,** *9*, 3745–3818.
36. Yaghi, O. M.; Li, H.; Groy, T. L. Construction of Porous Solids from Hydrogen-bonded metal Complexes of 1,3,5-Benzenetricarboxylic Acid. *J. Am. Chem. Soc.* **1996,** *118*, 9096–9101.
37. Escorihuela, J.; García-Bernabé, A.; Compañ, V., personal communication.

CHAPTER 10

CHEMICAL MESOSCOPICS FOR DESCRIPTION OF MAGNETIC METAL-CARBON MESOSCOPIC COMPOSITES SYNTHESIS

V. I. KODOLOV[1,2*], V. V. TRINEEVA[1,3], YU. V. PERSHIN[1], R. V. MUSTAKIMOV[1,2], D. K. ZHIROV[3], I. N. SHABANOVA[1,3], N. S. TEREBOVA[1,3], and T. M. MAKHNEVA[3]

[1]*Basic Research – High Educational Centre of Chemical Physics & Mesoscopics, Izhevsk, Russia*

[2]*Kalashnikov Izhevsk State Technical University, Izhevsk, Russia*

[3]*Udmurt Federal Research Centre, Ural Division, Russian Academy of Sciences, Izhevsk, Russia*

[*]*Corresponding author. E-mail: vkodol.av@mail.ru*

ABSTRACT

In this paper, the metal-carbon mesoscomposites synthesis stages at the growth of metal atomic magnetic moment are considered with a chemical mesoscopic position. The process occurs at the charge's quantization with the certain phase coherence and then with the chemical bonds' formation because of interference as well as in the reduction–oxidation processes possible annihilation takes place. The process of pair electron division and the shift of electrons on the high atomic levels for metal are explained by the annihilation origin. In this case, the metal atomic magnetic moment growth is observed in the dependence on the number of the electrons which participates in reduction–oxidation (redox) process. The production of metal-carbon mesoscopic composites is carried out with the use of mechanic chemical interaction between microscopic particles of metal oxides and macromolecules of polymers at the active medium presence. At the mesoscopic composites production, the sign variable loadings are applied. These loadings are appeared at the grinding with subsequent pressing through pores in stream of heat inert gas. This stream throws out through the nozzle into a vapor phase containing a protective polymer solution.

10.1 INTRODUCTION

The mesoscopic ideas may be successfully applied for the understanding of chemical particle reactivity at the conditions when the charge quantization and phase coherence take place. These phenomena are the reason for the interference and, possible, annihilation appearance. The transition from mesoscopic physics to mesoscopic chemistry and chemical mesoscopics is realized by means of the consideration of mesoscopic systems interaction with macroscopic systems. According to Imri[1] and Moskalets,[2] the mesoscopic system (ms) continually connects with alone or several macroscopic systems which are external medium for investigated sample (ms). Macroscopic systems are many more on size than a mesoscopic system. According to mesoscopics notions, the macroscopic systems are named as reservoirs, or banks, or contacts. These banks or reservoirs are sources and/or drains for particles and energy. The reservoirs are very big in order to that the transfer of energy between ms particles did not influence on their state. In other words, in any situation, the electron systems within reservoirs are found in an equilibrium state which is characterized by definite values of temperature (T) and chemical potential (μ).

The restrictions for processes defined by the limits of chemical mesoscopic are concluded in the following:

- The mesoscopic particle which may be presented as a big molecule or the linked molecules group is found in the active interaction with the medium.
- In this case, the size of phase coherency is located in limits up to 1000 nm
- Then the phenomena, such as interference, spectrum quantization, and charge quantization, are appeared.
- In other words, there is the source of quant radiation which activates the certain functional (active) groups in medium is formed when mesoscopic particles and also confined space take place.

Chemical mesoscopics explains these processes by the mesoscopic reactor (or nanoreactor) formation. In our case, the metal oxide cluster formed is coordinated with hydroxyl groups of polyvinyl alcohol (PVA) and promotes to its separation, as well as the hydrogen separation, from polymer chain.

10.2 WHAT ARE MESOSCOPIC REACTORS OR EARLY NAMED NANOREACTORS AND HOW THE CORRESPONDENT PROCESSES ARE REALIZED WITHIN MESOSCOPIC REACTORS?

Let us give the definition of this notion.

Mesoscopic reactors are defined as specific nanostructures and can be a nanosized cavity in polymeric matrices or space bounded part, in which reactive chemical particles are orientated with the creation of transitional state before predetermined product formation.

It is known[3,4] as the following classification of nanoreactors (now mesoscopic reactors):

One-dimensional nanoreactors (mesoscopic reactors) are considered as clearance between probe and surface, or crystal canals or complexes or crystal solvates, or macromolecules, or micelles, or vesicles, or pores.

For two-dimensional nanoreactors (mesoscopic reactors) the following classification can be presented: Double electrical layer, or monomolecular layers on the surface, or membranes, or interface layers (boundaries), or adsorption layers.

There are other cluster nanoreactors (mesoscopic reactors) as metal clusters.

Thus, the great variety of mesoscopic reactors, formed at different conditions, takes place. Then it is necessary to note that the formation of correspondent activated complex, which determines the process direction, is relieved within mesoscopic reactors. The charge quantization into medium promotes the process evolution during a certain time interval. Also, it is noted that the formation of nanostructures within mesoscopic reactors is determined by the nature of reactants, which participate in synthesis, and by the energetic and geometric characteristics of mesoscopic reactors.

The analysis of modern scientific data and our experiments show that the following peculiarities of nanostructures formation within nanoreactors may be noted:

1. The principal peculiarity—the decreasing of collateral parallel processes and the process direction to the special product side.
2. The low energetic expenses and the high rates of processes.
3. The dependence of obtained nanostructures properties from the energetic and geometric characteristics of nanoreactors.

In addition, it the absence of progressive and rotatory motions and the decreasing of vibration mode possibilities for the creation of processes realization conditions according to principles of chemical mesoscopics is necessary. When these conditions are secured, the transport "quantized electrons" takes place across the mesoparticles. In order to this process promotes to proceeding of redox processes. These conditions may be possible only at the correspondent sizes and forms of mesoscopic reactors and also at the interactions of mtsoscopic reactors walls with clusters created within these reactors.

What basic parameters and equations can be used for the nanostructures' formation within nanoreactors and what matrices are possible?

First, the coordination number of elements or elements group necessary to take into account the definition of process direction. In this case, it is possible the application of multiplate theory.

Second, the chemical potential difference is a basic factor for the self-organization process starts.

Third, the growth and form for nanostructures can be determined with the using of Kolmogorov–Avrami equations.[5–7]

The development of the nanostructures' self-organization process is estimated by means of the equation based on Kolmogorov–Avrami equations.[8,9] In this equation, the redox potentials are used because of the metal reduction occur in process of metal/carbon nanocomposite formation:

$$W = 1 - k \exp[-\tau^n \exp(zF\Delta\varphi/RT)],$$

where k is proportionality coefficient considering the temperature factor, τ is duration of process, n is fractal dimension, z is number of electrons participating in the process, $\Delta\varphi$ is difference of potentials on the border "nanoreactor wall – reaction mixture," F is Faraday number, R is the universal gas constant, and T is temperature. The calculations are made when the process duration changing takes with a half-hour interval. It was accepted for calculations that n equals 2 (two-dimensional growth), the potential of redox process during the metal reduction equals 0.34 V, temperature equals 473 K, Faraday number corresponds to 26.81 A·h/mol, universal gas constant (R) equals 2.31 W·h/mol·degree. The calculations are carried out for the thermal stage duration of the synthesis of copper–carbon mesocomposites.

Calculation results are given in Table 10.1.

TABLE 10.1 The Calculation Results of Process Duration for Copper/Carbon Nanocomposite Obtaining.

Duration (h)	0.5	1.0	1.5	2.0	2.5
Content of product, %	22.5	63.8	89.4	98.3	99.8

The calculation results practically coincide with experimental data.

Let us consider the processes of metal-carbon mesocomposites obtaining within mesoscopic reactors.

10.3 MESOSCOPIC REACTORS FOR METAL-CARBON MESOSCOPIC COMPOSITES SYNTHESIS

The aforesaid restrictions are at the metal/carbon mesoscopic composites obtaining. Let us consider the obtaining of metal-carbon mesoscopic composites from metal oxides (e.g., CuO) and PVA. The coordination number (CN) of copper maybe 2 or 4 in the dependence of metal oxidation state (for Cu^{+1} CN is equal to 2; for Cu^{+2} CN is equal to 4). Therefore, the reagents relation (CuO/PVA) can be equal to 4. At the grinding of reagents on the first stage, when the reagents relation is given as 1:4 [1 part—metal oxide, for example, copper oxide, and 4 parts—PVA], the decreasing of metal-containing phase sizes is explained by reduction–oxidation (redox) process. The formation of carbon shell from PVA chains starts off the coordination process of four chain fragments with the metal-containing cluster. Metals are reduced with the change of electron structure. When the dehydration and the dehydrogenization proceed, the connected fragments of polyacetylene and carbine appear. The fragments formed are coordinated on the metal. This is confirmed by spectroscopic and diffracted investigations. During the second (thermal) stage, which is realized at the temperature increasing to 400°C (no more), the result obtained on the first stage is fixed.[10–13]

At the common grinding of copper oxide particles with PVA particles (or concentrated water solution) the metallic phase clusters fall between macromolecules of polyvinyl alcohol or, in other words, into reservoir in which banks are PVA macromolecules. The difference of banks' chemical potentials stimulates the electron transport from bank with big potential to bank with small potential as is shown on the scheme.

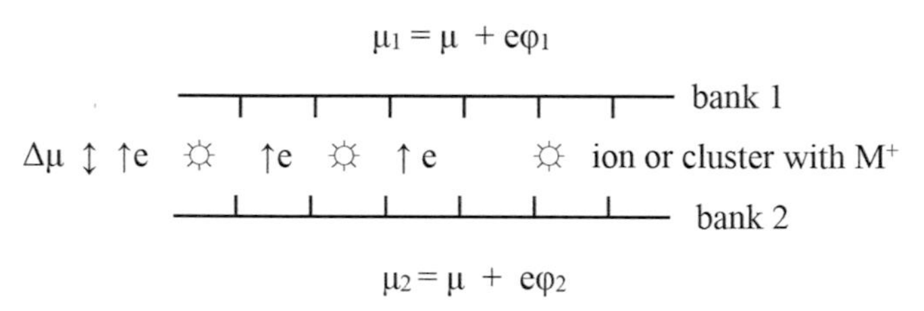

FIGURE 10.1 Possible scheme of nanostructures formation stages with using metal-containing cluster and polymer macromolecules orientated in the layer structure.

The metal (example, copper) within the cluster has a positive charge. Therefore, the negatively charged quants are directed to the positively charged atom. In our case, the negatively charged quants from polyvinyl alcohol acetate and hydroxyl groups are transferred to copper positive charged quants. As a result, the annihilation with the electromagnetic direct field formation takes place. In this process the acetic acid and water are formed, and also the banks (macromolecules) structures are changed: The polyacetylene and carbine fragments appear. There are unpaired electrons on joints of these fragments.

At the same time, the annihilation process stimulates the formation of d electrons to flow from metal to carbon shell of mesoscopic particle formed which is presented as metal (copper) carbon mesoparticle (or mesogranul). This process is accompanied by the growth of the metal atomic magnetic moment. The presented scheme of copper–carbon mesoscopic composites formation is confirmed by transition electron microscopy (TEM), X-ray photoelectron spectroscopy, roentgenograms, and also electron paramagnetic resonance. It is necessary to note that the mesoscopic composites obtained have magnetic characteristics. Owing to the delocalized electrons on carbon shell the obtained mesoscopic composites are active at the modification by the mechanic chemical methods.

The electron flow moving from nanoreactor walls across the metal cluster leads to the change of metal oxidation state. As a result of reduction–oxidation processes, the metal clusters are formed within carbon shells that contain fragments of polyacetylene and carbine chains. Such shells actively interact with d metals and initiate to d electrons transition of metals on higher energy levels that promote the magnetic characteristics increasing.

The structures of metal-carbon mesoscopic composites with active carbon shells are defined by means of the complex of methods including x-ray photoelectron spectroscopy, transition electron microscopy with high permission, electron microdiffraction, and also EPR spectroscopy (Figs. 10.2–10.5, Tables 10.1–10.3).

The image of copper/carbon mesoscopic composite structure obtained by TEM with high permission is shown in Figure 10.2.

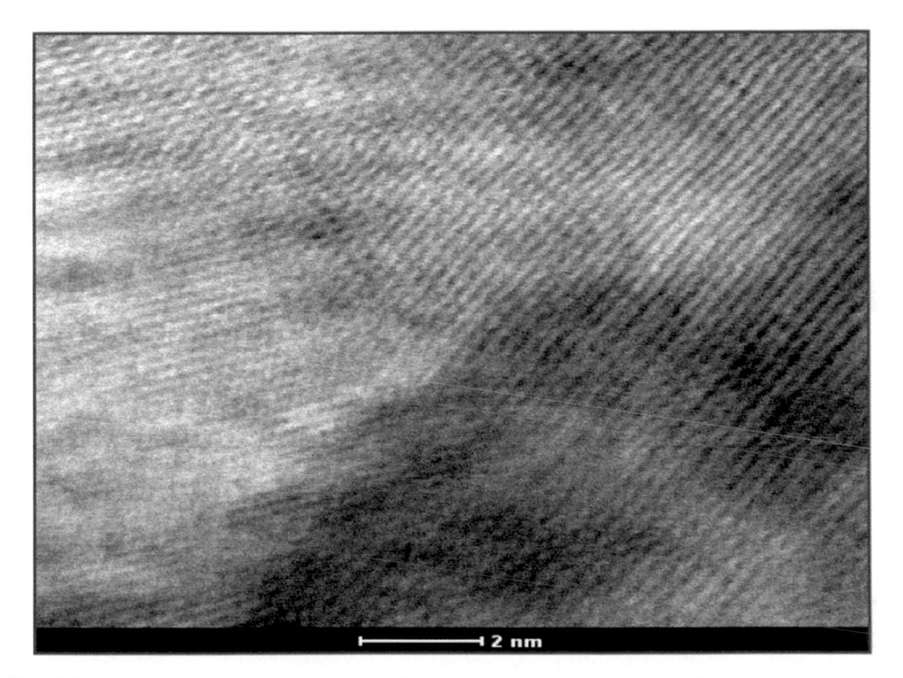

FIGURE 10.2 TEM microphotograph for copper/carbon nanocomposite.

Data of TEM with high solution testify that the carbon fiber consists of carbon atoms because of the correspondence of the fiber diameter and carbon atom diameter or C–H group diameter value.

According to Figure 10.2, the carbon structure shells for clusters of copper are presented as carbon fibers. This fact is confirmed by electron microdiffraction results. The carbon fibers formation is caused by the realization of reduction–oxidation process with the appearance of reduced copper and the carbonization of polymeric hydrocarbon chains.

The reduced copper formation at the redox process is confirmed by the diffract gram (Fig. 10.3).

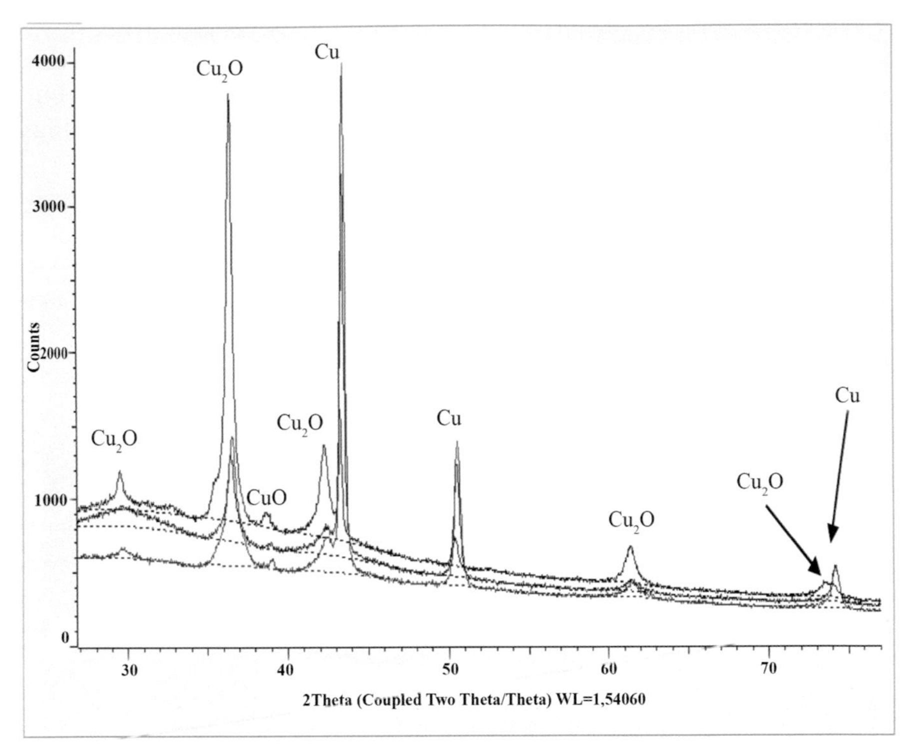

FIGURE 10.3 The diffract gram of copper–carbon mesoscopic composites obtained from copper oxide within matrices for different marks of PVA.

Cu/C NC (red): PVA mark BF-14, NC (blue): BF-17, NC (dark): BF-24.

Table 10.2 shows the dependence of the process results from the nature of the polymeric matrix or marks of PVA.

TABLE 10.2 The Influence of PVA Marks on Process Completion.

M Marks of PVA	The line color on diffract grams of copper–carbon mesoscopic composite	Process completion, %
PVA mark BF-17	Blue	77
PVA mark BF-14	Red	82
PVA mark BF-24	Dark	72

The correspondent marks of PVA contain the different quantity of acetic groups. The big number for PVA designates the most quantity of acetic groups in the polymer. However, the process completion at the use of this PVA mark is less at the comparison with PVA mark B-14. Therefore, the PVA mark B-14 is used for copper–carbon mesoscopic composite synthesis. The composition of metal-containing phase in this mesoscopic particle presents in Table 10.3.

TABLE 10.3 Composition of Metal Containing Phases in Copper–Carbon Mesocomposite.

Phase	Cu/C nanocomposite
CuO	1.17%
Cu_2O	5.19%
Cu	93.64%

Thus, the reduction–oxidation process accompanied by the copper formation takes place at the interaction of copper oxide and polyvinyl alcohol. At the same time in this process, the carbon fibers are formed.

In correspondence with C1s spectra (Fig. 10.4) the carbon fibers contain the carbine and polyacetylene fragments.

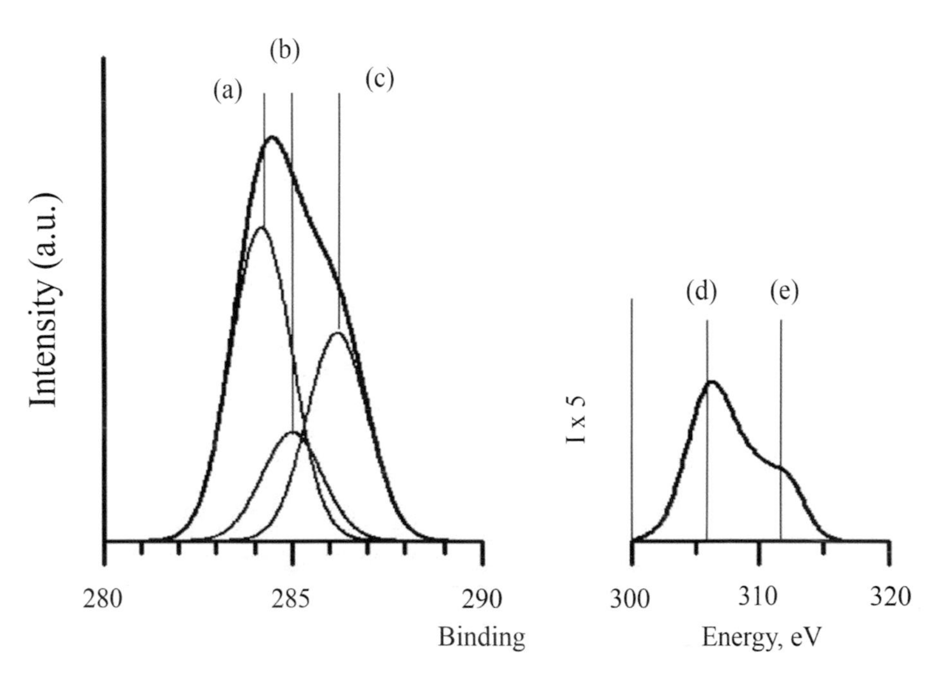

FIGURE 10.4 C1s spectrum of very Cu–C mesoscopic composite.

In C1s spectra, there are lines corresponding to the C1s energy for CH groups. A peak at 285 eV (C–H bond) can be concerned with polyacetylene fragment of carbon fiber, and addition to peak at 281–282 eV is concerned with Carbine fragment. According to C1s spectra, three types of satellites (sp, sp^2, sp^3 hybridization) with different relation between them are found. These satellites also take place in the C1s spectrum. The intensities relation as I_{sp}^{2}/I_{sp}^{3} for Cu/C nanocomposite corresponds to 1.7. The connection of these fragments is only possible at the unpaired electrons' formation on the joints of connections. Therefore, the investigations of EPR spectra are carried out. The results of these investigations have to Figure 10.5 and Table 10.4.

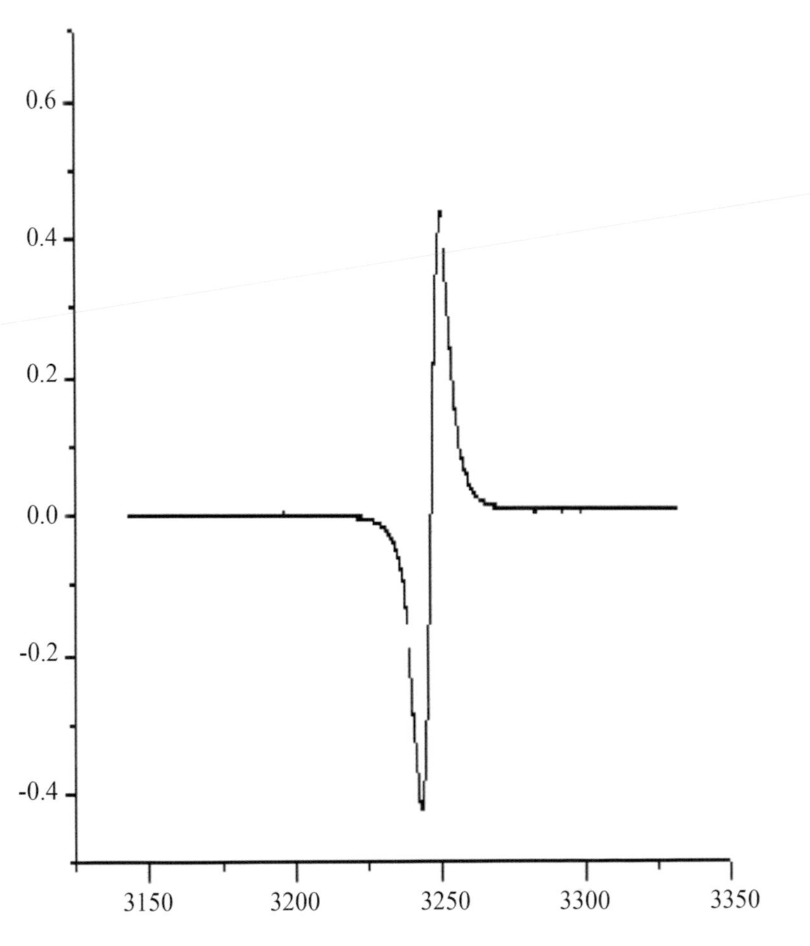

FIGURE 10.5 EPR spectrum of Cu–C mesoscopic composite carbon shell.

The coordination processes lead to the changes of metal electron structure with unpaired electrons formation. That is accompanied by the metal atomic magnetic moment increasing and also the appearance of unpaired electrons on the carbon shell surface (Table 10.4).

TABLE 10.4 Experimental EPR Data and Atomic Magnetic Moments (μ_B) for Copper–Carbon Mesoscopic Composite.

Copper–Carbon mesoscopic composite	g-Factor	Number of unpaired electrons, spin/g	Atomic magnetic moment, μ_B copper–carbon mesoscopic composite/massive sample
Copper–carbon mesoscopic composite	2.0036	1.2×10^{14}	1.3/–

In accordance with[15] the energetic characteristics for copper–carbon mesoscopic composite can be presented by the following data (Table 10.5).

TABLE 10.5 Energetic Characteristics of Copper–Carbon Mesoscopic Composite.

Relation of copper and carbon content, %	50/50
Density of copper–carbon mesocomposite, g/cm^3	1.71
Summary mass of copper–carbon mesocomposite, au	36.75
Middle size of copper–carbon mesocomposite, d, nm	25
Specific surface of copper–carbon mesocomposite, m^2/g	160
frequency of skeleton vibration of copper/carbon mesoscopic composite, s^{-1}	4×10^{11}
middle vibration energy of copper–carbon mesoscopic composite, erg	1.6×10^{13}

On our mind it's necessary to compare mesoparticles on the energetic parameters.

10.4 REACTIVITY OF COPPER–CARBON MESOSCOPIC COMPOSITE AT THE REACTIONS WITH OXIDIZERS

The presence of active double bonds and delocalized electrons in carbon shell of metal-carbon mesoscopic composites give the possibility for their modification by means of redox and addition processes. Therefore, the substances' interactions, in which elements (Si and P) have the highest oxidation state (+4 or +5), with such electron containing compounds as metal/carbon nanocomposites, are evident. The plain scheme of the reduction process is shown below.

$$НК\ (-\delta,\ or\ \sum e) \quad + \quad P^{+5} \quad \rightarrow \quad НК\ \text{--}\ P^{+3}\ (P^{+2},\ P^{0})$$

$$НК\ (-\delta,\ or\ \sum e) \quad + \quad Si^{+4} \quad \rightarrow \quad НК\ \text{--}Si^{0}\ (Si^{+2})$$

The mechanism is based on the chemical mesoscopics notions for electron transport across positive charged chemical particles. The proposed scheme is confirmed by x-ray photoelectron spectroscopy (P2p and Si2p spectra).

According to P2p spectrum, (Fig. 10.6) phosphorus in phosphorus-containing copper/carbon mesoscopic composite changes the oxidation state from +5 to zero. The binding energy P2p changes from 135 eV, corresponding to PO_4 group, to 129 eV for P^0. The process flows on 90%. It is possible the interaction copper and phosphorus in this case.[16] The X-ray pattern analysis of phosphorus-containing Cu/C nanocomposite shows the presence of peaks for groups Cu-C-P at θ equaled to 43°.

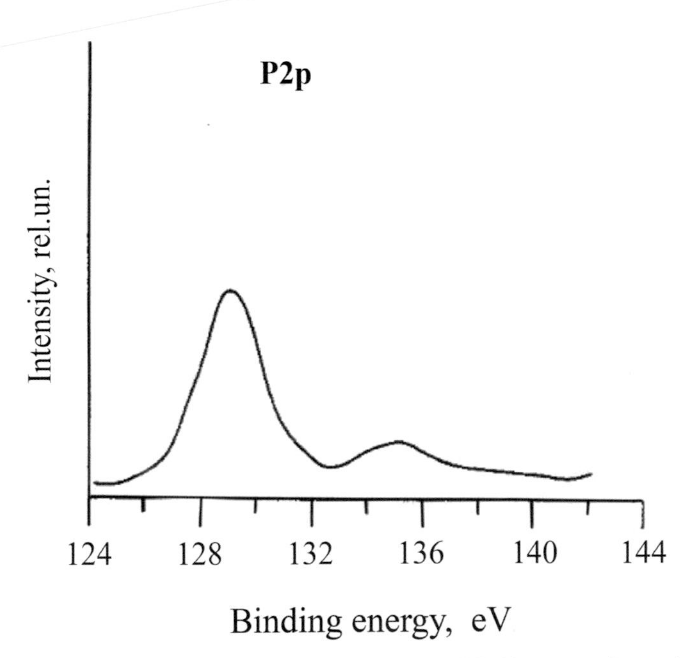

FIGURE 10.6 P2p spectrum of Cu/C mesocomposite modified by ammonium polyphosphate at the relation 1:1.

C1s spectrum of this mesoscopic composite (Fig. 10.7b) is distinguished from the C1s spectrum of nonmodified mesocomposite (Fig. 10.7a): The correspondent form C–H on 15% smaller than in spectrum of nonmodified mesocomposite. In turn, the relation of intensities for sp^2 and sp^3 hybridization is increased that can be linked with the mesoscopic granul increasing and approach its form to roundish.

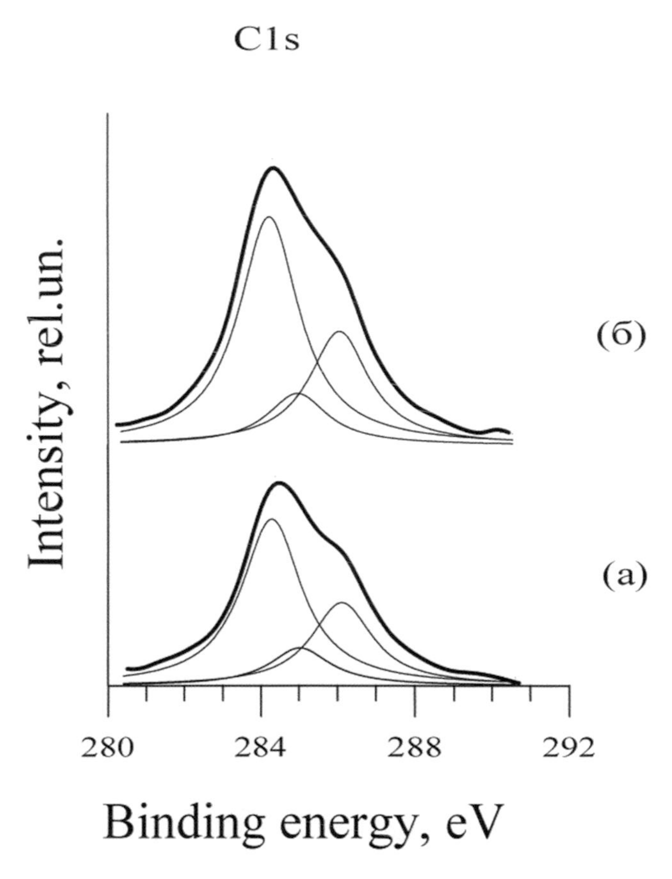

FIGURE 10.7 C1s spectra: (a) C1s spectrum of copper/carbon mesocomposite; (b) C1s spectrum of modified copper/carbon mesoscopic composite.

Analogous investigations are carried out[17,18] for copper–carbon mesoscopic composite modified by silicon-containing substances. Spectra Si2p and C1s for Cu/C nanocomposite (MC) modified by silica at the relation MC/silica = 1 leads to Figure 10.8.

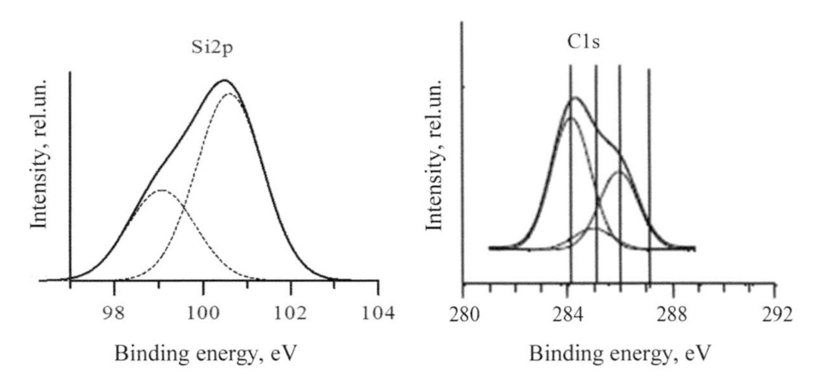

FIGURE 10.8 X-ray photoelectron spectra for modified Cu/C (Si) mesocomposite: (a) Si2p spectrum; (b) C1s spectrum.

According to the Si2p spectrum the relation of spectrum form intensities shows the reduction–oxidation process development on 51.4%. C–H intensity in the spectrum of modified Si-containing mesocomposite on 65% smaller the correspondent value in spectrum of initial mesocomposite. The thickness Si-containing shell for Cu/C (Si) mesocomposite in comparison with a shell of modified P containing mesoscopic composite is 4 times higher.

The investigation of modified metal/carbon mesoscopic composites by means of TEM of energy resolution shows that the shell from carbon fibers on mesoscopic granul surface is well preserved. For instance, the TEM image of phosphorus-containing Cu/C nanocomposite is presented below (Fig. 10.9).

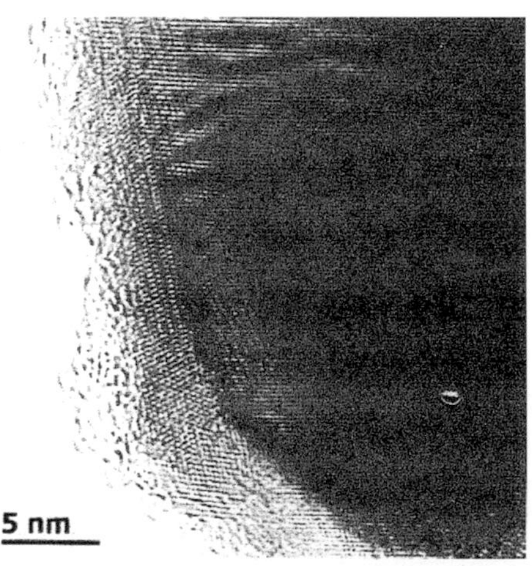

FIGURE 10.9 TEM image of phosphorus-containing copper/carbon mesoscopic composite.

At the reduction process, the copper atomic magnetic moment growth is observed (Table 10.6) which is accompanied by the increasing of unpaired electron values (Table 10.7).

TABLE 10.6 The Values of Copper Atomic Magnetic Moment in the Interaction Products for Systems Cu/C NC – APPh (or SiO_2)

Systems Cu/C NC – substances	μ_{cu}
Cu/C NC – silica	3.0
Cu/C NC – APPh	2.0
Cu/C NC – APPh, relation 1:0.5	4.2

TABLE 10.7 The Unpaired Electron Values (from EPR Spectra) for Systems "Cu/C NC – Silica" and "Cu/C NC – APPh" (relation 1:1) in Comparison with Mesoparticle Cu/C NC.

Substance	Quantity of unpaired electrons, spin/g
Cu/C nanocomposite	1.2×10^{17}
system "Cu/C NC – SiO_2"	3.4×10^{19}
system "Cu/C NC – APPh"	2.8×10^{18}

The results obtained cause natural questions concerning the reason for magnetic characteristics growth for metal atoms in the clusters within the nanocomposites granul. Answer on this question may be found in the following hypothesis:

The hypothesis is concluded in the formation of delocalized (unpaired) electrons because of the electrons shift on high energetic levels at the action electromagnetic radiation (field). It is possible the appearance of this radiation may be conditioned by the annihilation of quants of positive and negative charges which are appeared in the redox process at mechanochemical modification of metal/carbon nanocomposites. Certainly, this hypothesis is necessary to confirm by corresponding experiments.

It is interesting that the decripted processes are realized by the mechanochemical methods which can be perspective for the mesoscopic composites' production.

10.5 MECHANOCHEMICAL METHOD OF METAL-CARBON MESOCOMPOSITES OBTAINING

The magnetic mesoparticles obtaining with a possibility to the regulation of metal magnetic atomic moment is based on the combination of third variants of action on mesosystems formed on the boundary of reagents phases because of sign variable loadings (Fig. 10.10)

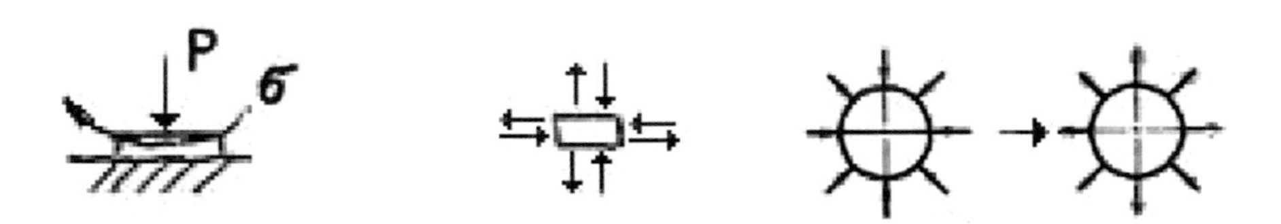

FIGURE 10.10 Schemes of mechanic loadings on reactive mesoscopic systems. (a) The combination of pressure with displacement loadings; (b) pulsation mechanical loadings; (c) the pressure increasing with next momentary decreasing.

The continuous process realization for the mesoscopic composites obtaining by mechanochemical method with thermal finishing is possible at following process steps:

1. The grinding of the mixture containing metallic and polymeric phases at the relation corresponding to the metal coordination number. It is necessary to remember about the possible change of coordination number at the metal oxidation state change during the process. The grinding is carried out according to the first scheme (Fig. 10.10a) in the reactor in which there is the bottom from porous ceramics through pores in which the mesoparticles are crushed.
2. The mesoparticles crushed through the micropores (may be nano-pores) proceeds into the warm inert gas stream. The gas is heated to the temperature which is necessary for the process continuation. Then the gas stream throws out through the nozzle into the reservoir with a vapor phase containing the protect polymer solution.
3. The mesoparticles formed in the process second stage touch in drops of polymer solution or substances from protecting the shell of nanostructures. Then the solvent is evaporated at vacuum. As a result, the mesoscopic product is obtained. This product consists of metal-carbon mesoparticle and shell protected from coagulation.

Proposed methods of metal-carbon mesoscopic composites synthesis[19] in mesoscopic reactors of polymeric matrices and obtained mesoscopic products have the following advantages:

On method:

- The technology originality which allows the application of metallurgical and polymeric raw for the synthesis of mesoscopic composites.
- The low energetic expenses and the absence of special requirements to catalysts and media.
- The absence of large discharges of heat, as well as volatile and other products, soiled medium.
- The possibilities of prognosis, control, and regulation of processes for the metal-carbon mesoscopic composites production.

On mesoscopic products:

- Most assortments of mesoscopic products determined by the production method.
- High activity of mesoscopic products in media and compositions owing to the metal and delocalized electrons' presence.
- Magnetic characteristics of metal-carbon mesoscopic composites.
- High effectiveness from the application of metal-carbon mesoscopic composites in media and compositions.

There are many examples of the metal-carbon mesoscopic composites modification, as well as the polymeric materials' modification, by these mesoscopic composites. So mesoscopic composites modified by the substances containing elements such as nitrogen, phosphorus, sulfur, silicon, and also metals, like aluminum, iron, nickel, copper, are known.[20] The modification of different materials by metal-carbon mesoscopic composites are also investigated.[21–23] The theoretical and experimental methods show that the mesoscopic composites electromagnetic field action on the medium molecules leads to the medium molecule self-organization and to the formation of nanostructured fragments (fractals) in the material composition. These changes can be accompanied by the processes of active particles' appearance. The changes in the media electron structure are possible under the mesoscopic composites minute quantities influence that it is confirmed by the X-ray photoelectron spectroscopic investigations. Below C1s spectra for films of PVA and polyvinyl alcohol, modified by means of the addition of 0.001 and 0.0001% of Cu–C mesoscopic composite into liquid PVA, in comparison with the spectrum of very Cu–C mesoscopic composite (Fig. 10.11).

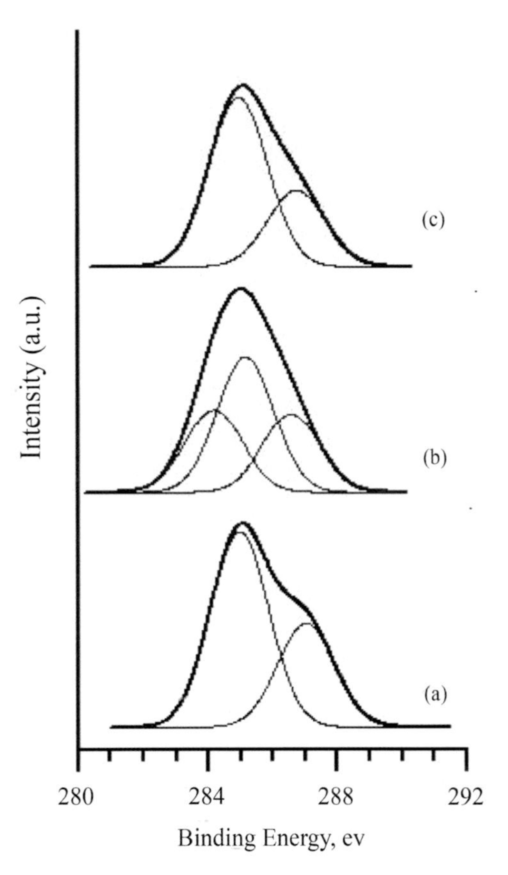

FIGURE 10.11 C1s spectra for polyvinyl alcohol modified by 0.001% (b) and 0.0001% (c). Cu–C mesocomposite in comparison with initial mesoscopic composite (a).

The mechanism of polymeric materials modification processes by minute quantities of metal-carbon mesoscopic composites is proposed. The estimation methods of self-organization in polymeric compositions and materials, modified by metal/carbon minute quantities are also proposed. The orientation processes, in which mesoscopic particles (nanocomposites) participate, lead to the changes of submolecular structures and properties of materials. Usually, the process of polymeric matrix self-organization stimulates the increase of material heat capacity, the material density, its thermal stability, and also the adhesive strength. The improvement of other properties of materials, modified by the super small quantities of metal-carbon mesoscopic composites, is also possible. Some results of material modifications by minute quantities of metal-carbon mesoscopic composites are represented, for instance, epoxy resins, conductive glues, glass-reinforced plastics, polyvinylchloride films, polycarbonate glasses. There is one of examples. The modification of industrial epoxy materials EZC–11 and ferric epoxy material with the use of 0.005% of Cu–C mesocomposite increases their adhesive strength by more than 60%. These materials contain fillers, the particles of which are organized within well-regulated net formed under the influence of mesoscopic composite. In this case, the formation of filler interacted particles' chains is possible.

10.6 CONCLUSION

The creation of reactive mesoscopic materials with regulated magnetic characteristics which can find the application as modifiers of materials properties, catalysts for different processes, effective inhibitors of corrosion, sorbents, stimulators of plant growth, is very topical. The present investigation has a fundamental character. It is based on the ideas concerning the change of metal-carbon mesoscopic composites reactivity. The investigations are dedicated to the mechanic chemical reduction–oxidation processes in which the electron transport from the mesoscopic composite cluster to carbon shell takes place. In this case, the electron delocalization is found. For the first time, the metal-carbon mesoscopic composite modification by the mechanic chemical process with the use of active substances including also bioactive systems is carried out. The activity of metal-carbon mesoscopic composites is caused by the structure and composition of correspondent composites, which contain the delocalized electrons and double bonds on the surface of the carbon shell.

Thus, at the mechanic chemical reduction/oxidation synthesis the changes of element oxidation states, as well as the increasing of the metal atomic moment for a cluster, can be appeared. At the same time, the modifiers' elements and functional groups are discovered in the carbon shell of mesoscopic composites modified. These facts open a new era for further investigations and development of metal-carbon mesoscopic composites application fields.

REFERENCES

1. Imri, I. *Introduction in Mesoscopic Physics*. M.: Physmatlit; 2009, p 304.
2. Moskalets, M. V. *Fundamentals of Mesoscopic Physics*. NTU KhPI: Khar'kov; 2010, p 186.

3. Kodolov, V. I.; Khokhriakov, N. V. *Chemical Physics of Formation and Transformation Processes of Nanostructures and Nanosystems*. IzhSACA: Izhevsk; 2009; Vol. 1, p. 361,Vol. 2, p 415.
4. Kodolov, V. I.; Trineeva, V. V. New Scientific Trend – Chemical Mesoscopics. *Chem. Phys. Mesos.* **2017,** *19* (3), 454–465.
5. Morozov, A.D. Introduction in Fractal Theory. –ICT: M. Izhevsk; 2002, p 160.
6. Kolmogorov, A. N.; Fomin, S. V. Introductory Real Analysis. Prentice Hall: Portland, USA, 2009, p 403.
7. Wunderlikh, B. Physics of Macromolecules. M.: Mir; 1979, Vol. 2, p. 422.
8. Kodolov, V. I.; Khokhriakov, N. V.; Trineeva, V. V.; Blagodatskikh, I. I. Nanostructure Activity and its Display in Nanoreactors of Polymeric Matrices and in Active Media. *Chem. Phys. Mesos.* **2008,** *10* (4), 448–460.
9. Kodolov V.I. The Addition to Previous Paper. *Chem. Phys. Mesos.* **2009,** *11* (1), 134–136.
10. Trineeva, V. V.; Vakhrushina, M. A.; Bulatov, D. I.; Kodolov, V. I. The Obtaining of Metal/Carbon Nanocomposites and Investigation of Their Structure Phenomena. *Nanotechnics* **2012,** *N4*, 50–55.
11. Trineeva, V. V.; Lyakkhovich, A. M.; Kodolov, V. I. Forecasting of the Formation Processes of Carbon Metal Containing Nanostructures Using the Method of Atomic Force Microscopy. *Nanotechnics* **2009,** *4* (20), 87–90.
12. Kodolov, V. I.; Blagodatskikh, I. I.; Lyakhovich, A. M. et al. Investigation of the Formation Processes of Metal Containing Carbon Nanostructures in Nanoreactors of Polyvinyl Alcohol at Early Stages. *Chem. Phys. Mesos* **2007,** *9* (4), 422–429.
13. Kodolov, V. I.; Trineeva, V. V.; Blagodatskikh, I. I.; Vasil'chenko, Yu. M.; Vakhrushina, M. A.; Bondar A. Yu. The Nanostructures Obtaining and the Synthesis of Metal/Carbon Nanocomposites in Nanoreactors. In *Nanostructure, Nanosystems and Nanostructured Materials: Theory, Production and Development*. Apple Academic Press: Toronto-New Jersey; 2013, pp 101–145
14. Kodolov, V. I.; Trineeva, V. V. Fundamental Definitions for Domain of Nanostructures and metal/Carbon Nanocomposites. In *Nanostructure, Nanosystems and Nanostructured Materials: Theory, Production and Development*. Apple Academic Press: Toronto-New Jersey; 2013, pp 1–42.
15. Kodolov, V. I.; Trineeva, V. V. Energetic Characteristics of Metal/Carbon Nanocomposites. *JCDNM* **2015,** *7* (2), 223–228.
16. Kodolov, V. I.; Trineeva, V. V.; Kopylova, A. A. et al. Mechanochemical Modification of Metal/Carbon Nanocomposites. *Chem. Phys. Mesos*. **2017,** *19* (4), 569–580.
17. Shabanova, I. N.; Kodolov, V. I.; Terebova, N. S.; Trineeva, V. V. *X Ray Electron Spectroscopy in Investigations of Metal/Carbon Nanosystems and Nanostructured Materials*. Udmurt Un iversity: M.-Izhevsk; 2012, p 252.
18. Kodolov, V. I.; Trineeva, V. V.; Terebova, N. S. et al. Changes of Electron Structure and Magnetic Characteristics of Modified Copper/Carbon Nanocomposites. *Chem. Phys. Mesos*. **2018,** *20* (1), 72–79.
19. Kopylova, A. A.; Kodolov, V. V. Investigation of Coper/Carbon Nanocomposite Interaction with Silicium Atoms from Silicon Compounds. *Chem. Phys. Mesos*. **2014,** *16* (4), 556–560.
20. Shabanova, I. N.; Terebova, N. S. Dependence of the Value of the Atomic Magnetic Moment of d metals on the Chemical Structure of Nanoforms. In *The Problems of*

Nanochemistry for the Creation of New Materials IEPMD: Torun, Poland; 2012, pp 123–131.

21. Kodolov, V. I.; Khokhriakov, N. V.; Kuznetsov, A. P. To the Issue of the Mechanism of the Influence of Nanostructures on Structurally Changing Media at the Formation of "Intellectual" Composites. Nanotechnics **2006,** *3* (7), 27–35.
22. Kodolov, V. I.; Khokhriakov, N. V.; Trineeva, V. V.; Blagodatskikh, I. I. Problems of Nanostructure Activity Estimation, Nanostructures Directed Production and Application. Nanomaterials Yearbook, 2009. *From Nanostructures, Nanomaterials and Nanotechnologies to Nanoindustry*. Nova Science Publishers, Inc.: NY; 2010, p 1–18.
23. Chashkin, M. A.; Kodolov, V. I.; Zakharov, A. I. et al. Metal/carbon Nanocomposites for Epoxy Compositions: Quantum-Chemical Investigations and Experimental Modeling. *Poly. Res. J.* **2011,** *5* (1), 5–19.

CHAPTER 11

ANALYSIS OF FUMES ARISING DURING FABRICATION OF HYBRID ADHESIVE-LASER JOINTS

MILAN MARONEK[1], IGOR NOVÁK[2], IVAN MICHALEC[1], JÁN MATYAŠOVSKÝ[3], PETER JURKOVIČ[3*], KATARÍNA VALACHOVÁ[4], and LADISLAV ŠOLTÉS[4]

[1]*Slovak University of Technology in Bratislava, Faculty of Materials Science and Technology, Trnava, Slovakia*

[2]*Polymer Institute, Slovak Academy of Sciences, Bratislava, Slovakia*

[3]*VIPO, Partizánske, Slovakia*

[4]*Centre of Experimental Medicine, Institute of Experimental Pharmacology and Toxicology, Bratislava, Slovakia*

[*]*Corresponding author. E-mail: pjurkovic@vipo.sk*

ABSTRACT

The chapter deals with the identification of gaseous effluents resulting from the preparation of hybrid adhesive-laser joints. CrNiMo AISI 316 1-mm thick stainless steel was used as the base material (adherent), which was joined by four types of Loctite two-component epoxy adhesives: Hysol 9466, Hysol 9455, Hysol 9492, and Hysol 9497 with Trumpf D70 welding head. The chemical composition of the adhesives was realized by Fourier transform infrared spectroscopy analysis; the effect of elevated temperature was analyzed by simultaneous thermal analysis and thermogravimetric analysis. The simulation of the thermal decomposition of the adhesive due to the temperature cycle of welding was performed by pyrolysis. The gaseous products were analyzed by gas chromatography and mass spectrometry. In the thermal degradation process, the cyclic hydrocarbons, such as phenol, bisphenol, benzene, unofuran, and many others, are created. In terms of the number of individual degradation products, it can be stated that fewer products have been found in Hysol 9466 and Hysol 9497, whereas the largest number of volatile thermal decomposition products has been found in Hysol 9492. The identified gaseous products are hazardous to the human body, so it is necessary to ensure consistent exhaustion of the resulting fumes when making hybrid joints.

11.1 INTRODUCTION

The use of hybrid-bonding technologies using a combination of adhesive bonding and mechanical bonding (riveting) or welding is currently used mainly in the aerospace industry but also penetrates the automotive and other industries. This is due to a number of features that make hybrid joints characteristic. It is primarily a more favorable distribution of stresses in the joint[1,2] at its load, higher toughness, increased joint stiffness, increased fatigue properties, increased corrosion resistance and others. Several methods for making hybrid joints are now known. The first is adhesion bonding in three possible modifications.[3] For spot welding of thin sheets, the weld through and flow in methods[3,4] are used. In addition, hybrid-bonding methods with subsequent plasma welding,[5,6] MIG welding,[7] laser welding,[8–10] and laser-TIG welding[11] were also tested. Thus, in the case of hybrid joints, thermal degradation of the used adhesive occurs, resulting in a number of fumes. The aim of this chapter is a qualitative analysis of the thermal degradation of the adhesives used in the preparation of hybrid joints and the assessment of possible risks arising from their formation.

11.2 EXPERIMENTAL

The basic material (adherent) was an austenitic, CrNiMo corrosion-resistant nonstabilized AISI 316 steel with a thickness of 1 mm. The exact chemical composition was determined by the SPECTRO Spectrocast optical emission spectrometer and is shown in Table 11.1.

TABLE 11.1 Chemical Composition of AISI 316 Used as Adherent.

C (%)	Mn (%)	Si (%)	P (%)	S (%)	Cr (%)	Ni (%)	Mo (%)	Al (%)	Co (%)	Cu (%)	Nb (%)	Ti (%)	V (%)
0.06	0.73	0.48	0.027	0.009	17.38	9.65	1.65	0.040	0.09	0.38	0.02	0.01	0.06

Four different types of two-component structural epoxy adhesives from Loctite were designed: Hysol 9466, Hysol 9455, Hysol 9492, and Hysol 9497. Hysol 9466 is a tough industrial structural epoxy adhesive with high shear strength and chemical and solvent resistance. Hysol 9455 is a fast curing epoxy adhesive with very low viscosity, high shear strength, and versatile use. Hysol 9492 excels in high-temperature resistance and Hysol 9497 in high thermal conductivity. The basic manufacturer-reported adhesive characteristics are shown in Table 11.2.

TABLE 11.2 Basic Characteristics of the Adhesives Used by the Manufacturer.

	Hysol 9466	Hysol 9455	Hysol 9492	Hysol 9497
Resin sort	Epoxy	Epoxy	Epoxy	Epoxy
Hardener	Amine	Methylene thiol	Amine	Amine
Volume mixing ratio (resin:hardener)	2:1	1:1	2:1	2:1

Fourier transform infrared spectroscopy (FTIR) was used to determine the chemical composition of the adhesives. Adhesive spectra were recorded on a NICOLET 8700™ Thermo Scientific FTIR spectrometer in the mid-infrared range (4000–650 cm^{-1}) on a disposable ATR insert using a germanium crystal. The number of scans ranged from 256 to 1024—the least scans were chosen for Hysol 9455 adhesive (particularly the larger material) and a resolution was 4 cm^{-1}. OMNIC 8.1™ spectroscopic software was used for both measurement and evaluation, and spectra were evaluated on a qualitative basis using a built-in spectrum library.

Simultaneous thermal analysis was performed on NETZSCH STA 409 C/CD to determine the behavior of adhesives during increasing temperatures. The heating was carried out in the range from 20 to 400°C and the heating rate was 10°C×min^{-1}. Helium with a purity of 99.999% was used as shielding gas.

Thermogravimetric analysis on a TA Instruments DMA Q800 was performed to heat the adhesives during heating up to 500°C. Pyrolysis was performed on adhesive samples to simulate the welding process and a combination of gas chromatography and mass spectrometry was used to identify the amount and composition of volatile thermal decomposition products formed during the pyrolysis. The CDS Pyroprobe Model 5150 was used for pyrolysis, the Agilent 7890A GC and the Agilent 5975C MSD were used as the gas chromatograph. The measurement conditions are shown in Table 11.3.

TABLE 11.3 Pyrolysis Parameters of Adhesives Used.

Sample weight (g)	1.5
Initial temperature (°C)	300
Heating to initial temperature ($°C \times min^{-1}$)	15
Maximum temperature	1000
Heating rate to maximum temperature ($°C \times s^{-1}$)	10
Endurance at maximum temperature (s)	5

Hybrid joints were created in two steps. First, adhesive joints were made. Prior to the adhesive bonding, the bonded surfaces were abraded with 100 and 240 μm abrasive paper. Subsequently, the bonded areas were degreased with Loctite 7061 alcohol-based and dried. Both adhesive components (resin and hardener) were extruded using a Loctite 96001 displacement gun with a static mixer to provide the necessary mixing of both components. To obtain maximum strength, a layer of adhesive was applied to both bonded areas, and the samples were then fixed in spring clips at 25°C for 48 h for sufficient cure of the adhesive. The thickness of the adhesive was approximately 50 μm. The dimensions of the test samples are shown in Figure 11.1.

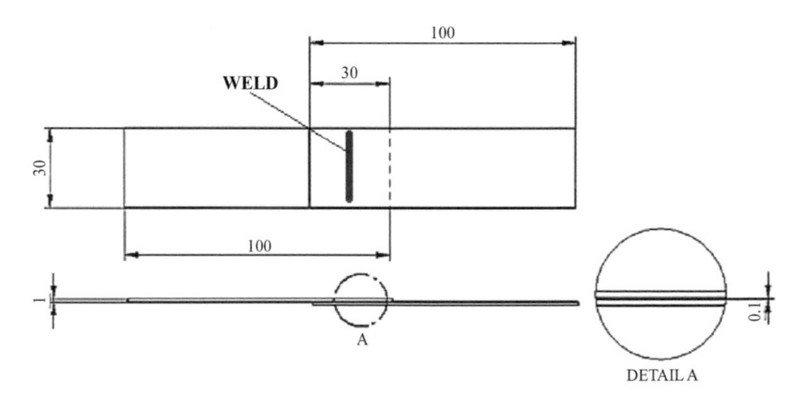

FIGURE 11.1 Dimensions of test samples.

In preparation of the hybrid joints, the weld joint was made along the vertical axis of the adhesive surface, the weld length being 28 mm. The TRUMPF TruDisk 4002 solid-state disk laser with D70 welding head was used to make hybrid joints. The collimation and focusing distance was 200 mm. Welded at 700 W, with a welding speed of 20 mm×min^{-1} at a laser beam focusing of 1 mm below the surface with a maximum output power of 2 kW. As a protective atmosphere, 99.996% purity (4N6) was used. The diameter of welding spot was 200 μm. Hybrid joint test specimens are shown in Figure 11.2, a typical hybrid joint macrostructure is shown in Figure 11.3.

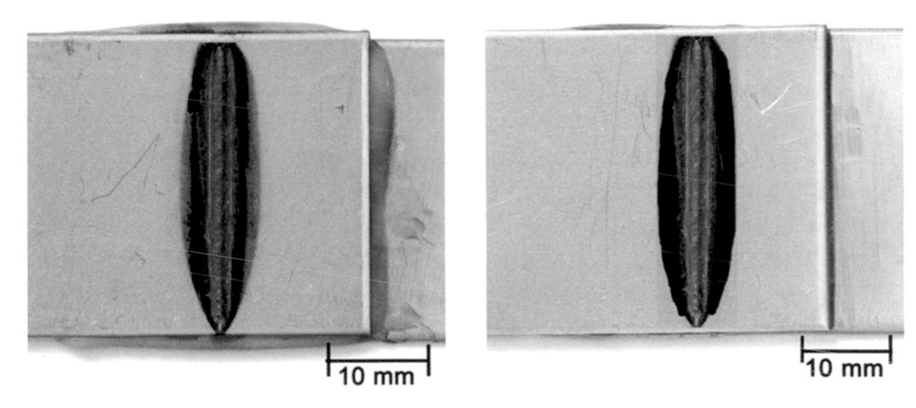

FIGURE 11.2 The view at the hybrid joints.

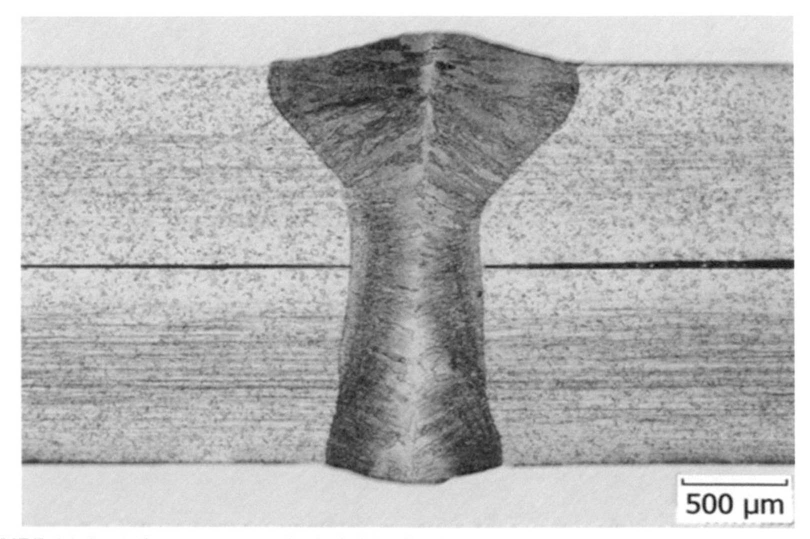

FIGURE 11.3 Microstructure of a hybrid adhesion-laser joint.

In the application of fusion welding processes, the thermal layer of the adhesive layer is completely thermally decomposed at the point of welding. The measurements have shown that at the boundary of the bonded materials at a distance of 2.5 mm from the weld axis, the temperature reaches about 300°C.

11.3 RESULTS AND DISCUSSION

11.3.1 CHEMICAL COMPOSITION OF ADHESIVES

The results of the individual wave numbers are shown in Figures 11.4 and 11.5. Function groups are assigned to each wave number in figures. The results show that the main component of the above adhesives is diglycidyl bisphenol A. All adhesives had very similar wavelength spectra, suggesting a similar chemical composition. The Hysol 9455 adhesive exhibited a more pronounced deviation from other adhesives in the wavelength range of 1050–1120, in which sulfur oxides (S=O) were cleaved. This finding confirms the presence of sulfur in the hardener (methylenethiol) in Hysol 9455 adhesive.

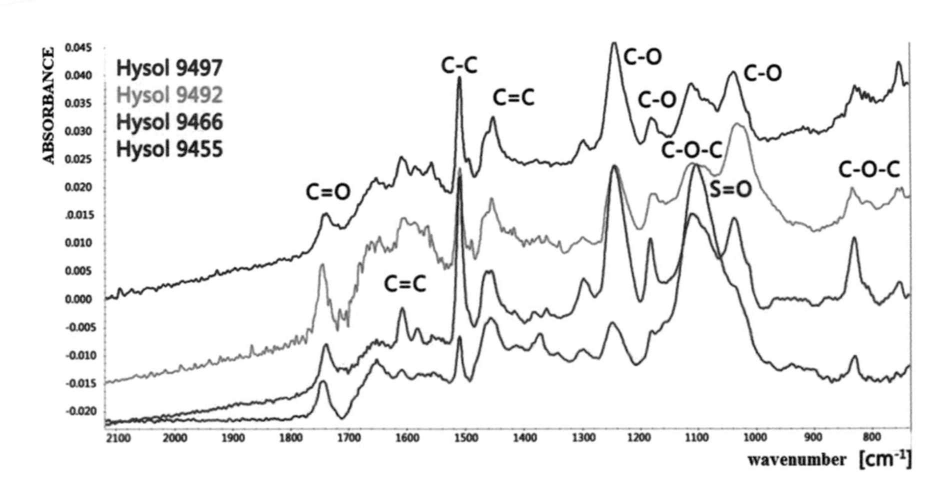

FIGURE 11.4 FTIR measurement results—wave numbers 2100–800.

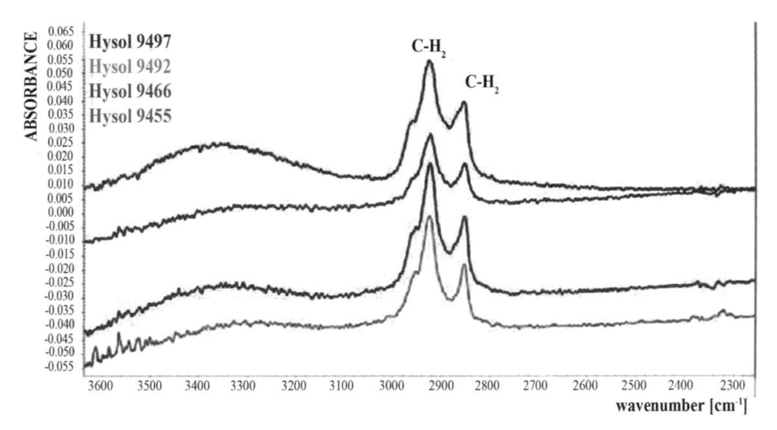

FIGURE 11.5 FTIR measurement results—wave numbers 2300–3600.

11.3.2 STA RESULTS

The Hysol 9466 STA adhesive is shown in Figure 11.6. The glass transition temperature was identified at 52.6°C. The start of melting of the adhesive was at 330°C, accompanied by a pronounced endothermic reaction, the peak of which was observed at 337.7°C. After the melting started, a continuous massive weight loss occurred, accompanied by another endothermic reaction, which in this case was the beginning of the evaporation of the adhesive. The evaporation temperature was set at 370°C, representing a difference of 42.3°C between the melting and evaporation temperatures. The first derivative of the functional thermogravimetric curve showed that the greatest mass loss (14.34%×min^{-1}) was observed at 379.4°C.

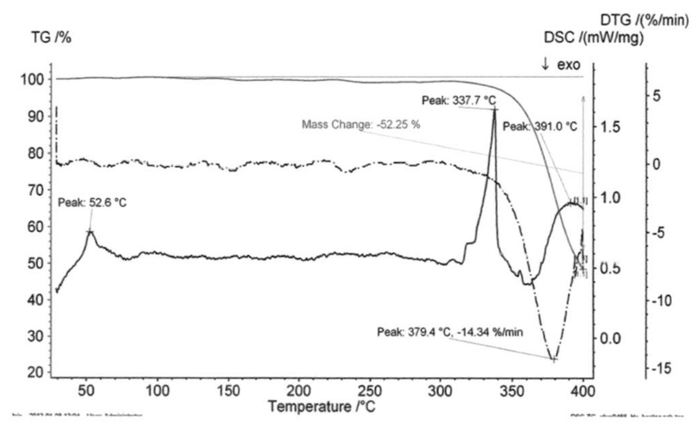

FIGURE 11.6 STA Hysol 9466—heating.

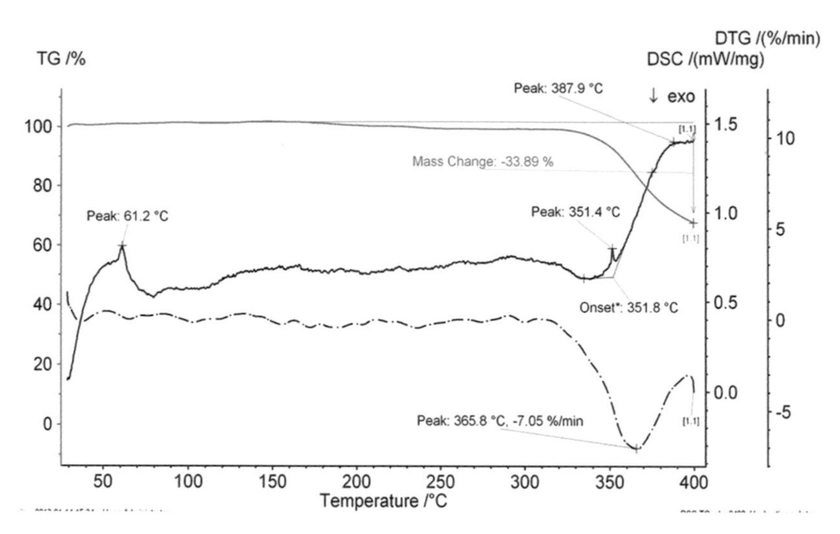

FIGURE 11.7 STA Hysol 9492—heating.

The Hysol 9497 STA adhesive is shown in Figure 11.8. The glass transition temperature was determined at 62.4°C. Melting of the adhesive was observed at 327°C and was accompanied by an endothermic reaction of 2 mW×mg^{-1}. As with Hysol 9492, evaporation of the adhesive also began immediately after melting. The greatest weight loss (3.52%×min^{-1}) occurred at 361.8°C.

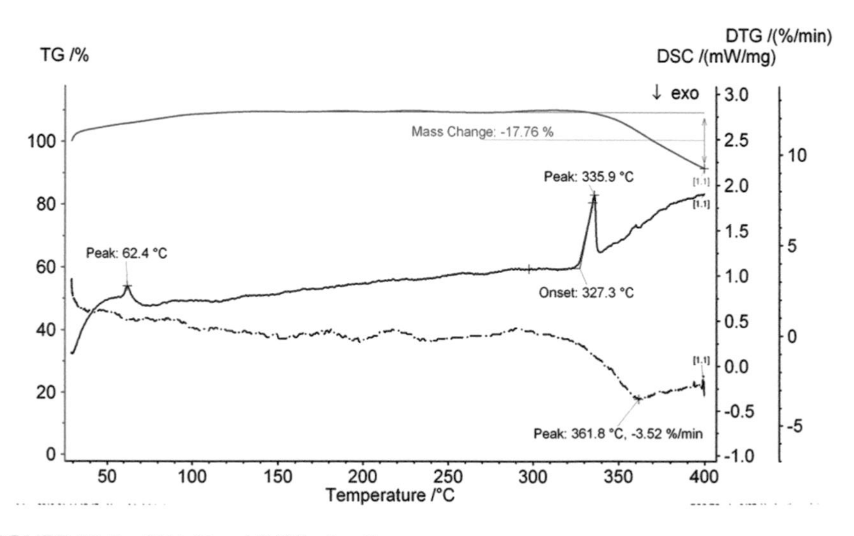

FIGURE 11.8 STA Hysol 9497—heating.

The Hysol 9455 STA adhesive is documented in Figure 11.9. The glass transition temperature was not detected as the measurement was made from 20°C for the instrument limitation. The melting point was set at 344.5°C, the evaporation occurred immediately after the start of melting. It can be seen from the graph that the heating of the adhesive was not accompanied by a constant or increasing positive (endothermic) reaction as with all previous adhesives, but rather a linear decrease in the area of exothermic reactions. The breakthrough occurred at the melting point (344.5°C) with the immediate evaporation of the adhesive, when the character of the reaction turned to endothermic. The greatest weight loss of material (20.78%×min^{-1}) occurred at 372.7°C.

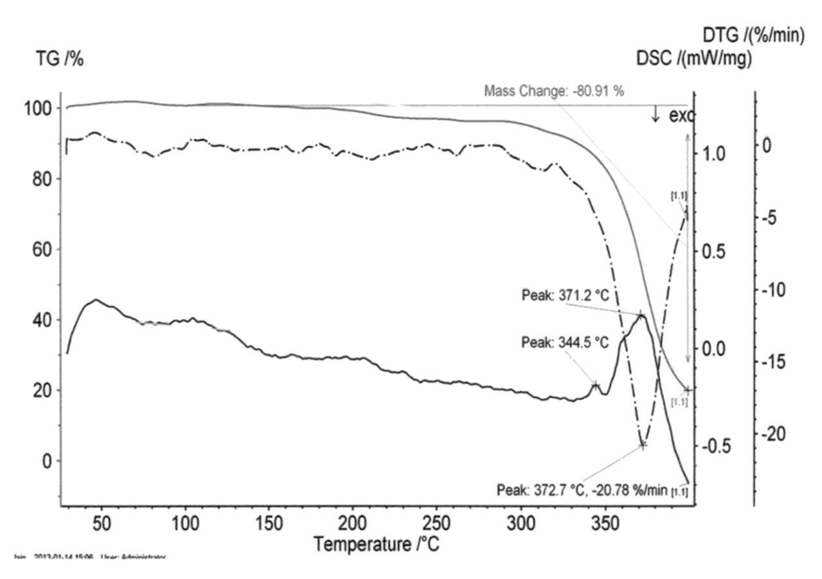

FIGURE 11.9 STA Hysol 9455—heating.

To determine the glass transition temperature of Hysol 9455 adhesive, a differential scanning calorimetry (DSC) analysis was performed on a Perkin-Elmer instrument, which allows to measure from temperatures below 0°C. The measurement (Fig. 11.10) started from -60°C and showed that the glass transition temperature of Hysol 9455 was 5.2°C. This makes it possible to explain the very fast curing of the adhesive.

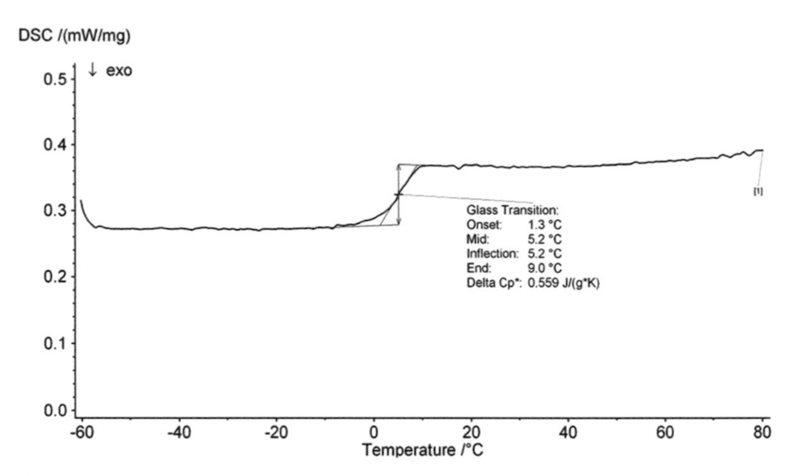

FIGURE 11.10 DSC analysis—Hysol 9455.

The results of the thermogravimetric analysis are shown in Figure 11.11. The greatest weight loss (up to 95% at 180–400°C) was observed on Hysol 9455. Hysol 9466 showed an 85% weight loss at 330–450°C. The Hysol 9492 adhesive degraded by 40 wt% in the temperature range 330–400°C and Hysol 9497 adhesive by 30 wt% at temperatures of 330–500°C. From the above-mentioned results, it can be stated that the tested adhesives are thermally stable from 300 to 350°C, above these temperatures, the thermal stability decreases significantly.

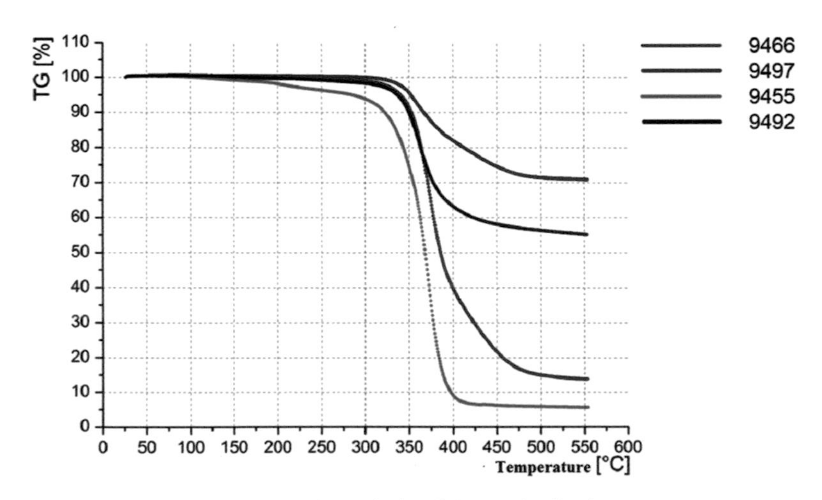

FIGURE 11.11 Thermogravimetric analysis of the used adhesives.

11.3.2.1 THERMAL DECOMPOSITION ANALYSIS OF ADHESIVES

The results of identifying volatile thermal decomposition products of Hysol 9466 are shown in Figure 11.12. Overall, 63 products were identified by gas chromatography captured and mass spectrometry. A full list of identified compounds is provided in Appendix A. The results show that due to rapid heating (pyrolysis), the evaporation of the adhesive occurs in the form of hydrocarbons. The most frequently observed product was phenol, which corresponds to the nature of the basic component of the epoxy adhesive—bisphenol A.

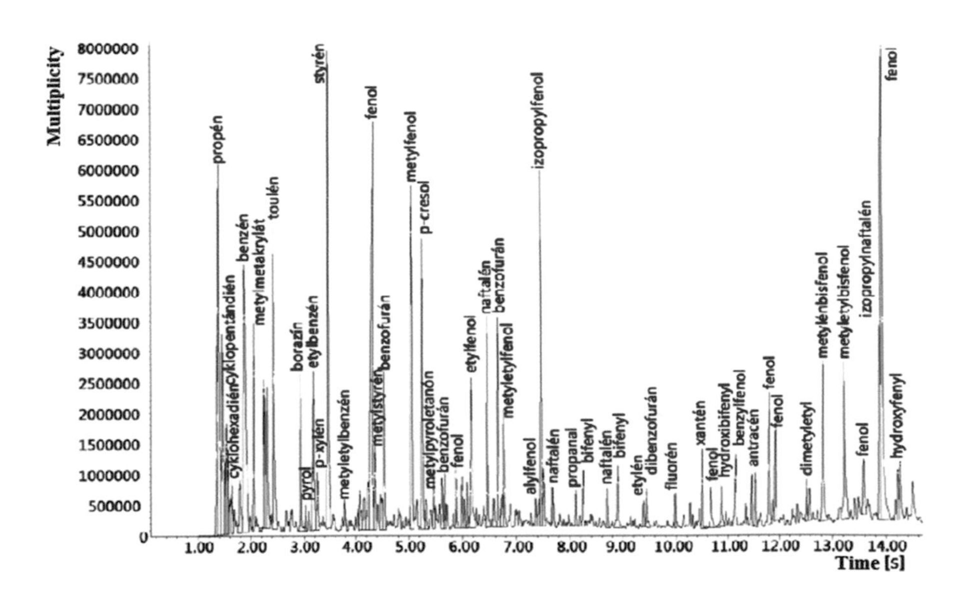

FIGURE 11.12 GC–MS Hysol 9466 adhesive record.

The results of the gas chromatography recording of Hysol 9455 are shown in Figure 11.13. A total of 59 compounds has been identified, which the full list is given in Appendix B. In particular, Hysol 9466 has identified cyclic and aromatic hydrocarbons. Most abundant products were recorded in the first five seconds of tracking.

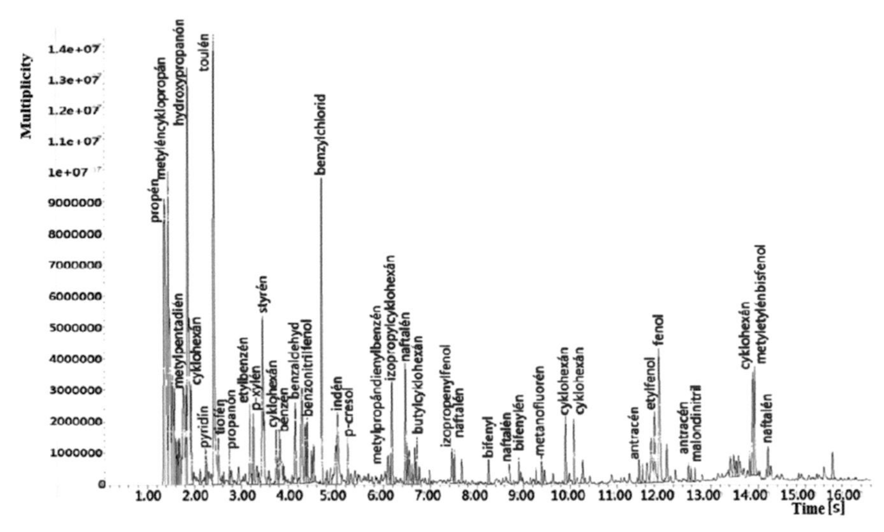

FIGURE 11.13 GC–MS Hysol 9455 adhesive record.

The gas chromatography results of Hysol 9492 are shown in Figure 11.14. Mass spectrometry identified a total of 63 compounds listed in Appendix C. Nitrogen-containing compounds were also identified with cyclic and aromatic hydrocarbons, whose compounds were in the hardener.

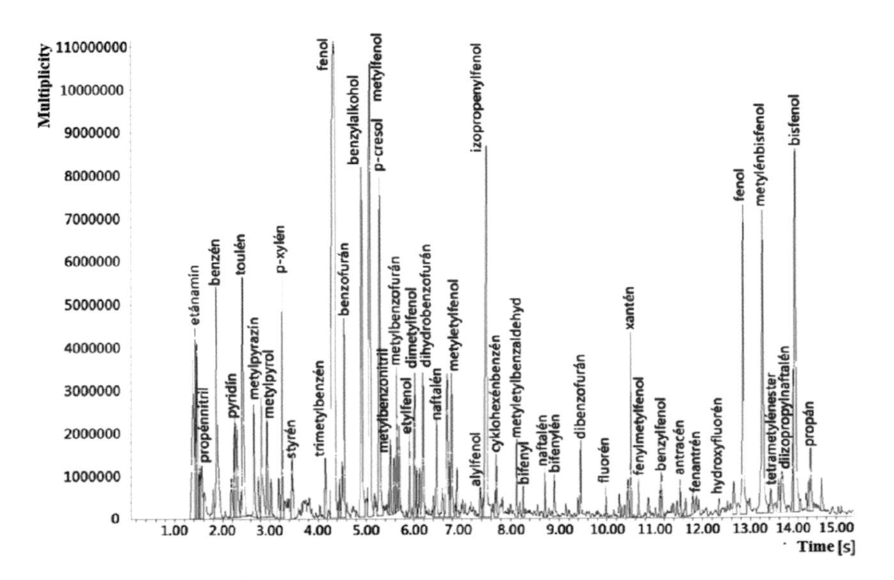

FIGURE 11.14 GC–MS Hysol 9492 adhesive record.

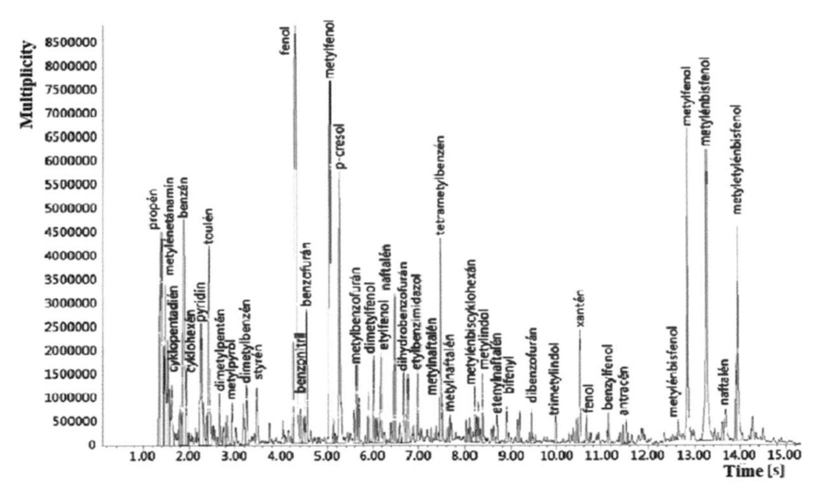

FIGURE 11.15 GC–MS Hysol 9497 adhesive record.

Based on the results of pyrolysis, it can be stated that from the thermal decomposition point of view all adhesives do not differ significantly in the type of compounds formed. From the point of view of the abundance of the individual compounds, it can be stated that the least abundances were found for Hysol 9466 and Hysol 9497, while the highest abundance of volatile thermal decomposition products was found for Hysol 9492.

11.4 CONCLUSIONS

The identified gaseous products from hybrid linkages of two-component epoxy adhesives pose a health risk to the human body, and subsequent research will aim to enhance their thermo-oxidative stability by applying nontoxic flame retardants or effective antioxidants. Simulation of thermal decomposition of the adhesive and the possibility to increase its thermo-oxidative stability is proposed by differential scanning calorimetry (DSC) under nonisothermal conditions. The obtained kinetic parameters allow determining the length of the induction period for any temperature.

ACKNOWLEDGMENT

This work was supported by the Slovak Research and Development Agency under the contract Nos. APVV-15-0124 and VEGA 2/0019/19 and the grants APVV-16-0177 and KEGA 001STU-4/2019.

KEYWORDS

- **hybrid joining**
- **laser beam welding**
- **epoxy resins**
- **fumes**
- **thermal degradation**

REFERENCES

1. Chang, B.; Shi, Y.; Dong, S. Comparative Studies on Stresses in Weld-Bonded, Spot-Welded and Adhesive-Bonded Joints. *J. Mater. Process. Technol.* **1999,** *87*, 230–236.
2. Al-Samhan, A.; Darwish, S. M. H. Finite Element Modeling of Weld-Bonded Joints. *J. Mater. Process. Technol.* **2003,** *142*, 587–598.
3. Da Silva, L. F. M.; Pirondi, A.; Ochsner, A. *Hybrid Adhesive Joints*; Springer: London, 2011; p 309. ISBN 978-3-642-16623-5.
4. Hassan, M. F.; Megahed, S. M. *Current Advances in Mechanical Design and Production VII*; Pergamon: Cairo, 2000; p 665. ISBN 0-08-043711-7.
5. Jiang, J.; Zhang, Z. The Study on the Plasma Arc Weld Bonding Process of Magnesium Alloy. *J. Alloys Compd.* **2008,** *466*, 368–372.
6. Liu, L; Jiang, J. The Effect of Adhesive Layer on Variable Polarity Plasma Arc Weld Bonding Process of Magnesium Alloy. *J. Mater. Process. Technol.* **2009,** *209*, 2862–2870.
7. Liu, L.; Ren, D. A Novel Weld-Bonding Hybrid Process for Joining Mg Alloy and Al Alloy. *Mater. Des.* **2011,** *32*, 3730–3735.
8. Liu, L.; Ren, D. Effect of Adhesive on Molten Pool Structure and Penetration in Laser Weld Bonding of Magnesium Alloy. *Opt. Lasers Eng.* **2010,** *48*, 882–887.
9. Liu, L.; Wang, H.; Song, G. Microstructure Characteristics and Mechanical Properties of Laser Weld Bonding of Magnesium Alloy to Aluminum Alloy. *J. Mater. Sci.* **2007,** *42*, 565–572.
10. Liu, L.; Ren, D.; Li, Y. Static Mechanics Analyses of Different Laser Weld Bonding Structures in Joining AZ61 Mg Alloy. *Int. J. Adhes. Adhes.* **2011,** *31* (7), 660–665.
11. Liu, L.; Jiang, J. The Effect of Adhesive on Arc Behaviors of Laser-TIG Hybrid Weld Bonding Process of Mg to Al Alloy. *IEEE Trans. Plasma Sci.* **2011,** *39* (1), 581–587.

APPENDIX A: LIST OF IDENTIFIED COMPOUNDS RESULTING FROM THE PYROLYSIS OF HYSOL 9466 ADHESIVE 9466.

No.	RT	Area	Compound name	Q	M_w	CAS
1	1.358	40,945,736	Propene	90	42.047	000115-07-1
2	1.507	13,212,036	1,3-Pentadiene, (*E*)-	95	68.063	002004-70-8
3	1.533	26,188,443	1,3-Cyclopentadiene	91	66.047	000542-92-7
4	1.792	23,623,790	1,4-Cyclohexadiene	81	80.063	000628-41-1
5	1.876	1,31E+08	Benzene	94	78.047	000071-43-2
6	2.058	36,224,847	Methyl methacrylate	90	100.052	000080-62-6
7	2.246	22,501,826	1*H*-Pyrrole, 1-methyl-	91	81.058	000096-54-8
8	2.265	33,507,917	Pyridine	89	79.042	000110-86-1
9	2.310	33,925,028	Pyrrole	72	67.042	000109-97-7
10	2.427	80,804,248	Toluene	95	92.063	000108-88-3
11	2.939	29,132,565	Borazine, 2-methyl-	64	95.100	021127-95-7
12	3.030	4623,395	1*H*-Pyrrole, 3-methyl-	91	81.058	000616-43-3
13	3.185	48,684,807	Ethylbenzene	81	106.078	000100-41-4
14	3.263	17,744,723	*p*-Xylene	97	106.078	000106-42-3
15	3.464	1,17E+08	Styrene	97	104.063	000100-42-5
16	3.788	4477,571	Benzene, (1-methylethyl)-	70	120.094	000098-82-8
17	4.170	13,443,046	Benzene, 1-ethyl-2-methyl-	87	120.094	000611-14-3
18	4.248	17,371,729	Ethanol, 2,2′-oxybis-	72	106.063	000111-46-6
19	4.319	1,86E+08	Phenol	91	94.042	000108-95-2
20	4.364	20,688,682	a-Methylstyrene	96	118.078	000098-83-9
21	4.481	13,438,939	Pyridine, 2,3-dimethyl-	76	107.073	000583-61-9
22	4.539	34,156,806	Benzofuran	93	118.042	000271-89-6
23	5.058	95,108,601	Phenol, 2-methyl-	97	108.058	000095-48-7
24	5.258	70,024,781	*p*-Cresol	97	108.058	000106-44-5
25	5.330	10,402,606	Ethanone, 1-(1-methyl-1*H*-pyrrol -2-yl)-	90	123.068	000932-16-1
26	5.634	12,509,085	Benzofuran, 2-methyl-	89	132.058	004265-25-2
27	5.686	11,213,712	Benzofuran, 2-methyl-	97	132.058	004265-25-2
28	5.900	14,596,194	Phenol, 2-ethyl-	97	122.073	000090-00-6
29	6.010	10,642,287	Phenol, 2,4-dimethyl-	94	122.073	000105-67-9
30	6.049	6806,774	Benzofuran, 2,3-dihydro-	83	120.058	000496-16-2
31	6.107	10,155,339	Benzene, 1-butynyl-	90	130.078	000622-76-4

No.	RT	Area	Compound name	Q	M_w	CAS
32	6.178	43,324,840	Phenol, 4-ethyl-	94	122.073	000123-07-9
33	6.476	43,689,645	Naphthalene	95	128.063	000091-20-3
34	6.587	8525,869	Phenol, 2,4,6-trimethyl-	97	136.089	000527-60-6
35	6.677	45,440,293	Benzofuran, 2,3-dihydro-	87	120.058	000496-16-2
36	6.768	25,096,488	Phenol, 4-(1-methylethyl)-	95	136.089	000099-89-8
37	7.371	6195,577	2-Allylphenol	86	134.073	001745-81-9
38	7.474	82,042,289	*p*-Isopropenylphenol	93	134.073	004286-23-1
39	7.513	12,198,840	Naphthalene, 1-methyl-	96	142.078	000090-12-0
40	7.675	5340,819	Naphthalene, 1-methyl-	96	142.078	000090-12-0
41	8.116	6946,493	Propanal, 2-methyl-3-phenyl-	94	148.089	1000131-87-6
42	8.258	12,568,120	Biphenyl	95	154.078	000092-52-4
43	8.705	9858,756	Naphthalene, 2-ethenyl-	92	154.078	000827-54-3
44	8.900	16,410,961	Biphenylene	91	152.063	000259-79-0
45	9.392	5960,163	Ethylene, 1,1-diphenyl-	93	180.094	000530-48-3
46	9.444	9820,683	Dibenzofuran	91	168.058	000132-64-9
47	9.982	6903,727	Fluorene	97	166.078	000086-73-7
48	10.500	16,045,466	9*H*-Xanthene	93	182.073	000092-83-1
49	10.662	13,044,087	Phenol, 2-(phenylmethyl)-	97	184.089	028994-41-4
50	10.876	13,312,956	*p*-Hydroxybiphenyl	95	170.073	000092-69-3
51	11.142	18,815,191	*p*-Benzylphenol	97	184.089	000101-53-1
52	11.472	10,889,094	Anthracene	95	178.078	000120-12-7
53	11.543	11,405,087	Anthracene	96	178.078	000120-12-7
54	11.809	33,386,603	Phenol, 4-(2-phenylethenyl)-	86	196.089	003839-46-1
55	11.926	19,987,028	Phenol, 4-(1-methyl-1 -phenylethyl)-	94	212.120	000599-64-4
56	12.528	10,870,015	1,1′-Biphenyl, 3,3′,4,4′-tetramethyl-	83	210.141	004920-95-0
57	12.586	8261,541	[1,1′-Biphenyl]-2-ol,5-(1,1-dimethylethyl)-	86	226.136	000577-92-4
58	12.839	46,322,677	Phenol,2-[(4-hydroxyphenyl)methyl]-	97	200.084	002467-03-0
59	13.241	51,237,578	Phenol,4,4′-methylenebis-	98	200.084	000620-92-8
60	13.591	17,932,677	2-(4′-Hydroxyphenyl)-2-(4′-methoxyphenyl)propane	95	242.131	016530-58-8
61	13.895	57,023,414	1,7-di-iso-propylnaphthalene	74	212.157	1000374-06-1

No.	RT	Area	Compound name	Q	M_w	CAS
62	13.941	1,65E+08	Phenol, 4,4′-(1-methylethylidene) bis-	98	228.115	000080-05-7
63	14.219	13,267,095	2-(4′-Hydroxyphenyl)-2-(4′-methoxyphenyl)propane	98	242.131	016530-58-8

APPENDIX B: LIST OF IDENTIFIED COMPOUNDS RESULTING FROM PYROLYSIS OF HYSOL 9455 ADHESIVE.

Pk	RT	Area	Compound name	Q	M_w	CAS
1	1.358	27,848,197	Propene	86	42.047	000115-07-1
2	1.384	108,575,152	Methylenecyclopropane	72	54.047	006142-73-0
3	1.539	19,076,610	1-Buten-3-yne, 2-methyl-	87	66.047	000078-80-8
4	1.578	11,482,839	Methacrolein	90	70.042	000078-85-3
5	1.688	31,817,968	1,3-Pentadiene, 3-methyl-, (*E*)-	81	82.078	002787-43-1
6	1.837	31,752,143	2-Propanone, 1-hydroxy-	86	74.037	000116-09-6
7	1.870	60,586,493	Benzene	95	78.047	000071-43-2
8	1.941	22,992,879	Cyclohexene	90	82.078	000110-83-8
9	2.272	23,955,935	Pyridine	76	79.042	000110-86-1
10	2.427	12,479,793	Toluene	95	92.063	000108-88-3
11	2.518	14,938,704	Thiophene, 3-methyl-	87	98.019	000616-44-4
12	2.745	120,707,585	2-Propanone,1-(1-methylethoxy)-	64	116.084	042781-12-4
13	2.939	5608,369	1*H*-Pyrrole, 2,5-dimethyl-	72	95.073	000625-84-3
14	3.185	10,307,915	Ethylbenzene	90	106.078	000100-41-4
15	3.256	50,871,983	*p*-Xylene	95	106.078	000106-42-3
16	3.464	23,815,729	Styrene	95	104.063	000100-42-5
17	3.742	15,723,490	Cyclohexane, (1-methylethyl)-	94	126.141	000696-29-7
18	3.781	44,696,760	Benzene, (1-methylethyl)-	90	120.094	000098-82-8
19	3.827	43,704,275	Cyclohexane, (1-methylethylidene)-	80	124.125	005749-72-4
20	4.151	15,047,741	Benzaldehyde	95	106.042	000100-52-7
21	4.293	8973,178	Phenol	91	94.042	000108-95-2
22	4.364	15,513,182	.a.-Methylstyrene	95	118.078	000098-83-9
23	4.403	18,647,988	Benzonitrile	94	103.042	000100-47-0
24	4.494	7674,903	Benzene,1-ethenyl-3-methyl-	96	118.078	000100-80-1

Pk	RT	Area	Compound name	Q	M_w	CAS
25	4.539	7587,641	Benzofuran	64	118.042	000271-89-6
26	4.708	17,271,825	Benzyl chloride	95	126.024	000100-44-7
27	4.818	11,652,785	Indane	64	118.078	000496-11-7
28	4.896	8875,467	Cyclopentane,1-methylene-3-(1-methylethylidene)-	64	122.110	073913-74-3
29	5.045	9326,979	Indene	92	116.063	000095-13-6
30	5.259	8282,652	*p*-Cresol	96	108.058	000106-44-5
31	6.101	8015,013	Benzene, 1-methyl-1,2-propadienyl-	90	130.078	022433-39-2
32	6.192	5959,738	4-Isopropylcyclohexanone	93	140.120	005432-85-9
33	6.477	11,221,430	Naphthalene	95	128.063	000091-20-3
34	6.554	7191,262	Benzo[*c*]thiophene	97	134.019	000270-82-6
35	6.619	26,073,238	Cyclopentanecarboxylic acid, 2-methylene-, methyl ester	78	140.084	110550-98-6
36	6.678	11,129,440	Benzofuran, 2,3-dihydro-	86	120.058	000496-16-2
37	6.723	9716,616	Cyclohexene, 1-butyl-	64	138.141	003282-53-9
38	6.768	37,156,878	Phenol, 4-(1-methylethyl)-	94	136.089	000099-89-8
39	6.982	9670,859	Quinoline	90	129.058	000091-22-5
40	7.468	7530,814	*p*-Isopropenylphenol	93	134.073	004286-23-1
41	7.513	12,337,385	Naphthalene, 2-methyl-	96	142.078	000091-57-6
42	7.675	12,145,094	Naphthalene, 2-methyl-	97	142.078	000091-57-6
43	8.258	11,629,241	Biphenyl	95	154.078	000092-52-4
44	8.706	53,510,788	Naphthalene, 2-ethenyl-	92	154.078	000827-54-3
45	8.900	12,661,514	Biphenylene	94	152.063	000259-79-0
46	9.269	27,848,197	Acetic acid, 4-isopropylidenecyclohexyl ester	72	182.131	1000186-82-7
47	9.399	108,575,152	4*a*,9*a*-Methano-9*H*-fluorene	97	180.094	019540-84-2
48	9.451	19,076,610	Bibenzyl	76	182.110	000103-29-7
49	9.904	11,482,839	Cyclohexane, 1,1′-(1-methylethylidene)bis-	83	208.219	054934-90-6
50	0.261	31,817,968	*N*-(4-Methoxyphenyl)-2-hydroxyimino-acetamide	90	194.069	1000143-61-3
51	1.472	31,752,143	Anthracene	96	178.078	000120-12-7
52	1.809	60,586,493	Phenol, 4-(2-phenylethenyl)-	86	196.089	003839-46-1
53	2.528	22,992,879	Anthracene, 1,2,3,4-tetrahydro-9,10-dimethyl-	90	210.141	094573-50-9

Pk	RT	Area	Compound name	Q	M_w	CAS
54	2.587	23,955,935	[1,1′-Biphenyl]-2-ol,5-(1,1-dimethylethyl)-	87	226.136	000577-92-4
55	13.500	12,479,793	Malonodinitrile, 2-(3-dimethylamino -1-phenyl-2-propenylideno)-	83	223.111	065996-13-6
56	3.591	14,938,704	4,4′-Ethylidenediphenol	64	214.099	002081-08-5
57	3.824	120,707,585	Cyclohexanone, 4,4′-(1-methylethylidene)bis-	93	236.178	007418-16-8
58	3.928	5608,369	Phenol, 4,4′-(1-methylethylidene) bis-	98	228.115	000080-05-7
59	4.213	10,307,915	Naphthalene, 2,6-bis(1,1-dimethylethyl)-	76	240.188	003905-64-4

APPENDIX C: LIST OF IDENTIFIED COMPOUNDS ARISING FROM PYROLYSIS OF HYSOL 9492 ADHESIVE.

Pk	RT	Area	Compound name	Q	M_w	CAS
1	1.436	46,054,817	Ethanamine, *N*-methylene-	78	57.058	043729-97-1
2	1.507	11,486,623	2-Propenenitrile	64	53.027	000107-13-1
3	1.798	16,426,092	2-Butenal	70	70.042	004170-30-3
4	1.883	127,061,221	Benzene	91	78.047	000071-43-2
5	2.187	15,347,458	Pyrazine	91	80.037	000290-37-9
6	2.245	21,474,237	1*H*-Pyrrole, 1-methyl-	91	81.058	000096-54-8
7	2.271	41,742,409	Pyridine	92	79.042	000110-86-1
8	2.388	7468,381	1*H*-Pyrrole-2,5-dione	64	97.016	000541-59-3
9	2.433	105,523,509	Toluene	95	92.063	000108-88-3
10	2.673	39,251,432	1*H*-Pyrazole, 4,5-dihydro-4,5-dimethyl-	64	98.084	028019-94-5
11	2.822	59,217,910	Pyrazine, methyl-	91	94.053	000109-08-0
12	2.945	32,366,654	1*H*-Pyrrole, 3-methyl-	64	81.058	000616-43-3
13	3.029	12,620,853	1*H*-Pyrrole, 3-methyl-	90	81.058	000616-43-3
14	3.191	17,033,616	Ethylbenzene	81	106.078	000100-41-4
15	3.256	78,945,137	*p*-Xylene	97	106.078	000106-42-3
16	3.464	25,085,967	Styrene	97	104.063	000100-42-5
17	4.157	23,516,484	Benzene, 1,2,4-trimethyl-	64	120.094	000095-63-6
18	4.338	521,885,284	Phenol	91	94.042	000108-95-2

Pk	RT	Area	Compound name	Q	M_w	CAS
19	4.455	9385,941	Benzonitrile	95	103.042	000100-47-0
20	4.507	20,885,895	Benzene, 1-ethenyl-3-methyl-	96	118.078	000100-80-1
21	4.552	69,688,566	Benzofuran	90	118.042	000271-89-6
22	4.915	170,527,761	Benzyl alcohol	96	108.058	000100-51-6
23	5.090	260,237,649	Phenol, 2-methyl-	97	108.058	000095-48-7
24	5.297	215,334,514	*p*-Cresol	97	108.058	000106-44-5
25	5.472	6621,225	Benzene, 1-methyl-4-(1-methylethenyl)-	97	132.094	001195-32-0
26	5.511	21,790,810	Benzonitrile, 2-methyl-	96	117.058	000529-19-1
27	5.589	15,177,001	Benzofuran, 7-methyl-	95	132.058	017059-52-8
28	5.641	41,851,612	Benzofuran, 2-methyl-	86	132.058	004265-25-2
29	5.686	21,489,487	Benzofuran, 2-methyl-	97	132.058	004265-25-2
30	5.906	26,364,141	Phenol, 2-ethyl-	95	122.073	000090-00-6
31	6.016	40,404,358	Phenol, 3,5-dimethyl-	97	122.073	000108-68-9
32	6.062	15,296,366	Benzofuran, 2,3-dihydro-	87	120.058	000496-16-2
33	6.191	51,262,592	Phenol, 4-ethyl-	95	122.073	000123-07-9
34	6.405	8627,022	Phenol, 2-(1-methylethyl)-	76	136.089	000088-69-7
35	6.476	36,446,031	Naphthalene	95	128.063	000091-20-3
36	6.690	70,442,551	Benzofuran, 2,3-dihydro-	86	120.058	000496-16-2
37	6.781	49,851,564	Phenol, 4-(1-methylethyl)-	95	136.089	000099-89-8
38	7.377	10,054,135	2-Allylphenol	68	134.073	001745-81-9
39	7.500	222,642,966	*p*-Isopropenylphenol	93	134.073	004286-23-1
40	7.682	8410,847	Naphthalene, 1-methyl-	90	142.078	000090-12-0
41	7.707	16,945,930	Benzene, 2-cyclohexen-1-yl-	68	158.110	015232-96-9
42	8.122	22,653,791	Benzaldehyde, 4-(1-methylethyl)-	93	148.089	000122-03-2
43	8.258	9644,098	Biphenyl	95	154.078	000092-52-4
44	8.705	18,935,497	Naphthalene, 2-ethenyl-	90	154.078	000827-54-3
45	8.900	16,732,835	Biphenylene	90	152.063	000259-79-0
46	9.444	18,799,002	Dibenzofuran	91	168.058	000132-64-9
47	9.982	9552,163	Fluorene	94	166.078	000086-73-7
48	10.442	12,502,415	Xanthene-9-carboxylic acid	72	226.063	000082-07-5
49	0.500	50,397,104	9*H*-Xanthene	91	182.073	000092-83-1
50	0.662	11,645,043	Phenol, 2-(phenylmethyl)-	97	184.089	028994-41-4

Pk	RT	Area	Compound name	Q	M_w	CAS
51	0.869	8271,500	*p*-Hydroxybiphenyl	95	170.073	000092-69-3
52	1.142	17,404,666	*p*-Benzylphenol	97	184.089	000101-53-1
53	1.543	10,289,894	Anthracene	95	178.078	000120-12-7
54	1.647	7047,273	2-Phenanthrenol	93	194.073	000605-55-0
55	1.802	7579,772	Phenol, 4-(2-phenylethenyl)-, (*E*)-	70	196.089	006554-98-9
56	1.880	4897,454	Benzene, 1-methyl-2-(4-methylphenoxy)-	83	198.104	003402-72-0
57	2.347	6682,470	2-Hydroxyfluorene	83	182.073	002443-58-5
58	2.658	18,759,202	Phenol, 2,2′-methylenebis-	95	200.084	002467-02-9
59	2.852	145,147,117	Phenol, 2-[(4-hydroxyphenyl) methyl]-	96	200.084	002467-03-0
60	3.260	176,136,783	Phenol, 4,4′-methylenebis-	95	200.084	000620-92-8
61	3.668	39,874,005	*n*-Heptanoic acid, methyl(tetramethylene)silyl ester	91	228.155	1000217-03-6
62	3.902	60,641,400	1,7-di-iso-propylnaphthalene	64	212.157	1000374-06-1
63	3.947	184,993,394	Phenol, 4,4′-(1-methylethylidene) bis-	98	228.115	000080-05-7

APPENDIX D: LIST OF IDENTIFIED COMPOUNDS RESULTING FROM PYROLYSIS OF HYSOL 9497 ADHESIVE.

Pk	RT	Area	Compound name	Q	M_w	CAS
1	1.358	45,648,176	Propene	90	42.047	000115-07-1
2	1.436	20,376,255	Ethanamine, *N*-methylene-	72	57.058	043729-97-1
3	1.533	38,771,344	1,3-Cyclopentadiene	90	66.047	000542-92-7
4	1.792	17,355,475	1,3-Cyclohexadiene	70	80.063	000592-57-4
5	1.876	144,685,302	Benzene	91	78.047	000071-43-2
6	1.941	26,094,005	Cyclohexene	70	82.078	000110-83-8
7	2.265	48,375,826	Pyridine	64	79.042	000110-86-1
8	2.388	8329,716	1,3-Cycloheptadiene	83	94.078	004054-38-0
9	2.433	85,135,111	Toluene	95	92.063	000108-88-3

Pk	RT	Area	Compound name	Q	M_w	CAS
10	2.673	16,755,575	2-Pentene, 4,4-dimethyl-, (*Z*)-	72	98.110	000762-63-0
11	2.822	12,769,370	Fampridine	90	94.053	000504-24-5
12	2.945	13,417,869	1*H*-Pyrrole, 3-methyl-	80	81.058	000616-43-3
13	3.263	24,913,918	Benzene, 1,3-dimethyl-	97	106.078	000108-38-3
14	3.464	26,541,731	Styrene	97	104.063	000100-42-5
15	4.319	293,078,423	Phenol	91	94.042	000108-95-2
16	4.429	14,967,024	Benzonitrile	94	103.042	000100-47-0
17	4.500	8983,417	Benzene, 1-propenyl-	93	118.078	000637-50-3
18	4.546	39,035,214	Benzofuran	91	118.042	000271-89-6
19	5.070	147,848,402	Phenol, 2-methyl-	97	108.058	000095-48-7
20	5.278	132,350,313	*p*-Cresol	97	108.058	000106-44-5
21	5.582	8591,163	Benzofuran, 7-methyl-	95	132.058	017059-52-8
22	5.634	22,756,810	Benzofuran, 2-methyl-	83	132.058	004265-25-2
23	5.686	12,305,756	Benzofuran, 2-methyl-	97	132.058	004265-25-2
24	5.900	13,789,936	Phenol, 2-ethyl-	97	122.073	000090-00-6
25	6.010	21,964,748	Phenol, 3,5-dimethyl-	97	122.073	000108-68-9
26	6.055	7145,137	Benzofuran, 2,3-dihydro-	74	120.058	000496-16-2
27	6.107	6904,544	Cycloprop[*a*]indene, 1,1*a*,6,6*a*-tetrahydro-	93	130.078	015677-15-3
28	6.185	29,826,777	Phenol, 4-ethyl-	93	122.073	000123-07-9
29	6.476	41,292,073	Naphthalene	95	128.063	000091-20-3
30	6.587	7199,025	Phenol, 2,4,6-trimethyl-	94	136.089	000527-60-6
31	6.684	26,594,133	Benzofuran, 2,3-dihydro-	87	120.058	000496-16-2
32	6.742	8983,007	1*H*-Benzimidazole, 2-ethyl-	64	146.084	001848-84-6
33	6.774	23,602,532	Phenol, 3-(1-methylethyl)-	90	136.089	000618-45-1
34	6.988	21,666,009	Quinoline	95	129.058	000091-22-5
35	7.481	71,678,424	Benzene, 1,2,3,4-tetramethyl-	90	134.110	000488-23-3
36	7.513	13,516,895	Naphthalene, 2-methyl-	96	142.078	000091-57-6
37	7.682	6748,550	Naphthalene, 1-methyl-	95	142.078	000090-12-0
38	8.038	5551,767	Quinoline, 8-methyl-	96	143.073	000611-32-5
39	8.219	16,766,818	Cyclohexane, 1,1′-methylenebis-	95	180.188	003178-23-2
40	8.258	8037,528	Biphenyl	94	154.078	000092-52-4
41	8.304	10,507,936	1*H*-Indole, 6-methyl-	93	131.073	003420-02-8
42	8.705	10,014,226	Naphthalene, 2-ethenyl-	95	154.078	000827-54-3

Pk	RT	Area	Compound name	Q	M_w	CAS
43	8.900	11,180,217	Biphenylene	90	152.063	000259-79-0
44	9.126	7714,951	1*H*-Indole, 2,5-dimethyl-	95	145.089	001196-79-8
45	9.185	7258,189	1*H*-Indole, 2,3-dimethyl-	96	145.089	000091-55-4
46	9.444	11,209,534	Dibenzofuran	91	168.058	000132-64-9
47	9.975	10,570,556	1*H*-Indole, 2,3,5-trimethyl-	92	159.105	021296-92-4
48	10.442	7851,402	9*H*-Carbazole, 2-methyl-	62	181.089	003652-91-3
49	10.500	31,223,168	9*H*-Xanthene	91	182.073	000092-83-1
50	10.669	8758,690	Phenol, 2-(phenylmethyl)-	96	184.089	028994-41-4
51	11.142	10,847,348	*p*-Benzylphenol	97	184.089	000101-53-1
52	11.472	5294,419	Anthracene	93	178.078	000120-12-7
53	11.543	5924,819	1,10*b*(2*H*)-Dihydropyrano [3,4,5-ik]fluorene	87	208.089	1000217-20-8
54	12.658	7646,216	Phenol, 2,2′-methylenebis-	97	200.084	002467-02-9
55	12.852	123,208,483	Phenol, 2-[(4-hydroxyphenyl) methyl] -	96	200.084	002467-03-0
56	13.260	141,139,733	Phenol, 4,4′-methylenebis-	96	200.084	000620-92-8
57	13.895	27,857,687	Naphthalene, 6-methoxy-2-(1-buten-3-yl)-	72	212.120	101327-54-2
58	13.934	100,005,318	Phenol, 4,4′-(1-methylethylidene)bis-	98	228.115	000080-05-7

CHAPTER 12

APPLICATION OF GREEN TECHNOLOGY FOR ENERGY CONSERVATION AND SUSTAINABLE DEVELOPMENT

E. P. APARNA[1*] and K. S. DEVAKY[2]

[1]*Research Scholar, School of Chemical Sciences, Mahatma Gandhi University, Kottayam, Kerala, India*

[2]*Professor, School of Chemical Sciences, Mahatma Gandhi University, Kottayam, Kerala, India*

[*]*Corresponding author. E-mail: aparnaep9@gmail.com*

ABSTRACT

In the modern era of technology, energy conservation and environmental protection is the most burning global issue, and it has a major role in our everyday life. Minimal use of fossil energy can be attained through energy preservation and the novel ideas and practical use of renewable energy like sunlight, wind, water, tides, geothermal heat, and so on. As Einstein said, "Neither energy can be created nor destroyed, rather we can alternate." The present chapter discusses the potentials of green technology to solve problems related to conventional energy sources in the environment.

12.1 INTRODUCTION

The demand for power has been increasing globally though utilization is faster than its production. For instance, the most common energy resources are coal, oil, natural gas, and so on, and are being formed through thousands of years of natural process.[1,2] Most of these energy resources are limited and could not be reused or renewed. A proper conception of energy helps to decrease its requirement, and thereby saving one unit of energy is equivalent to the creation of two units. In fact, every energy production, one way or another, adversely affects the nature; hence, the reduced production helps to save our environment.[3,4]

Energy preservation is a best exercise to lessen energy usage resourcefully or by reducing the utilization of energy service. It is the process that reduces pollution by utilizing natural energy. It also reduces energy cost and saves the conventional energy resources by using renewable resources.[4–6] Renewable resources are natural resources which are refilled by natural production or other frequent processes in a fine time frame. Energy can be preserved either by decreasing the wastage and loss or by improving the effectiveness through advanced technologies, operation, and maintenance.[7,8] The current chapter describes the different dimensions of energy conservation.

12.2 USE OF ENERGY FROM RENEWABLE RESOURCES

12.2.1 SUN: AS A SOURCE OF ENERGY

Sunlight is an important renewable energy source. There are two groups of light energy: active and passive. Photovoltaic cell is used in active technology, while harvesting solar light in passive energy systems.[9] The amount of energy obtained from all the nonrenewable resources such as coal, oil, natural gas, and so on is about half the amount of solar energy reaching the surface.[10] Electricity-generating photovoltaic cell displaces all other energy sources and reduces the release of potentially harmful emissions and toxic substances into the environment. The first photovoltaic cell, Si-based, was produced in 1970s.[8] These cells were produced using doping technology.

An efficiency of 32% is shown by silicon photovoltaic cells in terms of total energy conversion. They can only take up energy from sunlight. Thin-film cells are the second-generation solar cells. They contain thin films of various semiconductor materials placed onto substrates, yielding more efficient, thinner, and flexible cells. Amorphous silicon, cadmium telluride, and copper indium gallium selenide are the key materials used in commercial thin-film photovoltaic cells.[11]

The third-generation solar cells are more efficient, safer, and cheaper to produce. Examples include perovskite solar cells based on hybrid perovskite structured crystals such as methylammonium lead iodide and other compound halides.[12] These materials can be produced at lower temperatures and absorb light well. These are also cost-effective since they are made from commonly available industrial chemicals. Gallium arsenide is another material used to improve solar cell efficiency. Gallium arsenide solar panels are primarily used on spacecraft.

Nanomaterial-based solar technologies are also developing. Based on the particle size, the bandgap of nanoparticle can be changed, which greatly affects the properties of solar cells. Examples of these semiconductor materials are cadmium sulfide, cadmium selenide, lead sulfide, and others. Conversion of infrared radiation into visible light can be achieved by upconverting phosphor nanoparticles doped with rare-earth elements. Transparent bifacial cells, tandem cells, and solar thermophotovoltaic devices are emerging technologies to boost the efficiency of photovoltaic cells.[11]

12.2.2 WIND ENERGY

Electricity can be produced by using wind. Turbines are driven by the wind and thereby ensure the proper working of generator that produces electricity, which is considered as a dependable source of energy, and it has less adverse effect on the environment. The wind is an intermittent energy source which cannot release on demand. It varies with time, climate, and so on. As the quantity of wind power in a region increases, more conservative power sources are needed to back it up. Power management methods like releasable power sources, importing and exporting power to neighboring areas, hydroelectric power, energy storage, or reducing demand when wind production is low can in many cases overcome these problems.[13–16]

Wind power does not need fuels and fuel costs. The price of wind power is much more stable than the prices of fossil fuel sources. The marginal cost required for the construction of wind station can be reduced by improved turbine technology. The length and weight of the turbine blade improve the performance and increase the production of power in a lesser capital investment and maintenance. Using fossil fuels release toxic elements into the environment on a larger scale compared to wind energy. Habitat disturbances by the turbine generator sound are the available, reported disadvantage of wind energy.[17]

12.2.3 HYDRO ENERGY

Hydro energy is also called hydroelectric power, which is produced from water. According to statistical calculations, the amount of electricity generated by hydropower is expected to grow about 3.1% by each year for the next 25 years.[18,19] This method of producing electricity is relatively cheap compared to gas or coal plants. Depending on changing energy demands, the dam and reservoir help to produce the required amount of electricity by using the stored water; by this way, the flexibility of electricity production can be achieved. Here the start-up time of turbines is lesser than any other electricity plant.

The main benefit of the conventional plant is to store water at a cheaper rate for release later as electricity. Hydroelectric stations have long economic lives. Operating labor cost is frequently low. Since hydroelectric dams do not require fuel, power generation does not produce any toxic gases. The same amount of electricity produced by hydroelectric projects and through this 3 billion tons of CO_2 emissions instead of making electricity through fossil fuels. The low greenhouse gas impact of hydroelectricity is found chiefly in temperate climates. These hydroelectric power stations work by using the potential energy of water stored in a specific height; hence, they do not require fuels and chemicals, and they do not release any kind of toxic elements into the environment, thus reducing the emission of CO_2, methane, and any other kind of greenhouse gases. Environmental impact of hydroelectric stations is lesser than any other electric plants.[20]

The major disadvantages associated with hydroelectric projects are as follows: ecosystem damage and loss of land, water loss by evaporation, siltation and flow shortage, methane emissions, and relocation of the people living where the reservoirs are placed. The construction of hydroelectric power stations requires large areas of land but nuclear stations required a small area. The main disadvantage of the hydroelectric station is that the dam may act as a water bomb which cannot be controlled, whereas the nuclear power station failure can reduce quickly; meanwhile, the cost difference varies on a larger scale. Hydroelectricity along with other sources can achieve an adequate amount of electricity in all seasons. The failure of a hydroelectric power station is much danger than the failure of a nuclear power station. However, nuclear power can reduce its output reasonably quickly. Since the cost of nuclear power is higher due to high infrastructure costs, the cost per unit energy goes up significantly with low production. But hydroelectricity can supply power at a much lower cost. Thus, hydroelectricity is a complement to other sources of energy. Compared with wind power, the easily regulated character of hydroelectricity is used to compensate for the intermittent nature of wind power. In some cases, wind power can be used as additional to water for later use in dry seasons.[21,22]

12.2.4 TIDAL ENERGY

Tidal power is produced by the conversion of energy from tides to electricity. Among various types of renewable energy, tidal energy has a comparatively high cost and limited availability of sites with adequately high tidal ranges. By new technologies, the availability of tidal power could be increased, and economic and environmental costs are decreased. The energy associated with the tidal flow can be easily converted into electricity with the aid of a tidal generator. The selection of a site for tidal power generation mainly depends on the tidal variation and tidal current velocities present at that particular place.

The following are the main challenges of tidal power. Tidal power affects the marine life. The turbines may adversely affect the swimming life in the sea. The advanced application like switching off the turbine immediately after detecting any animals entering the turbine causes energy loss. To ensure the security of marine animals is essential while placing the tidal generator in water. The breeding stream of migrating fishes is also disturbed by the tidal energy generator. The same acoustic concerns apply to tidal barrages. Environmental concerns are the damages by the turbine blade to marine animals and its sound, which are limited to a small area and do not affect the entire bay.[23]

Another challenge is the corrosion of metal parts due to saltwater. Nickel alloys, titanium, and stainless steel materials prevent corrosion. Mechanical fluids such as lubricants can be leak out into the sea, which is harmful to the marine life. Proper maintenance can minimize the leakage of harmful chemicals. Tidal energy is not used as a popular source of energy due to its initial setting up cost. Researches are going on to reduce the price of tidal energy. One such approach is the application of the orthogonal turbine, a simplified design that offers considerable cost savings. The water is 800 times denser than air, and the unsurprising and regular nature of tides makes a predominantly striking source of energy. Regular screening is the key step for reducing the cost.

12.2.5 GEOTHERMAL ENERGY

The power obtained from geothermal energy is known as geothermal power. Major technologies involved in geothermal energy production are flash steam power stations, dry steam power stations, and binary cycle power stations. The heat extraction in geothermal energy production is small compared with the heat produced from earth and thus considered as a sustainable and reusable form of energy. Geothermal electric stations emit lesser greenhouse gases compared with conventional coal-fired plants. These power stations are analogous to other thermal power stations as heat is used for heating water or another working fluid. The steam produced from the working fluid turns the turbine of a generator, thereby producing electricity. The various types of geothermal power stations include dry steam stations, flash steam stations, binary cycle power stations, and so on.

Dry steam stations: Dry steam stations are the simplest and oldest. A large amount of dry steam is required for the proper working of this type of power station. Therefore, dry steam station is not found very often. Such power stations are simple but very resourceful. Geothermal steam (temperature ≥ 150 °C) is the main fuel in such power stations. After rotating the turbine, the steam is emitted to a condenser. The steam then is reformed as liquid form and chills the water. Then the cold water carries condensate into deep wells through the pipe.

Flash steam stations: Highly pressured, well heated, water injects into the tank with lower pressure which helps to drive the turbines. The temperature of fluid is around 180 °C. Flash steam stations use groundwater having temperatures greater than 182 °C. This high temperature is sufficient to generate adequate pressure required for the upward flow of water through the wells. The upward flow reduces the pressure, and a certain amount of steam is produced by hot water. The steam is used for power generation. Both existing water and compressed steam sent back to the reservoir, completing the circle sustainably.

Binary cycle power stations: Binary cycle power stations are recently developed. The fluid used here has a low temperature of about 57 °C. The secondary fluid vaporized by this fluid drives the turbines. These kinds of power stations are existing commonly. Both organic Rankine and Kalina cycles are used in binary cycle power stations.[24] An efficiency of 10–13% is shown by this type of power station.

The main environmental impacts of geothermal power are as follows: The deep earth fluids contain certain gases like carbon dioxide, methane, hydrogen sulfide, radon, and ammonia. The release of those gases might cause global warming and acid rain. From a theoretical point of view, the geothermal stations release back these gases into earth's atmosphere by C capture and storage. In addition to these gases, hot water from geothermal sources may contain toxic chemicals like boron, mercury, arsenic, antimony, and so on. Release of these can cause environmental damage. Geothermal station building can harmfully affect land stability. Geothermal stations can generate earthquakes due to water injection. Minimum use of land and water is the major advantage of geothermal stations. It does not require fuel. The capital cost of the geothermal station is high.[25]

12.3 USE OF ALTERNATIVE ENERGY SOURCES

12.3.1 BIOENERGY

Bioenergy is the energy in the form of electricity or gas that is produced from organic matter called biomass. This biomass includes plant parts, agricultural waste, and food waste even to sewage. The term bioenergy also covers transport fuels made from organic matter. The biomass used for the production of bioenergy ranges by region. Heat and fuel are simultaneously obtained from bioenergy. Since the energy in the biomass is acquired from the sun during photosynthesis, it can be considered as a renewable source.

The biomass is converted into energy commonly by three processes, namely, chemical, thermal, and biochemical. In chemical process, biomass is first broken down by chemical agents and then to liquid fuel, for example, ethanol obtained from corn by its fermentation. In thermal process, heat is used to convert biomass to energy by combustion or gasification. In biochemical conversion, bacteria or other organisms are used to convert biomass to energy by composting or fermentation. Bioenergy is important as it can dramatically reduce greenhouse emissions and dangerous gases that are the main reason for global warming and major changes in climate. Due to environmental impacts, large-scale use of forest biomass in energy production is opposed by many environmental groups.[26]

12.3.1.1 BIOMASS

Natural substances like wood waste, sawdust, and combustible agricultural wastes can be used as energy sources (stored energy from the sun) with far fewer greenhouse gas emissions compared with petroleum-based fuel sources. These materials are known as biomass. A large amount of CO_2 is released by the combustion of biomass, even though it belongs to a renewable energy source because by photosynthesis the produced CO_2 could return to new crops.[27]

12.3.1.2 BIOFUELS

Fuels derived from biomass are known as biofuel. Since biomass can be directly used as a fuel, the terms biomass and biofuel are interchangeable. Fluid or gaseous fuels used for transportation are commonly known as biofuel. The biofuel is normally considered as renewable energy since the biomass used for the manufacture of biofuel can regrow quickly. The example of biofuels includes bioethanol and biodiesel. Bioethanol, an alcohol-based biofuel, is mainly prepared from carbohydrates by fermentation. Transesterification of oils or fats also produces biodiesel.

Biofuels are commonly classified as follows:

(1) first-generation biofuels,
(2) second-generation biofuels,
(3) third-gencration biofuels, and
(4) fourth-generation biofuels.

The first-generation or conventional biofuels are prepared from food crops on arable land, for example, biodiesel or ethanol produced by the transesterification of sugar, starch, or vegetable oil. Second-generation biofuels are fuels produced from biomass of plant and animal origin. Third-generation biofuels are produced via simple economical reactions from algae in ionic liquids. Nonarable land biomass is used for the synthesis of fourth-generation biofuels. Examples of biofuels include biogas, syngas, green diesel, ethanol, biodiesel, straight unmodified edible vegetable oil, bioethers, and so on. Biofuels also produce air pollution. Carbon dioxide, carbon monoxide, nitrous oxides, airborne carbon particulates, and so on are the major pollutants produced from biofuels.[28]

12.3.2 HYDROGEN

Hydrogen is also known as a zero-waste production fuel. Hydrogen is commonly used as a fuel in fuel cells as well as internal combustion engines.

It has used as a commercial fuel cell in vehicles and spacecraft propulsion. A combination of hydrogen (H_2) and oxygen (O_2) produces water (H_2O) and releases energy. This energy facilitates hydrogen to act as a fuel.

$$2H_2\ (g) + O_2\ (g) \rightarrow 2H_2O\ (g) + \text{Energy}$$

Hydrogen is generally considered an energy carrier. Since hydrogen is nonobtainable on earth naturally and is made from a principal energy source on an industrial scale, hydrogen fuel can be produced mainly by partial oxidation of methane by steam reforming of fossil fuels, and so on. Biomass gasification or electrolysis of water is the other technique used for the synthesis of hydrogen, but small quantities of hydrogen can be achieved by all these techniques.[29]

12.4 TECHNIQUES TO IMPROVE ENERGY EFFICIENCY

12.4.1 EFFICIENT ENERGY USE

Efficient energy use also called energy efficiency is the global goal to reduce the amount of energy required to provide products and services. For example, traditional lights can be replaced with LED lights, fluorescent lights, or natural skylight windows. The emission of greenhouse gases can be reduced by reducing the energy use. Energy efficiency and renewable energy coexist as twin pillars of sustainable energy. From the economical point of view, energy efficiency is also favorable by reducing power purchase. The other major advantages achieved by energy efficiency techniques are low impact on climate change, reduced air pollution rate and corresponding improvement in health, improved energy security, reduction of the price risk for energy consumers, and so on.[30]

12.4.2 ENERGY STORAGE

The energy produced at one time can be utilized when it needs, with the aid of energy storage. An accumulator or battery is used for such purposes. The main intention of energy storage is to transfer one energy form that is not easy to store in a convenient manner. Some common examples are rechargeable battery, hydroelectric dam, ice storage tanks, fossil fuels such as coal and gasoline, and so on. The types of energy storage techniques are variable nowadays. Examples are as follows: fossil fuel storage, mechanical storage, electrical and electromagnetic storage, biological storage, electrochemical storage, and so on. The technique of energy storage is adequate to store the power of certain megawatt-hour. The most favorable boundary of an energy storage device mainly depends on locality and cost.[31]

12.4.3 GREEN BUILDING

Sustainable and energy competent buildings can be developed by the use of green technology. Green buildings have a reduced carbon footprint and a lower impact on the environment. In the present life, green building has an important role. Every phase of the building and the methods used to maintain building procedures are sustainable and energy resourceful as possible. The major production systems used in green building are as follows: solar power, biodegradable materials, green insulation, use of smart appliances, cool roofs, sustainable resource sourcing, low-energy house and zero-energy building design, electronic glass systems, sustainable indoor technologies, water competent technologies, self-powered constructions, and so on.[32]

12.4.4 MICROGENERATION

Microgeneration is the small-scale production of heat and electric power by individuals such as small businesses and communities to satisfy their own needs, as alternatives or in addition to the traditional power supply. Microgeneration is greatly encouraged by environmental concern due to the zero-carbon footprints and low cost. They are practically very useful in conditions such as the electrical grid is in long distance or unreliable grid power and so on. Compared with large-scale generation, it is less economic and has low environmental impact.[33]

12.4.5 PASSIVE SOLAR BUILDING DESIGN

The passive solar building design is mainly focused on collecting and saving the solar energy. In winter, the saved solar energy is given out as heat, and in summer, the solar power avoids solar heat. Mechanical or electrical devices are not commonly used in passive solar building designs. Cultivation of solar energy in passive solar building design is mainly by the use of heat resistance and shadowing. In new buildings, the passive solar design can be applied easily. The modification of existing buildings can also be achieved with the passive solar building design. The passive solar design utilizes sunlight excluding active mechanical systems. By making use of external energy supply, passive solar building design converts sunlight into usable heat and causes air circulation for ventilating.

The passive solar device uses solar power either directly or indirectly for keeping space heating, water heating, slowing indoor air temperature swings, and earth sheltering. A solar furnace is the most important part of a passive solar design. The furnace concentrates its receivers by using an outer energy source. The passive solar design has developed scientifically as a combination of climatology, thermodynamics, fluid mechanics, and human thermal comfort.[34]

12.4.6 COMBINED HEAT AND POWER (CHP)

Cogeneration is another term commonly used for representing CHP. In CHP electricity generation, heat capturing took place simultaneously. Cogeneration technology can be arranged cost-effectively with few geographic limitations. The advantage of the design of cogeneration is to make use of both renewable and nonrenewable energy sources. In certain cases, the electrical energy and thermal energy are required simultaneously; in such cases, the technology of cogeneration is applied. In the usual electricity generation, a large amount of energy is wasted in the form of heat. Using this heat cogeneration technology can accomplish efficiencies of more than 80% compared with electricity generation in the conventional method. The technology of cogeneration is applied in various areas such as commercial buildings, residential, institutions, municipal, manufacturers, and so on.[35]

12.5 CONCLUSION

Conservation and protection of nature is the main concern of every government. Energy conservation plays a significant role in lowering climate change and making the environment safe. Energy conservation is attained by using the energy more efficiently or by reducing the utilization of energy service. By using green technologies, the nonrenewable resources are replaced by renewable resources. High energy efficiency can be achieved with the aid of green technologies. The application of green technology also reduces the environmental pollution.

KEYWORDS

- **energy conservation**
- **renewable energy sources**
- **alternative energy sources**
- **energy efficiency improving techniques**

REFERENCES

1. Zehner, O. Unintended Consequences of Green Technologies. In *Green Technology*; Sage: London, 2011; pp 427–432.
2. Solangi, K.; Islam, M.; Saidur, R.; Rahim, N.; Fayaz, H. A Review on Global Solar Energy Policy. *Renewable Sustainable Energy Rev.* **2011,** *15*, 2149–2163.
3. Ahmad, S.; Ab Kadir, M. Z. A.; Shafie, S. Current Perspective of the Renewable Energy Development in Malaysia. *Renewable Sustainable Energy Rev.* **2011,** *15*, 897–904.
4. Twidell, J.; Weir, T. *Renewable Energy Resources*; Routledge: Abingdon, UK, 2015.
5. Chattopadhyay, R. *Green Tribology, Green Surface Engineering, and Global Warming*; ASM International: Cleveland, OH, 2014.
6. Allcott, H. Social Norms and Energy Conservation. *J. Public Econ.* **2011,** *95*, 1082–1095.
7. Hassett, K. A.; Metcalf, G. E. Energy Conservation Investment: Do Consumers Discount the Future Correctly? *Energy Policy* **1993,** *21*, 710–716.
8. Vettriselvan, R.; Ruben Anto, M.; Jesu Rajan, F. Rural Lighting for Energy Conservations and Sustainable Development. *Int. J. Mech. Eng. Tech* **2018,** *9* (7), 604–611.
9. Chen, K. K. Assessing the Effects of Customer Innovativeness, Environmental Value and Ecological Lifestyles on Residential Solar Power Systems Install Intention. *Energy Policy* **2014,** *67*, 951–961.
10. Valero, A.; Valero, A.; Calvo, G.; Ortego, A. Material Bottlenecks in the Future Development of Green Technologies. *Renewable Sustainable Energy Rev.* **2018,** *93*, 178–200.
11. Bagher, A. M.; Vahid, M. M. A.; Mohsen, M. Types of Solar Cells and Application. *Am. J. Opt. Photonics* **2015,** *3*, 94–113.
12. Yan, J.; Saunders, B. R. Third-Generation Solar Cells: A Review and Comparison of Polymer: Fullerene, Hybrid Polymer and Perovskite Solar Cells. RSC Adv. **2014,** *4*, 43286–43314.
13. Walwyn, D. R.; Brent, A. C. Renewable Energy Gathers Steam in South Africa. *Renewable Sustainable Energy Rev.* **2015,** *41*, 390–401.

14. Jones, N. F.; Pejchar, L.; Kiesecker, J. M. The Energy Footprint: How Oil, Natural Gas, and Wind Energy Affect Land for Biodiversity and the Flow of Ecosystem Services. *Bioscience* **2015,** *65*, 290–301.
15. Holttinen, H.; Meibom, P.; Ensslin, C.; Hofmann, L.; Mccann, J.; Pierik, J. *Design and Operation of Power Systems with Large Amounts of Wind Power*; VTT Research Notes 2493, Citeseer, 2009.
16. Holttinen, H.; Lemstrom, B.; Meibom, P.; Bindner, H.; Orths, A.; Van Hulle, F.; Ensslin, C.; Tiedemann, A.; Hofmann, L.; Winter, W. *Design and Operation of Power Systems with Large Amounts of Wind Power*; State of the Art Report, VTT, 2007.
17. Bakker, R. H.; Pedersen, E.; Van Den Berg, G. P.; Stewart, R. E.; Lok, W.; Bouma, J. Impact of Wind Turbine Sound on Annoyance, Self-Reported Sleep Disturbance and Psychological Distress. *Sci. Total Environ.* **2012,** *425*, 42–51.
18. Sørensen, B. *Renewable Energy: Its Physics, Engineering, Environmental Impacts, Economics and Planning*, 2nd Lightly Updated Printing of 3rd Edition; Academic Press: Burlington, MA, 2004.
19. Wehrli, B. Climate Science: Renewable but Not Carbon-Free. *Nat. Geosci.* **2011,** *4*, 585.
20. Zhang, C.; Anadon, L. D. Life Cycle Water Use of Energy Production and Its Environmental Impacts in China. *Environ. Sci. Technol.* **2013,** *47*, 14459–14467.
21. Macknick, J.; Newmark, R.; Heath, G.; Hallett, K. *Review of Operational Water Consumption and Withdrawal Factors for Electricity Generating Technologies*; National Renewable Energy Lab. (NREL): Golden, CO, 2011.
22. Dodder, R. S. A Review of Water Use in the US Electric Power Sector: Insights from Systems-Level Perspectives. *Curr. Opin. Chem. Eng.* **2014,** *5*, 7–14.
23. Williams, G. E. Geological Constraints on the Precambrian History of Earth's Rotation and the Moon's Orbit. *Rev. Geophys.* **2000,** *38*, 37–59.
24. Marcuccilli, F.; Zouaghi, S. Radial Inflow Turbines for Kalina and Organic Rankine Cycles. *System* **2007,** *2*, 1.
25. Fridleifsson, I. B.; Bertani, R.; Huenges, E.; Lund, J. W.; Ragnarsson, A.; Rybach, L. The Possible Role and Contribution of Geothermal Energy to the Mitigation of Climate Change. In *IPCC Scoping Meeting on Renewable Energy Sources*; IPCC: Luebeck, Germany, Citeseer, 2008; pp 59–80.
26. Khanal, S. K. *Anaerobic Biotechnology for Bioenergy Production: Principles and Applications*; John Wiley & Sons: Hoboken, NJ, 2011.
27. Klass, D. L. *Biomass for Renewable Energy, Fuels, and Chemicals*; Elsevier: Cambridge, MA, 1998.
28. Knothe, G. Biodiesel and Renewable Diesel: A Comparison. *Prog. Energy Combust. Sci.* **2010,** *36*, 364–373.
29. Momirlan, M.; Veziroglu, T. N. The Properties of Hydrogen as Fuel Tomorrow in Sustainable Energy System for a Cleaner Planet. *Int. J. Hydrogen Energy* **2005,** *30*, 795–802.
30. Forsström, J.; Lahti, P.; Pursiheimo, E.; Rämä, M.; Shemeikka, J.; Sipilä, K.; Tuominen, P.; Wahlgren, I. *Measuring Energy Efficiency, Indicators and Potentials in Buildings, Communities and Energy Systems*; VTT Technical Research Centre of Finland, VTT Research Notes 2581, 2011.
31. Chen, H.; Cong, T. N.; Yang, W.; Tan, C.; Li, Y.; Ding, Y. Progress in Electrical Energy Storage System: A Critical Review. *Prog. Nat. Sci.* **2009,** *19*, 291–312.

32. Kats, G.; Alevantis, L.; Berman, A.; Mills, E.; Perlman, J. *The Costs and Financial Benefits of Green Buildings*; A Report to California's Sustainable Building Task Force, 2003; p 134.
33. Fritsch, A. J.; Gallimore, P. *Healing Appalachia: Sustainable Living through Appropriate Technology*; University Press of Kentucky: Lexington, KY, 2007.
34. Norton, B. Solar Water Heaters: A Review of Systems Research and Design Innovation. *Green* **2011,** *1*, 189–207.
35. Guo, T.; Henwood, M. I.; Van Ooijen, M. An Algorithm for Combined Heat and Power Economic Dispatch. *IEEE Trans. Power Syst.* **1996,** *11*, 1778–1784.

CHAPTER 13

IMMUNOMODULATORY MOLECULES FROM HIMALAYAN MEDICINAL PLANTS

FRANCISCO TORRENS[1*] and GLORIA CASTELLANO[2]

[1]*Institut Universitari de Ciència Molecular, Universitat de València, Edifici d'Instituts de Paterna, PO Box 22085, E-46071 València, Spain*

[2]*Departamento de Ciencias Experimentales y Matemáticas, Facultad de Veterinaria y Ciencias Experimentales, Universidad Católica de Valencia San Vicente Mártir, Guillem de Castro-94, E-46001 València, Spain*

[*]*Corresponding author. E-mail: torrens@uv.es*

ABSTRACT

Plant-derived compounds that modulate the immune responses emerge as frontline treatment agents versus cancer, infectious diseases, and autoimmunity. Forty phytochemicals are isolated from five Bhutanese *Sowa Rigpa* medicinal plants (*Aconitum laciniatum*, *Ajania nubegina*, *Corydalis crispa*, *Corydalis dubia*, and *Pleurospermum amabile*) and 14 purified compounds, tested for their immunomodulatory properties via a murine dendritic cell line, and cytotoxicity versus a human cholangiocyte cell line, via xCELLigence real-time cell monitoring. The five medicinal plants, which grow in extreme Himalayan mountain ecology, are used in the scholarly Bhutanese traditional medicines for treating various disorders, which bear relevance to modern disease pathologies (e.g., inflammation, tumor, and infections). The study not only identifies potential immunomodulatory compounds from five Bhutanese medicinal plants but also provides molecular and immunological data to support their reported efficacy. A vast reservoir of selected species remains untapped in terms of phytochemical constituents and pharmacology, which results in the research gap for future studies. Further investigations of endemic plants, and their phytochemical constituents, are needed to understand completely the molecular mechanisms of their action in vivo and in vitro and to assure the plant extracts are safe for human use.

13.1 INTRODUCTION

Setting the scene: Defined small molecules produced by Himalayan medicinal plants (MPs) display immunomodulatory properties. Five MPs (*Aconitum laciniatum*, *Ajania nubegina*, *Corydalis crispa*, *Corydalis dubia*, and *Pleurospermum amabile*), which grow in extreme Himalayan mountain ecology, are used in the scholarly Bhutanese traditional medicines (TMs) for treating disorders, which bore relevance to modern disease pathologies, for example, inflammation, tumor, and infections. Inspired by the strong bioactivities of their crude extracts, 40 phytochemicals were isolated, and 14 of them were tested for their capacity to modulate dendritic cell activity and 10, for cytotoxicity. Herbs of TM ethnopharmacological evaluation and antioxidant (AO) activity (AOA) were informed.[1] Cannabinoids and pain were reviewed with new insights from old molecules.[2] Genuine and sequestered natural products (NPs) from genus *Orobanche* were revised.[3] Herbal extracts and NPs were discussed in alleviating nonalcoholic fatty liver disease.[4] Ethnomedicinal, phytochemical, and pharmacological *Perilla frutescens* investigations were analyzed.[5] Myanmar-MP phytochemistry, medicine, and pharmacological activities were examined.[6] *Lannea schweinfurthii* ethnomedicine, phytochemistry, and pharmacology were reviewed.[7] Chinese herbal medicine for Parkinson's disease was systematically revised.[8]

The TM was practiced virtually in all cultures, expanded globally and gained popularity.[9] A Q'eqchi' communities' ethnopharmacological field study in Guatemala was reported.[10] Chamomile, parsley, and celery-extract botanical therapeutics was published.[11] *Azorella glabra* was informed as an NP source with antiproliferative and cytotoxic activity.[12] Herbal-metabolomics, net-pharmacology, and experiment-validation strategies were investigated in frankincense.[13] A study was reported on *Hagenia abyssinica*-essential oil (EO) antibacterial effects.[14] Kenyan-MPs, sea-algae, and medicinal-wild-mushrooms AOA, total phenolics, and flavonoids were published.[15] *Rosa cymosa*-fruits ethanolic-extract AO activates phosphatase and tensin homolog.[16] Human–platelet-aggregation inhibition by *Premna foetida* was informed.[17] *Lycium barbarum*-extract chemical constituents and bioactivities were studied.[18] Anticandidal potential of stem bark extract from *Schima superba* was reported.[19] *Ajuga genevensis*/*A. reptans*-extract phytochemical profile and AOA were published.[20] Constituents from *Carpesium divaricatum* aerial parts and bioactivity were informed.[21] Ecuadorian Amazon-rainforest-EOs chemical composition and bioactivity were reported.[22] Iberian-thyme-EOs antifungal potential and effect on *Candida albicans* were published.[23] *Lavandula angustifolia* EO alleviates neuropathic pain in mice with spared nerve injury.[24]

Earlier publications in *Nereis*, and so on, classified yams,[25] lactic acid bacteria,[26] fruits,[27] food spices,[28] and oil legumes[29] by principal component (PCA), cluster (CA), and meta-analysis (MA). The molecular classifications of 33 phenolic compounds derived from the cinnamic and benzoic acids from *Posidonia oceanica*,[30] 74 flavonoids,[31] 66 stilbenoids,[32] 71 triterpenoids and steroids from *Ganoderma*,[33] 17 isoflavonoids from *Dalbergia parviflora*,[34] 31 sesquiterpene lactones (STLs),[35,36] and STL artemisinin derivatives[37] were informed. A tool for interrogation of macromolecular structure was reported.[38] Mucoadhesive polymer hyaluronan favors transdermal penetration absorption of caffeine.[39,40] Polyphenolic phytochemicals in cancer prevention and therapy, bioavailability, and bioefficacy were reviewed.[41] From Asia to Mediterranean, soya bean, Spanish legumes, and commercial *soya bean* PCA, CA, and MA were published.[42] The NP AOs from herbs and spices improved the oxidative stability and frying performance of vegetable oils.[43] The relationship between vegetable oil composition and oxidative stability was revealed via a multifactorial approach.[44] It was informed chemical and biological screening approaches to phytopharmaceuticals,[45] cultural interbreeding in indigenous and scientific ethnopharmacology,[46] ethnobotanical studies of MPs, underutilized wild edible plants, food, medicine,[47] biodiversity as a source of drugs, *Cordia*, *Echinacea*, *Tabernaemontana*, *Aloe*,[48] phylogenesis by information entropy, avian birds and 1918 influenza virus.[49] The aim of this chapter is to review defined small molecules produced by Himalayan MPs, displaying immunomodulatory properties.

13.2 IMMUNOMODULATORY MOLECULES FROM HIMALAYAN MEDICINAL PLANTS

Five MPs: *A. laciniatum*, *A. nubegina*, *C. crispa*, *C. dubia*, and *P. amabile* were collected for their roots, aerial parts, and whole plant materials from the high-altitude Himalayan alpine mountains of Bhutan (3500–4900 m above sea level).[50] Forty compounds were isolated, belonging to alkaloids, flavonoids, terpenoids, phenylpropanoids, and furanocoumarins. Of the 40 isolated compounds, 14 of them, namely, pseudaconitine (**1**), 14-veratryolpseudaconitine (**2**), 14-*O*-acetylneoline (**3**, cf. Fig. 13.1), linalool oxide acetate (**4**), (*E*)-spiroether (**5**), luteolin (**6**), luteolin-7-*O*-β-D-glucopyranoside (**7**, cf. Fig. 13.2), protopine (**8**), ochrobirine (**9**, cf. Fig. 13.3), scoulerine (**10**), capnoidine (**11**, cf. Fig. 13.4), isomyristicin (**12**), bergapten (**13**), and isoimperatorin (**14**, cf. Fig. 13.5), were obtained in quantities sufficient for bioactivity screening. Compounds **1–3** belong to C_{19}-diterpenoid-alkaloid phytochemical class. Compounds **4–7** belong to terpenoids and flavonoids classes. Compounds **8–11** belong to benzylisoquinoline-type alkaloids. Compounds **12–14** belong to phenylpropanoids and furanocoumarins classes.

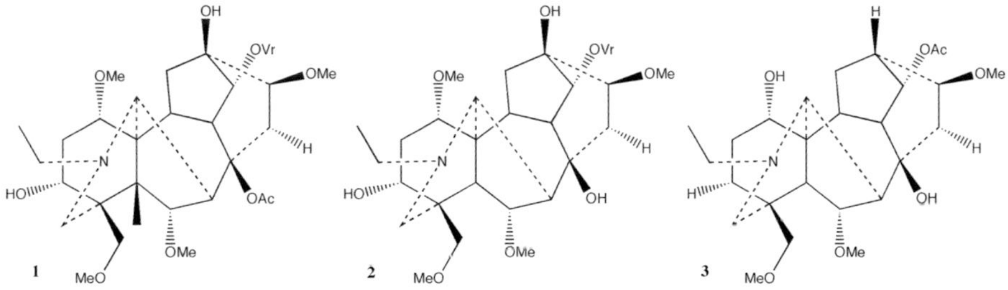

FIGURE 13.1 Structures of the compounds isolated from Bhutanese medicinal plant *Aconitum laciniatum*.

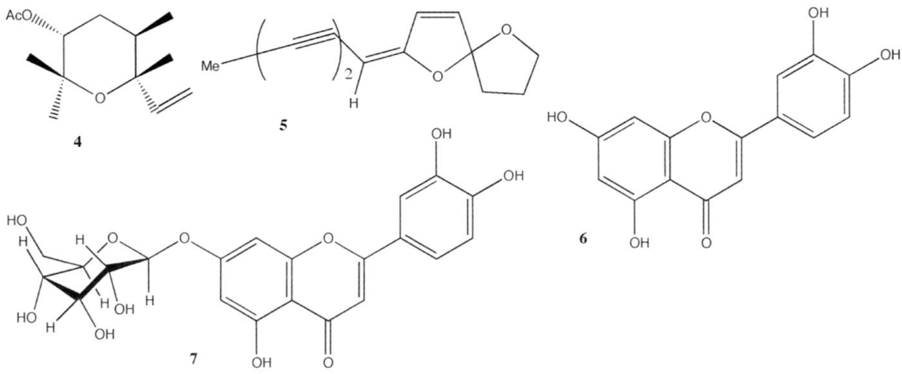

FIGURE 13.2 Structures of the compounds isolated from Bhutanese medicinal plant *Ajania nubegina*.

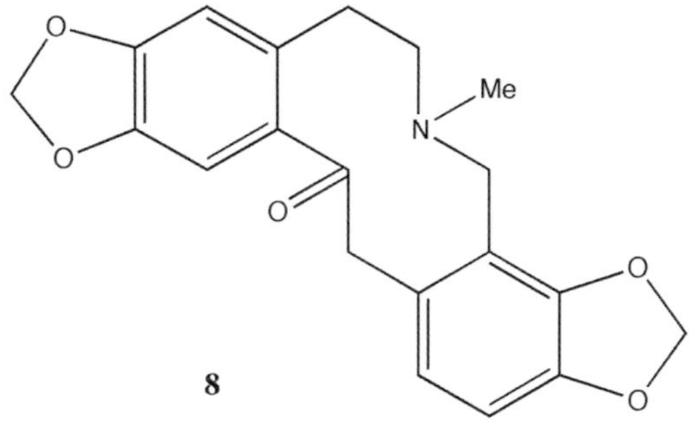

8

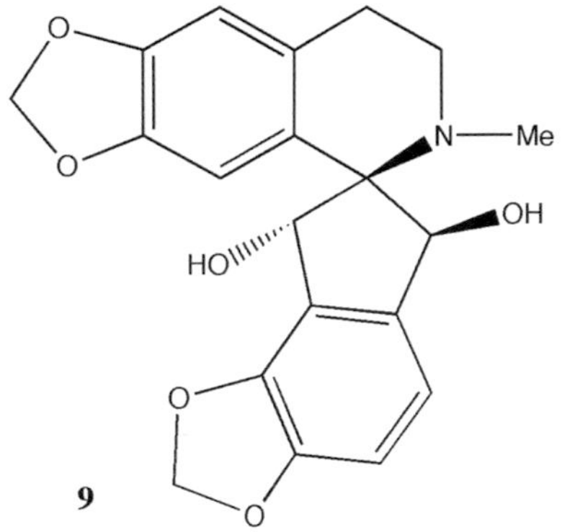

9

FIGURE 13.3 Structures of the compounds isolated from Bhutanese medicinal plant *Corydalis crispa*.

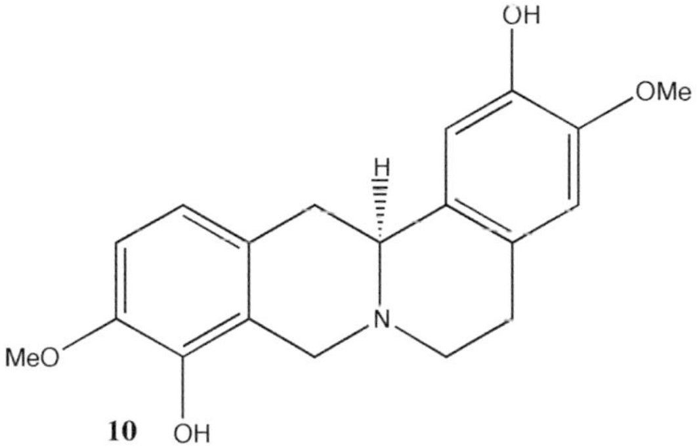

10

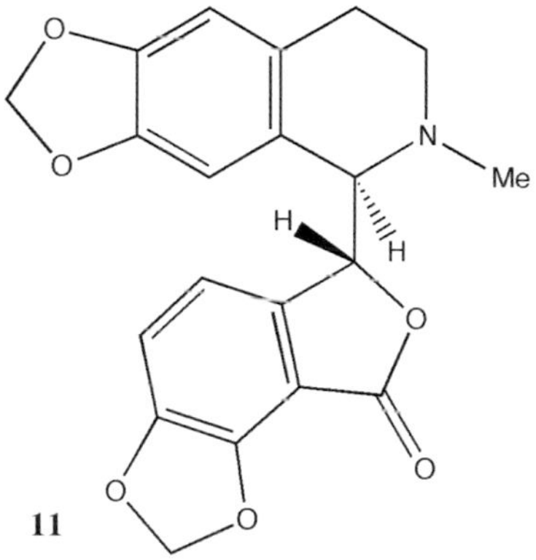

11

FIGURE 13.4 Structures of the compounds isolated from Bhutanese medicinal plant *Corydalis dubia*.

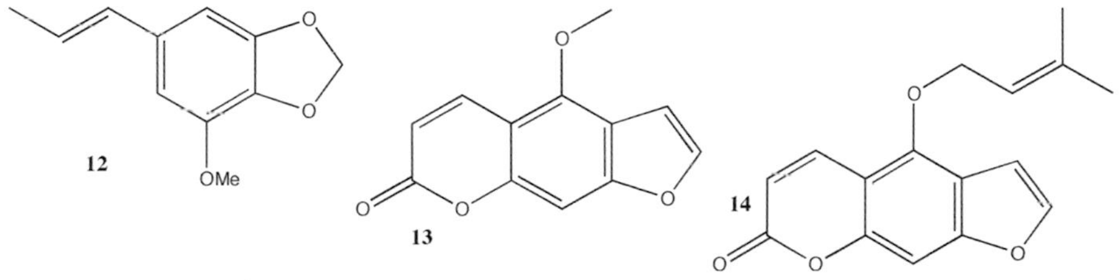

12

13

14

FIGURE 13.5 Structures of compounds isolated from Bhutanese medicinal plant *Pleurospermum amabile*.

13.3 DISCUSSION

The NPs, especially those derived from MPs, were used to help mankind sustain human health since medicine dawn. The TM existed since time immemorial and was accepted and utilized by the people throughout history. Since ancient times, MPs were a TM exemplary source. Plant-derived medicinal NPs attracted scientists' attention around the world for many years, because of their minimum side and positive effects on human health. In the pharmaceutical landscape, MPs with an ethnomedicinal long use history are a rich substance source for the treatment of various ailments and infectious diseases. The MPs are considered a numerous bioactive-compound-type repository, possessing varied therapeutic properties. Vast therapeutic-effect array associated with MPs includes anti-inflammatory, antiviral, antitumor, antimalarial, and analgesic properties. According to the World Health Organization, a variety of drugs are obtained from different MPs, and 80% of the world's developing population depends on TMs for their primary healthcare necessities. The MPs were used since antiquity to treat a number of health problems, and 80% of the world's developing population rely on TM to meet their healthcare needs. Widespread TM use is attributed to cultural acceptability, perceived efficacy versus certain disease types, physical accessibility, and affordability as compared to modern medicine.

Herbal TM is attracting rising attention and acceptance in the world because of its special contribution to chronic-diseases treatment. Plant extracts attracted attention mainly concentrated on their role in preventing diseases. Oxidative stress, which releases free oxygen radicals (FRs) in the body, is involved in a number of disorders, for example, cardiovascular diseases, diabetes, and cancer, which can be prevented by phytochemicals in plant extracts. Oxidation caused by FRs sets reduced capabilities to combat cancer, kidney damage, atherosclerosis and heart diseases. Different phytochemicals in MP NPs are safer than synthetic medicine and beneficial in FR-caused-diseases treatment. Their multiple bioeffects were described, for example, AOs, cellular signals, cardioprotective effects, antibiotics, anti-inflammation, anti-allergic, anticoagulation, antineoplastic, antimutagenesis, and anticarcinogenesis. Plant-extract bioactivities are mainly because of their polyphenols, flavonoids, and terpenoids contents. Numerous studies showed that compound-classes possess AO, anti-inflammatory and anti-cancer properties.

The MPs were an important modern-drug source for many centuries and are receiving rising attention worldwide with advancement in drug discovery techniques and technologies. Plants produce various secondary metabolites that belong to different major phytochemical classes, for example, flavonoids, tannins, terpenoids, saponins, triterpenoid saponins, alkaloids, phytosterols, carotenoids, fatty acids, and EOs, which are used for their defense and protection versus predators and herbivores. Based on the rationale, five MPs were selected (*A. laciniatum*, *A. nubigena*, *C. crispa*, *C. dubia*, and *P. amabile*), which are used in the scholarly Bhutanese TMs for treating various disorders, for example, fever, inflammation, malaria, tumor, leprosy, gout, mumps, abscess, liver, and heart infections. Since mankind's history, plants not only provided shelter, fuel, and food but also they were always a vital TM source imparting health benefits. Because of the high pharmaceutical-medicine cost and safety, renewed interest exists in plants and herbs used as a food and TM. The medicinal NPs obtained from a number of MPs are considered safer. Eighty percent of the world's population, especially in Africa and South Asia, depends on TM, mainly MP-derived NPs and phytomedicines, to accomplishing their basic healthcare needs. Many food and MPs, especially aromatic herbs and spices, were evaluated for nutrapharmaceutical potential, because of containing a wide NP bioactive compound and new chemical entities array. It is important to establish a strategy for realizing the investigation of the effects of processing procedures, comparing the global metabolites, absorption, and pharmacological properties.

13.4 FINAL REMARKS

From the previous results and discussion, the following final remarks can be drawn:

(1) Five MPs, which grow in extreme Himalayan mountain ecology, are used in the scholarly Bhutanese TMs for treating disorders, which bore relevance to modern disease pathologies, for example, inflammation, tumor, and infections.
(2) Inspired by the strong bioactivities of their crude extracts, 40 phytochemicals were isolated, and 14 of them were tested for their capacity to modulate dendritic cell activity and 10 for cytotoxicity. The study not only identified potential immunomodulatory compounds from five Bhutanese MPs but also provided molecular and immunological data to support their reported efficacy.
(3) A vast reservoir of selected species remains untapped in terms of phytochemical constituents and pharmacology, which results in the

research gap for future studies. Further investigations of endemic plants, and their phytochemical constituents, are needed to understand completely the molecular mechanisms of their action in vivo/vitro and to assure the plant extracts are safe for human use.

(4) Chemical and pharmaceutical industries must work together with researchers and traditional MP collectors, to use the maximum benefit of the plant under study.

(5) Herbal medicine extracts and NPs are effective in treating many diseases, although the mechanisms are still under exploration. Studies provided valuable information on the role of herbal medicine extracts and NPs via activating autophagy, which inducing agents are confirmed to be beneficial in the treatment of diseases in animal models and cell lines.

(6) Medicinal plants were used since ancient times and are still used as a primary source of medical treatment in developing countries. Plant-derived substances present advantages (e.g., low cost, rapid speed of drug discovery); their main disadvantage is the absence of common international standards for evaluating their quality, efficacy, and safety.

ACKNOWLEDGMENTS

The authors thank the support from Generalitat Valenciana (Project No. PROMETEO/2016/094) and Universidad Católica de Valencia *San Vicente Mártir* (Project No. UCV.PRO.17-18.AIV.03).

KEYWORDS

- **phytochemical**
- **scoulerine**
- **bergapten**
- **immunomodulator**
- **adjuvant**
- **cytoxicity**
- **dendritic cell**

REFERENCES

1. Rashid, S.; Ahmad, M.; Zafar, M.; Anwar, A.; Sultana, S.; Tabassum, S.; Ahmed, S. N. Ethnopharmacological Evaluation and Antioxidant Activity of Some Important Herbs Used in Traditional Medicines. *J. Tradit. Chin. Med.* **2016,** *36*, 689–694.
2. Vargas, J. M.; Andrade-Cetto, A. Ethnopharmacological Field Study of Three Q'eqchi Communities in Guatemala. *Front. Pharmacol.* **2018,** *9*, 1246-1–14.
3. Scharenberg, F.; Zidorn, C. Genuine and Sequestered Natural Products from the Genus *Orobanche* (Orobanchaceae, Lamiales). *Molecules* **2018,** *23*, 2821-1–30.
4. Zhang, L.; Yao, Z.; Ji, G. Herbal Extracts and Natural Products in Alleviating Non-alcoholic Fatty Liver Disease via Activating Autophagy. *Front. Pharmacol.* **2018,** *9*, 1459-1–7.
5. Ahmed, H. M. Ethnomedicinal, Phytochemical and Pharmacological Investigations of *Perilla frutescens* (L.) Britt. *Molecules* **2019,** *24*, 102-1–23.
6. Aye, M. M.; Aung, H. T.; Sein, C.; Armijos, C. A Review on the Phytochemistry, Medicinal Properties and Pharmacological Activities of 15 Selected Myanmar Medicinal Plants. *Molecules* **2019,** *24*, 293-1–34.
7. Maroyi, A. Review of Ethnomedicinal, Phytochemical and Pharmacological Properties of *Lannea schweinfurthii* (Engl.) Engl. *Molecules* **2019,** *24*, 732-1–23.
8. Jin, X. C.; Zhang, L.; Wang, Y.; Cai, H. B.; Bao, X. J.; Jin, Y. Y.; Zheng, G. Q. An Overview of Systematic Reviews of Chinese Herbal Medicine for Parkinson's Disease. *Front. Pharmacol.* **2019,** *10*, 155-1–13.
9. Galen, E. Traditional Herbal Medicines Worldwide, from Reappraisal to Assessment in Europe. *J. Ethnopharmacol.* **2014,** *158*, 498–502.
10. Vuckovic, S.; Srebro, D.; Vujovic, K. S.; Vucetic, C.; Prostran, M. Cannabinoids and Pain: New Insights from Old Molecules. *Front. Pharmacol.* **2018,** *9*, 1259-1–19.
11. Danciu, C.; Zupko, I.; Bor, A.; Schwiebs, A.; Radeke, H.; Hancianu, M.; Cioanca, O.; Alexa, E.; Oprean, C.; Bojin, F.; Soica, C.; Paunescu, V.; Dehelean, C. A. Botanical Therapeutics: Phytochemical Screening and Biological Assessment of Chamomile, Parsley and Celery Extracts against A375 Human Melanoma and Dendritic Cells. *Int. J. Mol. Sci.* **2018,** *19*, 3624-1–20.
12. Lamorte, D.; Faraone, I.; Laurenzana, I.; Milella, L.; Trino, S.; De Luca, L.; Del Vecchio, L.; Armentano, M. F.; Sinisgalli, C.; Chiummiento, L.; Russo, D.; Bisaccia, F.; Musto, P.; Caivano, A. Future in the Past: *Azorella glabra* Wedd. as a Source of New Natural Compounds with Antiproliferative and Cytotoxic Activity on Multiple Myeloma Cells. *Int. J. Mol. Sci.* **2018,** *19*, 3348-1–18.
13. Ning, Z.; Wang, C.; Liu, Y.; Song, Z.; Ma, X.; Liang, D.; Liu, Z.; Lu, A. Integrating Strategies of Herbal Metabolomics, Network Pharmacology, and Experiment Validation to Investigate Frankincense Processing Effects. *Front. Pharmacol.* **2018,** *9*, 1482-1–18.
14. Abebe, G.; Birhanu, G. A Comparative Study on Antibacterial Effects of *Hagenia abyssinica* Oil Extracted from Different Parts of the Plant using Different Solvents against Two Selected and Standardized Human Pathogens. *Afr. J. Microbiol. Res.* **2019,** *13*, 99–105.
15. Siangu, B. N.; Sauda, S.; John, M. K.; Njue, W. M. Antioxidant Activity, Total Phenolic and Flavonoid Content of Selected Kenyan Medicinal Plants, Sea Algae and Medicinal Wild Mushrooms. *Afr. J. Pure Appl. Chem.* **2019,** *13*, 43–48.

16. Wang, K. C.; Liu, Y. C.; El-Shazly, M.; Shih, S. P.; Du, Y. C.; Hsu, Y. M.; Lin, H. Y.; Chen, Y. C.; Wu, Y. C.; Yang, S. C.; Lu, M. C. The Antioxidant from Ethanolic Extract of *Rosa cymosa* Fruits Activates Phosphatase and Tensin Homolog *In Vitro* and *In Vivo*: A New Insight on its Antileukemic Effect. *Int. J. Mol. Sci.* **2019,** *20*, 1935-1–23.
17. Dianita, R.; Jantan, I. Inhibition of Human Platelet Aggregation and Low-Density Lipoprotein Oxidation by *Premna foetida* Extract and its Major Compounds. *Molecules* **2019,** *24*, 1469-1–14.
18. Xiao, X.; Ren, W.; Zhang, N.; Bing, T.; Liu, X.; Zhao, Z.; Shangguan, D. Comparative Study of the Chemical Constituents and Bioactivities of the Extracts from Fruits, Leaves and Root Barks of *Lycium barbarum*. *Molecules* **2019,** *24*, 1585-1–22.
19. Wu, C.; Wu, H. T.; Wang, Q.; Wang, G. H.; Yi, X.; Chen, Y. P.; Zhou, G. X. Anticandidal Potential of Stem Bark Extract from *Schima superba* and the Identification of its Major Anticandidal Compound. *Molecules* **2019,** *24*, 1587-1–16.
20. Toiu, A.; Mocan, A.; Vlase, L.; Pârvu, A. E.; Vodnar, D. C.; Gheldiu, A. M.; Moldovan, C.; Oniga, I. Comparative Phytochemical Profile, Antioxidant, Antimicrobial and *In Vivo* Anti-inflammatory Activity of Different Extracts of Traditionally Used Romanian *Ajuga genevensis* L. and *A. reptans* L. (Lamiaceae). *Molecules* **2019,** *24*, 1597-1–21.
21. Kleczek, N.; Michalak, B.; Malarz, J.; Kiss, A. K.; Stojakowska, A. *Carpesium divaricatum* Sieb. & Zucc. Revisited: Newly Identified Constituents from Aerial Parts of the Plant and Their Possible Contribution to the Biological Activity of the Plant. *Molecules* **2019,** *24*, 1614-1–12.
22. Noriega, P.; Guerrini, A.; Sacchetti, G.; Grandini, A.; Ankuash, E.; Manfredini, S. Chemical Composition and Biological Activity of Five Essential Oils from the Ecuadorian Amazon Rain Forest. *Molecules* **2019,** *24*, 1637-1–12.
23. Alves, M.; Gonçalves, M. J.; Zuzarte, M.; Alves-Silva, J. M.; Cavaleiro, C.; Cruz, M. T.; Salgueiro, L. Unveiling the Antifungal Potential of Two Iberian Thyme Essential Oils: Effect on *C. albicans* Germ Tube and Preformed Biofilms. *Front. Pharmacol.* **2019,** *10*, 446-1–11.
24. Sanna, M. D.; Les, F.; Lopez, V.; Galeotti, N. Lavender (*Lavandula angustifolia* Mill.) Essential Oil Alleviates Neuropathic Pain in Mice with Spared Nerve Injury. *Front. Pharmacol.* **2019,** *10*, 472-1–13.
25. Torrens-Zaragozá, F. Molecular Categorization of Yams by Principal Component and Cluster Analyses. *Nereis* **2013,** *2013* (5), 41–51.
26. Torrens-Zaragozá, F. Classification of Lactic Acid Bacteria against Cytokine Immune Modulation. *Nereis* **2014,** *2014* (6), 27–37.
27. Torrens-Zaragozá, F. Classification of Fruits Proximate and Mineral Content: Principal Component, Cluster, Meta-analyses. *Nereis* **2015,** *2015* (7), 39–50.
28. Torrens-Zaragozá, F. Classification of Food Spices by Proximate Content: Principal Component, Cluster, Meta-Analyses, *Nereis* **2016,** *2016* (8), 23–33.
29. Torrens, F.; Castellano, G. From Asia to Mediterranean: Soya Bean, Spanish Legumes and Commercial *Soya Bean* Principal Component, Cluster and Meta-Analyses. *J. Nutr. Food Sci.* **2014,** *4* (5), 98–98.
30. Castellano, G.; Tena, J.; Torrens, F. Classification of Polyphenolic Compounds by Chemical Structural Indicators and its Relation to Antioxidant Properties of *Posidonia oceanica* (L.) Delile. *MATCH Commun. Math. Comput. Chem.* **2012,** *67*, 231–250.

31. Castellano, G.; González-Santander, J. L.; Lara, A.; Torrens, F. Classification of Flavonoid Compounds by Using Entropy of Information Theory. *Phytochemistry* **2013,** *93*, 182–191.
32. Castellano, G.; Lara, A.; Torrens, F. Classification of Stilbenoid Compounds by Entropy of Artificial Intelligence. *Phytochemistry* **2014,** *97*, 62–69.
33. Castellano, G.; Torrens, F. Information Entropy-Based Classification of Triterpenoids and Steroids from *Ganoderma*. *Phytochemistry* **2015,** *116*, 305–313.
34. Castellano, G.; Torrens, F. Quantitative Structure–Antioxidant Activity Models of Isoflavonoids: A Theoretical Study. *Int. J. Mol. Sci.* **2015,** *16*, 12891–12906.
35. Castellano, G.; Redondo, L.; Torrens, F. QSAR of Natural Sesquiterpene Lactones as Inhibitors of Myb-Dependent Gene Expression. *Curr. Top. Med. Chem.* **2017,** *17*, 3256–3268.
36. Torrens, F.; Redondo, L.; León, A.; Castellano, G. Structure–Activity Relationships of Cytotoxic Lactones as Inhibitors and Mechanisms of Action. *Curr. Drug Discov. Technol.* submitted for publication.
37. Torrens, F.; Redondo, L.; Castellano, G. Artemisinin: Tentative Mechanism of Action and Resistance. *Pharmaceuticals* **2017,** *10*, 20-4–4.
38. Torrens, F.; Castellano, G. A Tool for Interrogation of Macromolecular Structure. *J. Mater. Sci. Eng. B* **2014,** *4* (2), 55–63.
39. Torrens, F.; Castellano, G. Mucoadhesive Polymer Hyaluronan as Biodegradable Cationic/Zwitterionic-Drug Delivery Vehicle. *ADMET DMPK* **2014,** *2*, 235–247.
40. Torrens, F.; Castellano, G. Computational Study of Nanosized Drug Delivery from *Cyclo*dextrins, Crown Ethers and Hyaluronan in Pharmaceutical Formulations. *Curr. Top. Med. Chem.* **2015,** *15*, 1901–1913.
41. Estrela, J. M.; Mena, S.; Obrador, E.; Benlloch, M.; Castellano, G.; Salvador, R.; Dellinger, R. W. Polyphenolic Phytochemicals in Cancer Prevention and Therapy: Bioavailability *versus* Bioefficacy. *J. Med. Chem.* **2017,** *60*, 9413 9436.
42. Torrens, F.; Castellano, G. From Asia to Mediterranean: Soya Bean, Spanish Legumes and Commercial *Soya Bean* Principal Component, Cluster and Meta-analyses. *J. Nutr. Food Sci.* **2014,** *4* (5), 98–98.
43. Redondo-Cuevas, L.; Castellano, G.; Raikos, V. Natural Antioxidants from Herbs and Spices Improve the Oxidative Stability and Frying Performance of Vegetable Oils. *Int. J. Food Sci. Technol.* **2017,** *52*, 2422–2428.
44. Redondo-Cuevas, L.; Castellano, G.; Torrens, F.; Raikos, V. Revealing the Relationship between Vegetable Oil Composition and Oxidative Stability: A Multifactorial Approach. *J. Food Compos. Anal.* **2018,** *66*, 221–229.
45. Torrens, F.; Castellano, G. Chemical/Biological Screening Approaches to Phytopharmaceuticals. In *Research Methods and Applications in Chemical and Biological Engineering*; Pourhashemi, A., Deka, S. C., Haghi, A. K., Eds.; Apple Academic–CRC: Waretown, NJ, in press.
46. Torrens, F.; Castellano, G. Cultural Interbreeding in Indigenous/Scientific Ethnopharmacology. In *Research Methods and Applications in Chemical and Biological Engineering*; Pourhashemi, A., Deka, S. C., Haghi, A. K., Eds.; Apple Academic–CRC: Waretown, NJ, in press.
47. Torrens, F.; Castellano, G. Ethnobotanical Studies of Medicinal Plants: Underutilized Wild Edible Plants, Food and Medicine. In *Innovations in Physical Chemistry*; Haghi, A. K., Ed.; Apple Academic–CRC: Waretown, NJ, in press.

48. Torrens, F.; Castellano, G. Biodiversity as a Source of Drugs: Cordia, Echinacea, Tabernaemontana and Aloe. In *Green Chemistry and Biodiversity: Principles, Techniques, and Correlations*; Aguilar, C. N., Ameta, S. C., Haghi, A. K., Eds.; Apple Academic–CRC: Waretown, NJ, in press.
49. Torrens, F.; Castellano, G. Phylogenesis by Information Entropy: Avian Birds and 1918 Influenza Virus. In *Modelling & Simulation*; Turner, S., Yunus, J., Eds.; IEEE: London, 2008; Vol 2, pp 1–2.
50. Wangchuk, P.; Apte, S. H.; Smout, M. J.; Groves, P. L.; Loukas, A.; Doolan, D. L. Defined Small Molecules Produced by Himalayan Medicinal Plants Display Immunomodulatory Properties. *Int. J. Mol. Sci*. **2018,** *19*, 3490-1–21.

CHAPTER 14

SUPERCONDUCTORS, MAGNETISM, QUANTUM METROLOGY, AND COMPUTING

FRANCISCO TORRENS[1*] and GLORIA CASTELLANO[2]

[1]*Institut Universitari de Ciència Molecular, Universitat de València, Edifici d'Instituts de Paterna, PO Box 22085, E-46071 València, Spain*

[2]*Departamento de Ciencias Experimentales y Matemáticas, Facultad de Veterinaria y Ciencias Experimentales, Universidad Católica de Valencia San Vicente Mártir, Guillem de Castro-94, E-46001 València, Spain*

**Corresponding author. E-mail: torrens@uv.es*

ABSTRACT

Increasingly useful in materials research and development, molecular modeling method combines computational chemistry techniques with molecular infography, for simulating and predicting the structure, chemical processes, and properties of materials. Molecular modeling techniques, in materials science, explore the impact of using molecular modeling for a number of simulations in industrial settings. They provide an overview of commonly used methods in atomic simulation of a broad range of materials, for example, superconductors, and so on. Molecular modeling techniques in materials science present the background and tools, for chemists and physicists, to perform in-silico experiments to understand relationships between the properties of materials and the underlying atomic structure, which insights result in more accurate data for designing application-specific materials that withstand real process conditions, for example, hot temperatures and high pressures.

14.1 INTRODUCTION

Setting the scene: superconductors (SCs), structure, O vacancies, O diffusion, time-reversal symmetry (TRS) breaking in SCs via loop Josephson-current (LJC) order, measurement of magnetic susceptibility (MS), explanation and definition of terms, analogy between the London equations of SC and Meissner effect, investigating quantum metrology in noisy channels, high-fidelity (Hi-Fi) spin, optical control of single Si-vacancy centers in SiC, and proposing the use of magnetic molecules for quantum computing.

Increasingly useful in materials research and development, molecular modeling is a method that combines computational chemistry techniques with molecular infography (graphics visualization), for simulating and predicting the structure, chemical processes, and properties of materials. Molecular modeling techniques, in materials science, explore the impact of using molecular modeling for a number of simulations in industrial settings. They provide an overview of commonly used methods in atomic simulation of a broad range of materials, for example, SCs, and so on. Molecular modeling techniques in materials science present the background and tools, for chemists and physicists, to perform in-silico experiments to understand relationships between the properties of materials and the underlying atomic structure, which insights result in more accurate data for designing application-specific materials that withstand real process conditions, for example, hot temperatures and high pressures. Even at the molecular scale, different theoretical approaches exist, corresponding to the level of detail at which the electrons in the system must be described. It is beyond the scope of this work to go into the detail of every theory. Notwithstanding, the study attempted to include important ideas so that the reader could gauge the approximations and errors implicit in a given simulation method. In this laboratory, Coronado group reviewed molecular magnetism from molecular assemblies to devices.[1]

In earlier publications, the periodic table of the elements (PTE),[2–4] quantum simulators,[5–13] science, ethics of developing sustainability via nanosystems, devices,[14] *green nanotechnology* as an approach towards environment safety,[15] molecular devices, machines as hybrid organic-inorganic structures,[16] PTE, quantum biting its tail, sustainable chemistry,[17] quantum molecular *spintronics*, nanoscience, and GRs[18] were reported. Cancer, its hypotheses,[19] precision personalized medicine from theory to practice, cancer,[20] how human immunodeficiency virus/acquired immunodeficiency syndrome (HIV/AIDS) destroy immune defenses, hypothesis,[21] 2014 emergence, spread, uncontrolled Ebola outbreak,[22,23] Ebola virus disease, questions, ideas, hypotheses, models,[24] metaphors that made history, reflections on philosophy, science and deoxyribonucleic acid,[25] scientific integrity and ethics, science communication, psychology,[26] theory, simulation, and the present and future of quantum technologies[27] were informed. In the present report, SCs, structure, O vacancies, O diffusion, TRS breaking in SCs via LJC order, measurement of MS, explanation and definition of terms, analogy between the London equations of SC and Meissner effect, investigating quantum metrology in noisy channels, Hi-Fi spin, optical control of single Si vacancies in SiC, and proposing the use of magnetic molecules for quantum computing are reviewed.

14.2 SUPERCONDUCTORS

The SCs are materials that present the ability to conduct electricity without resistance (*ballistic conduction*), below a critical temperature (CT),[28] which is larger than absolute zero, which phenomenon was first seen (1911) by Onnes in Hg at $He_{(l)}$ temperatures.[29] *Bednorz and Müller discovery (*1986) that certain ceramic-like materials exhibit SC at CTs, greater than 30 K, gave rise to an interest and activity in the area.[30] The excitement rose when Chu group (1987) found Y–Ba–Cu oxide to result SC above $N_{2(l)}$ temperatures.[31] A number of publications devoted to SC exist and some simulation aspects are discussed in this work. After extensive studies on Cu oxide SCs, the following are known: (1) SC occurs in two-dimensional CuO_2 arrays, which are formed by square-planar corner-sharing O atoms; (2) via addition of either cations or anions (hole, electron) created by chemical doping, the intermediate spacer layers between SC planes act as charge reservoirs; (3) because of an energy match between O-2p and Cu 3d orbitals, Fermi level is highly hybridized; (4) in all cases so far, the original antiferromagnetic insulator becomes metallic on doping to the correct level. The actual electronic SC mechanism is under debate, and no conclusive evidence exists on whether the pairing occurs because of magnetic or electronic reasons. Notwithstanding, electronic structure and classical molecular mechanics calculations provided insight into complex atomic geometries, charge distributions, and how they change with varying dopant types and amounts. Diffusion of O atoms in SCs was studied via model potentials. In the next sections, we deal with a few examples, which highlight atomistic simulation use applied to understanding static and dynamic SC properties.

14.3 STRUCTURE

Doped lanthanoid La and Nd Cu oxide systems were studied. Both present planar CuO_2 units, which is an essential feature for high-temperature (HT) SCs. Atoms of Cu are six-fold coordinated in La_2CuO_4, while they are four-fold coordinated in Nd_2CuO_4. For ionic oxides, for example, these, atomistic simulations were performed via empirical ion pair potentials (IPPs). Defect structures, because of O vacancies created by substitution of La^{3+} by divalent cations (e.g., alkaline earth metal Ca^{2+}, Sr^{2+}, Ba^{2+}), were simulated via IPPs. Empirical IPPs were used to simulate La_2CuO_4 crystal structure. Tetragonal to orthorhombic La_2CuO_4 distortion and HT SC $La_{2-x}M_xCuO_4$ (M = Ba, Sr) were studied. Crystal La_2CuO_4 structure is tetragonal above 533 K and orthorhombic below 533 K, which distortion occurs by planar-CuO_4 units buckling because of ionic interaction involving La^{3+}. The IPPs consist of a Coulomb V_c, a nonbond (NB) repulsion V_{nr} and an NB dispersion term V_v expressed as

$$V_c = \frac{q_i q_j}{r_{ij}} \tag{14.1}$$

$$V_{nr} = B_{ij} \exp\left(\frac{-r_{ij}}{\rho_{ij}}\right) \tag{14.2}$$

$$V_v = \frac{-C_{ij}}{r_{ij}^6} \tag{14.3}$$

where q_i and q_j are the ionic charges, r_{ij} is the separation distance, B_{ij}, C_{ij}, and ρ_{ij} are adjustable parameters to be fitted via the experimental data.

$$\frac{1}{\rho_{ij}} = \frac{\left(1/\rho_{ii} + 1/\rho_{jj}\right)}{2} \tag{14.4}$$

and

$$C_{ii} = \frac{3 I_i P_i^2}{4} \tag{14.5}$$

where I_i is the ionization potential and P_i, the polarizability of ion *i*. The strategy is to fit B_{ij} and ρ_{ij} of $Cu^{2+}...Cu^{2+}$, $O^{2-}...O^{2-}$, and $La^{3+}...La^{3+}$ pairs via experimental CuO and La_2O_3 unit cell (UC), which adjustable parameters are changed until the simulated lattice constant and atom positions match the experimental data, which parameters were used in La_2CuO_4-structure simulations. In order to simulate $La_{2-x}M_xCuO_4$ (M = Ba, Sr) structure, the strategy was to increase $La^{3+}...La^{3+}$ B_{ij} value and decrease that for $Cu^{2+}...Cu^{2+}$, which was justified since Ba^{2+} and Sr^{2+} are much larger than La^{3+} and, in the doped system, Cu atoms are usually in a higher oxidation state to compensate for excess charge. Based on the assumptions, the tetragonal to orthorhombic structural distortions were simulated for $La_{1.85}Ba_{0.15}CuO_4$ and $La_{1.85}Sr_{0.15}CuO_4$, which resulted in agreement with experiments.

14.4 OXYGEN VACANCIES

In Tl–Ba–Ca–Cu–O SCs case, two phases exist: one in which the lattice is primarily tetragonal with the ideal chemical formula $TlBa_2Ca_{n-1}Cu_nO_{2n+2.5}$ (n = 1, 2, 3, 4) and a second phase, which belongs to a body-centered tetragonal lattice, with an ideal chemical formula $Tl_2Ba_2Ca_{n-1}Cu_nO_{2n+4}$ (n = 1, 2, 3, 4). In the former, every UC contains 4.5 O atoms if n = 1, that is, half an O-vacancy exists per UC, in which simulation, for simplicity randomness is omitted and the vacancy can be repeated periodically. Building two UCs and omitting one O in the supercell, it is possible to construct a periodic defect model in $TlBa_2CuO_{4.5}$, which work was performed via the extended Hückel tight-binding (TB) method. No geometry optimization around the vacancy was performed. Different O-atom-vacancy positions were considered and, for every one, the net charge (NC) on every atom and the electric field gradient (EFG) were calculated. One of the positions for O-atom vacancy results in 2.4 and 0.62 for NC of the two Cu atoms, which is similar to $YBa_2Cu_3O_{7-x}$ where Cu^{3+} and Cu^{+} exist. The EFG calculated for O-atom vacancy is the largest, indicating that the electron and hole movement are most facilitated in the structure.

14.5 OXYGEN DIFFUSION

Several workers experimentally and theoretically studied O diffusion. Notwithstanding, until recently O-diffusion-paths nature in SCs was controversial. Structure of SC $YBa_2Cu_3O_7$ was computed. Calculating the mobile-ion defect energy along the diffusion path, the most favorable energy O diffusion resulted O1 to O4 to O1. On the other hand, O diffusion occurs via an O1 O jump to an O5 empty site and moves along empty O5 sites. Diffusion of O in $YBa_2Cu_3O_{7-x}$ (x = 0.09–0.27) was studied via molecular dynamics (MD) simulations (MDSs). All work was based on IPP, which parameters were derived fitting a rigid ion model to reproduce the calculated structure in 5% of the experimental structure. Calculations of MD at constant volume were performed on compositions $YBa_2Cu_3O_{6.91}$, $YBa_2Cu_3O_{6.82}$, and $YBa_2Cu_3O_{6.73}$ for 100 ps every one at temperatures in 1350–1500 K, which HTs are necessary in simulations to obtain accurate diffusion coefficients, which MDSs revealed that no O5 to O5 jump was present but the paths O1 to O5, O1 to O4 and O4 to O5 were possible. At lower temperatures, O1–O4–O1 path is the main contributor, while at HTs, O1–O5–O1 path becomes important.

14.6 SCS TIME-REVERSAL SYMMETRY BREAKING: LOOP JOSEPHSON-CURRENT ORDER

Recent muon (μ)-spin relaxation experiments found broken TRS in a number of SCs, which, from other viewpoints (e.g., specific heat, penetration depth, impurities sensitivity, etc.), appear to be conventional. Ghosh group proposed a novel SC ground state (GS), where Josephson currents (JCs) flow spontaneously between distinct but symmetry-related sites in a UC.[32] Such LJC state breaks TRS without the need for triplet, intersite or interorbital pairing, that is, they are compatible with a conventional Bardeen–Cooper–Schrieffer's type pairing mechanism. The JCs result from a nontrivial phase difference between the on-site pairing potentials on different sites, appearing spontaneously at SC CT. They explicitly showed how such instability emerged in Ginzburg–Landau's theory of a simple toy model. They estimated resulting-spontaneous-magnetization size that resulted consistent with many existing experiments. They discussed the crystal-symmetry requirements and applied their theory to the recently discovered TRS-breaking family but, otherwise, seemingly conventional SC family Re_6X (X = Zr, Hf, Ti), showing an LJC-instability possibility.

14.7 MAGNETIC-SUSCEPTIBILITY MEASUREMENT: TERMS EXPLANATION/DEFINITION

The intensity of a magnetic field (MF) is expressed in *oersteds*, although the word *gauss* is usually used in the same sense.[33] A field of one oersted (or one gauss) is of such intensity that a unit magnetic pole placed in it is acted on by a force of one dyne. If a substance is placed in an MF of a certain intensity, then MF intensity in the substance may be either smaller or larger than the intensity in the surrounding space (cf. Fig. 14.1). In the first, the substance is called *diamagnetic* (DM); in the second, *paramagnetic* (PM). *Ferromagnetism* (FM) case exists, in which MF intensity in the substance may be risen a million-fold or more. However, FM, although of technological importance, is rare in nature. It occurs in only a few metals, alloys, and compounds. The PM is common in nature, especially among the transition group elements. The DM is a universal property of matter. All substances, even though PM, present at least an underlying DM that must be corrected for in precise determination of the permanent magnetic moment. A substance may be DM and PM but, generally, whenever PM is present, it is much larger that it hides DM.

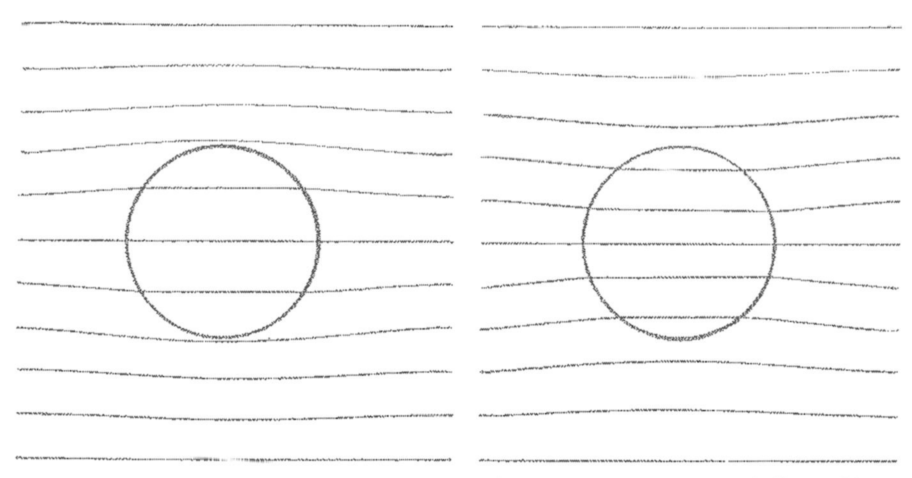

FIGURE 14.1 DM bodies (*left*) are less permeable than a vacuum to magnetic lines of force. PM bodies are more permeable than a vacuum.
Source: Ref. [33].

If a substance is placed in an MF of intensity H, the intensity in the substance is given by B, where:

$$B = H + 4\pi\mathfrak{I} \tag{14.6}$$

The quantity $\mathfrak{I}$ is called the intensity of magnetization, and $\mathfrak{I}/H = \kappa$ is MS per unit volume. The MS per unit mass is obtained dividing κ by the density. The symbol χ is used for MS per gram. The molar MS, χ_M, is MS per mol. In general, DM-substance susceptibility is independent of temperature and MF strength. The PM-substance MS is usually inversely proportional to the absolute temperature and is independent of MF strength. The FM-substance MS is dependent on temperature and MF strength in a rather complicated way. Many methods exist available for MS measurement.

14.8 LONDON EQUATIONS OF SUPERCONDUCTIVITY—MEISSNER EFFECT ANALOGY

Bilya reviewed literature, making a case for a theoretical London brothers SC laws—Meissner effect analogy.[34] Mathematical approach was involved in deriving equations that actually prove that an external applied MF, via SC materials, actually penetrate up to the vortices length or penetration depth in SC material, called London penetration depth, before MF decays completely (exponentially, cf. Fig. 14.2), disagreeing with Meissner theories that state that for a field via SCs, a complete DM effect takes place.

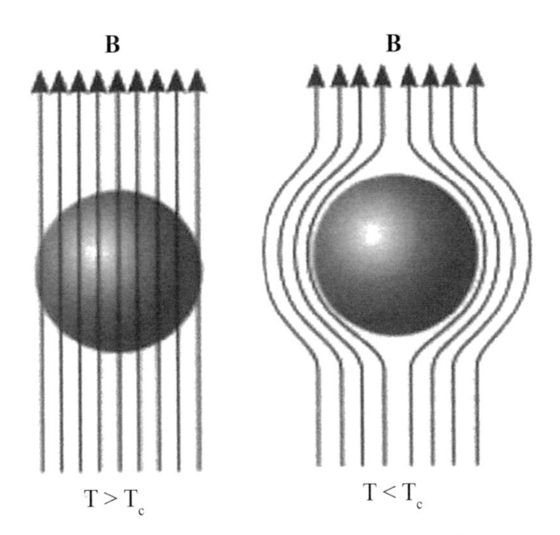

FIGURE 14.2 Magnetic field through normal and superconducting metals, respectively. *Source*: Ref. [34].

14.9 INVESTIGATING QUANTUM METROLOGY IN NOISY CHANNELS

Quantum *entanglement* lies at quantum-information and -metrology heart, in which, with a colossal amount of quantum Fisher information (QFI), entangled systems were ameliorated to result a better resource scheme. Notwithstanding, noisy channels eject QFI substantially. Falaye group investigated how N-quantum binary digit (*bit*) (*qubit*) Greenberger–Horne–Zeilinger's (GHZ) state QFI is ejected, when subjected to *decoherence* channels: bit-phase flip and generalized amplitude damping, which were experimentally induced.[35] They determined the evolution under the channels, deduced the eigenvalues and derived QFI. They found that when no interaction with the environment exists, Heisenberg's limit was achieved via rotations along z-direction. They showed that N-qubit-GHZ-state maximal mean QFI (F_{max}) dwindled as decoherence (d) rose, because of information low from the system to the environment, until $d = 0.5$, then revived to form a symmetry around $d = 0.5$, which revival was as a consequence of environment memory eject, which led to information back-ow from the environment to the system. The $d > 0.5$ leads to a situation where more noise yields more efficiency. They showed that at finite temperature, QFIs decayed more rapidly than at infinite one. They revealed that QFI could be enhanced by adjusting environmental temperature.

14.10 HIGH-FIDELITY SPIN/OPTICAL CONTROL OF SINGLE SI-VACANCY CENTERS IN SIC

Scalable quantum networking requires quantum systems with quantum processing capabilities, in which regard solid-state spin systems with reliable spin–optical interfaces result a leading hardware. Notwithstanding, available systems suffer from large electron–phonon interaction or fast spin dephasing. Kaiser group showed that the anionic Si vacancy in SiC results immune to both.[36] Thanks to its $^4A_2$ symmetry in GS and excited states, optical resonances result stable with near-Fourier-transform-limited line-widths, allowing optical-transition spin-selectivity exploitation. In combination with millisecond-long spin *coherence* times originating from the high-purity crystal, they showed Hi-Fi optical initialization and coherent spin control, which they exploited to show 1-kHz-resolution coherent coupling to single nuclear spins. Their findings made the defect a prime candidate for realizing memory-assisted quantum net applications, via semiconductor-based spin-to-photon interfaces and coherently coupled nuclear spins.

14.11 PROPOSING THE USE OF MAGNETIC MOLECULES FOR QUANTUM COMPUTING

Spins in solids or molecules present discrete energy levels, and the associated quantum states were tuned and coherently manipulated via external electromagnetic fields.[37] Spins provide one of the simplest platforms to encode a qubit, future-quantum computer elementary unit. Performing any useful computation demands much more than realizing a robust qubit (one needs a large number of qubits and a reliable manner with which to integrate them into a complex circuitry, which stores and processes information, and implement quantum algorithms), which *scalability* is arguably one of the challenges for which a chemistry-based *bottom-up* approach is best suited. Molecules, being much more versatile than atoms and yet microscopic, are the quantum objects with the highest capacity to form nontrivial ordered states at the nanoscale, and be replicated in large numbers via chemical tools.

14.12 FINAL REMARKS

From the present results and discussion, the following final remarks can be drawn:

(1) Increasingly useful in materials research and development, molecular modeling is a method that combines computational chemistry techniques with molecular infography, for simulating and predicting the structure, chemical processes, and properties of materials.
(2) Molecular modeling techniques, in materials science, explore the impact of using molecular modeling for a number of simulations in industrial settings. They provide an overview of commonly used methods in atomic simulation of a broad range of materials, for example, superconductors, and so on.
(3) Molecular modeling techniques in materials science present the background and tools, for chemists and physicists, to perform in-silico experiments to understand relationships between the properties of materials and the underlying atomic structure, which insights result in more accurate data for designing application-specific materials

that withstand real process conditions, for example, hot temperatures and high pressures.

(4) Even at the molecular scale, different theoretical approaches exist, corresponding to the level of detail at which the electrons in the system must be described. It is beyond the scope of this work to go into the detail of every theory. Notwithstanding, the study attempted to include important ideas so that the reader could gauge the approximations and errors implicit in a given simulation method.

ACKNOWLEDGMENTS

The authors thank the support from Fundacion Universidad Catolica de Valencia San VicenteMartir (Project No. 2019-217-001UCV).

KEYWORDS

- **structure**
- **oxygen vacancy**
- **oxygen diffusion**
- **time-reversal symmetry breaking**
- **loop Josephson-current order**
- **magnetochemistry**
- **magnetic-susceptibility measurement**

REFERENCES

1. Coronado, E.; Delhaès, P.; Gatteschi, D.; Miller, J., Eds. *Molecular Magnetism: From Molecular Assemblies to the Devices*; Springer: Heidelberg, Germany, 1996.
2. Torrens, F.; Castellano, G. Reflections on the Nature of the Periodic Table of the Elements: Implications in Chemical Education. In *Synthetic Organic Chemistry*; Seijas, J. A., Vázquez Tato, M. P., Lin, S. K., Eds.; MDPI: Basel, Switherland, 2015; Vol 18, pp 1–15.
3. Torrens, F.; Castellano, G. Nanoscience: From a Two-Dimensional to a Three-Dimensional Periodic Table of the Elements. In *Methodologies and Applications for Analytical and Physical Chemistry*; Haghi, A. K., Thomas, S., Palit, S., Main, P., Eds.; Apple Academic–CRC: Waretown, NJ, 2018; pp 3–26.

4. Torrens, F.; Castellano, G. Periodic Table. In *New Frontiers in Nanochemistry: Concepts, Theories, and Trends*; Putz, M. V., Ed.; Apple Academic–CRC: Waretown, NJ, in press.
5. Torrens, F.; Castellano, G. Ideas in the History of Nano/Miniaturization and (Quantum) Simulators: Feynman, Education and Research Reorientation in Translational Science. In *Synthetic Organic Chemistry*; Seijas, J. A., Vázquez Tato, M. P., Lin, S. K., Eds.; MDPI: Basel, Switzerland, 2015; Vol 19, pp 1–16.
6. Torrens, F.; Castellano, G. Reflections on the Cultural History of Nanominiaturization and Quantum Simulators (Computers). In *Sensors and Molecular Recognition*; Laguarda Miró, N., Masot Peris, R., Brun Sánchez, E., Eds.; Universidad Politécnica de Valencia: València, Spain, 2015; Vol 9, pp 1–7.
7. Torrens, F.; Castellano, G. Nanominiaturization and Quantum Computing. In *Sensors and Molecular Recognition*; Costero Nieto, A. M., Parra Álvarez, M., Gaviña Costero, P., Gil Grau, S., Eds.; Universitat de València: València, Spain, 2016; Vol 10, pp 31-1–5.
8. Torrens, F.; Castellano, G. Nanominiaturization, Classical/Quantum Computers/ Simulators, Superconductivity, and Universe. In *Methodologies and Applications for Analytical and Physical Chemistry*; Haghi, A. K., Thomas, S., Palit, S., Main, P., Eds.; Apple Academic–CRC: Waretown, NJ, 2018; pp 27–44.
9. Torrens, F.; Castellano, G. Superconductors, Superconductivity, BCS Theory and Entangled Photons for Quantum Computing. In *Physical Chemistry for Engineering and Applied Sciences: Theoretical and Methodological Implication*; Haghi, A. K., Aguilar, C. N., Thomas, S., Praveen, K. M., Eds.; Apple Academic–CRC: Waretown, NJ, 2018; pp 379–387.
10. Torrens, F.; Castellano, G. EPR Paradox, Quantum Decoherence, Qubits, Goals and Opportunities in Quantum Simulation. In *Theoretical Models and Experimental Approaches in Physical Chemistry: Research Methodology and Practical Methods*; Haghi, A. K., Ed.; Apple Academic–CRC: Waretown, NJ, 2018; Vol 5, pp 317–334.
11. Torrens, F.; Castellano, G. Nanomaterials, Molecular Ion Magnets, Ultrastrong and Spin–Orbit Couplings in Quantum Materials. In *Physical Chemistry for Chemists and Chemical Engineers: Multidisciplinary Research Perspectives*; Vakhrushev, A. V., Haghi, R., de Julián-Ortiz, J. V., Allahyari, E., Eds.; Apple Academic–CRC: Waretown, NJ, in press.
12. Torrens, F.; Castellano, G. *Nanodevices and Organization of Single Ion Magnets and Spin Qubits*. In *Chemical Science and Engineering Technology: Perspectives on Interdisciplinary Research*; Balköse, D., Ribeiro, A. C. F., Haghi, A. K., Ameta, S. C., Chakraborty, T., Eds.; Apple Academic–CRC: Waretown, NJ, in press.
13. Torrens, F.; Castellano, G. Superconductivity and Quantum Computing via Magnetic Molecules. In *New Insights in Chemical Engineering and Computational Chemistry*; Haghi, A. K., Ed.; Apple Academic–CRC: Waretown, NJ, in press.
14. Torrens, F.; Castellano, G. Developing Sustainability via Nanosystems and Devices: Science-Ethics. In *Chemical Science and Engineering Technology: Perspectives on Interdisciplinary Research*; Balköse, D., Ribeiro, A. C. F., Haghi, A. K., Ameta, S. C., Chakraborty, T., Eds.; Apple Academic–CRC: Waretown, NJ, in press.
15. Torrens, F.; Castellano, G. Green Nanotechnology: An Approach towards Environment Safety. In *Advances in Nanotechnology and the Environmental Sciences: Applications, Innovations, and Visions for the Future*; Vakhrushev, A. V.; Ameta, S. C.; Susanto, H., Haghi, A. K., Eds.; Apple Academic–CRC: Waretown, NJ, in press.

16. Torrens, F.; Castellano, G. Molecular Devices/Machines: Hybrid Organic-inorganic Structures. In *Research Methods and Applications in Chemical and Biological Engineering*; Pourhashemi, A., Deka, S. C., Haghi, A. K., Eds.; Apple Academic–CRC: Waretown, NJ, in press.
17. Torrens, F.; Castellano, G. The Periodic Table, Quantum Biting its Tail, and Sustainable Chemistry. In *Chemical Nanoscience and Nanotechnology: New Materials and Modern Techniques*; Torrens, F., Haghi, A. K., Chakraborty, T., Eds.; Apple Academic–CRC: Waretown, NJ, in press.
18. Torrens, F.; Castellano, G. Quantum Molecular Spintronics, Nanoscience and Graphenes. In *Molecular Physical Chemistry*; Haghi, A. K., Ed.; Apple Academic–CRC: Waretown, NJ, in press.
19. Torrens, F.; Castellano, G. Cancer and Hypotheses on Cancer. In *Molecular Chemistry and Biomolecular Engineering: Integrating Theory and Research with Practice*; Pogliani, L., Torrens, F., Haghi, A. K., Eds.; Apple Academic–CRC: Waretown, NJ, in press.
20. Torrens, F.; Castellano, G. Precision Personalized Medicine from Theory to Practice: Cancer. In *Molecular Physical Chemistry*; Haghi, A. K., Ed.; Apple Academic–CRC: Waretown, NJ, in press.
21. Torrens, F.; Castellano, G. AIDS Destroys Immune Defences: Hypothesis. *New Front. Chem.* **2014,** *23*, 11–20.
22. Torrens-Zaragozá, F.; Castellano-Estornell, G. Emergence, spread and uncontrolled Ebola outbreak. *Basic Clin. Pharmacol. Toxicol.* **2015,** *117* (Suppl. 2), 38.
23. Torrens, F.; Castellano, G. 2014 Spread/Uncontrolled Ebola Outbreak. *New Front. Chem.* **2015,** *24*, 81–91.
24. Torrens, F.; Castellano, G. Ebola Virus Disease: Questions, Ideas, Hypotheses and Models. *Pharmaceuticals* **2016,** *9*, 14-6.
25. Torrens, F.; Castellano, G. Metaphors That Made History: Reflections on Philosophy/Science/DNA. In *Molecular Physical Chemistry*; Haghi, A. K., Ed.; Apple Academic–CRC: Waretown, NJ, in press.
26. Torrens, F.; Castellano, G. Scientific Integrity Ethics: Science Communication and Psychology. In *Molecular Physical Chemistry*; Haghi, A. K., Ed.; Apple Academic–CRC: Waretown, NJ, in press.
27. Torrens, F.; Castellano, G. Theory and Simulation: The Present and Future of Quantum Technologies. In *Physical Biochemistry, Biophysics, and Molecular Chemistry: Applied Research and Interactions*; Torrens, F., Mahapatra, D. K., Haghi, A. K., Eds.; Apple Academic–CRC: Waretown, NJ, in press.
28. Hill, J. R.; Subramanian, L.; Maiti, A. *Molecular Modeling Techniques in Material Sciences*; CRC: Boca Raton, FL, 2005.
29. Onnes, H. K. Further Experiments with Liquid Helium. G. On the Electrical Resistance of Pure Metals, etc. VI. On the Sudden Change in the Rate at which the Resistance of Mercury Disappears. In *Through Measurement to Knowledge*; Gavroglu, K., Goudaroulis, Y., Eds.; *Boston Studies in the Philosophy of Science No. 124*; Springer: Dordrecht, The Netherlands, 1991; pp 267–272.
30. Bednorz, J. G.; Müller, A. K. Possible High T_c Superconductivity in the Ba–La–Cu–O System. *Z. Phys. B* **1986,** *64, 189–193.*

31. Wu, M. K.; Ashburn, J. R.; Torng, C. J.; Hor, P. H.; Meng, R. L.; Gao, L.; Huang, Z. J.; Wang, Y. Q.; Chu, C. W. Superconductivity at 93-K in a New Mixed-Phase Y–Ba–Cu–O Compound System at Ambient Pressure. *Phys. Rev. Lett.* **1987,** *58, 908–910.*
32. Ghosh, S. K.; Annett, J. F.; Quintanilla, J. Time-Reversal Symmetry Breaking in Superconductors through Loop Josephson-Current Order. *Int. J. Phys. Stud. Res.* **2018,** *1* (2), 33.
33. Selwood, P. W. *Magnetochemistry*; Interscience: New York, NY, 1943.
34. Bilya, M. A. Analogy between the London Equations of Superconductivity and Meissner Effect. *Int. J. Phys. Stud. Res.* **2018,** *1* (2), 49–49.
35. Falaye, B. J.; Adepoju, A. G.; Aliyu, A. S.; Melchor, M. M.; Liman, M. S.; Gonzalez Ramrez, M. D.; Oyewumi, K. J. Investigating Quantum Metrology in Noisy Channels. *Int. J. Phys. Stud. Res.* **2018,** *1* (2), 47–47.
36. Nagy, R.; Niethammer, M.; Widmann, M.; Chen, Y. C.; Udvarhelyi, P.; Bonato, C.; Hassan, J. U.; Karhu, R.; Ivanov, I. G.; Son, N. T.; Maze, J. R.; Ohshima, T.; Soykal, Ö. O.; Gali, Á.; Lee, S. Y.; Kaiser, F.; Wrachtrup, J. High-Fidelity Spin and Optical Control of Single Silicon-Vacancy Centres in Silicon Carbide. *Nat. Commun.* **2019,** *10*, 1954-1–8.
37. Gaita-Ariño, A.; Luis, F.; Hill, S.; Coronado, E. Molecular Spins for Quantum Computation. *Nat. Chem.* **2019,** *11*, 301–309.

CHAPTER 15

DRUG-DELIVERY SYSTEMS: STUDY OF QUATERNARY SYSTEMS METHYLXANTHINE + CYCLODEXTRIN + KCL + WATER AT (298.15 AND 310.15) K

CECÍLIA I. A. V. SANTOS[1,2*], ANA C. F. RIBEIRO[1], and MIGUEL A. ESTESO[2]

[1]*Department of Chemistry, Coimbra University Centre, University of Coimbra, 3004-535 Coimbra, Portugal*

[2]*U.D. Química Física, Universidad de Alcalá, 28871 Alcalá de Henares, Madrid, Spain*

Corresponding author. E-mail: ceciliasantos@qui.uc.pt

ABSTRACT

This chapter focus on the study of the effects of biologically relevant ions (KCl) on the stability of controlled release systems formed between a host cyclodextrin (β-CD and HP-β-CD) and a guest drug molecule (methylxanthine) aiming to provide accurate information for ulterior drug formulations.

Taylor dispersion technique was used to measure diffusion coefficients for both aqueous ternary mixtures of methylxantine + KCl + water and quaternary mixtures of CD+ methylxantine + KCl + water at temperatures of 298.15 and 310.15 K.

The changes on the behavior diffusion of these multicomponent systems and the coupled flows occurring in the solution, caused by the introduction of ions, were analyzed, and the new association constants for these solutes were estimated.

An analysis of the molecular interactions is done, explaining the new associations and competition for the CD cavity between the drug molecule and the ions present, leading us to a better understanding of the structure of these systems in aqueous solution and in conditions closer to the in-vivo environment.

15.1 INTRODUCTION

The modification of the physicochemical properties of a drug molecule achieved through its complexation with a cyclodextrin molecule may have considerable pharmaceutical potential. The stability constant and the stoichiometry of the inclusion complexes are useful indexes for estimating the stability of the complex and the changes in the physicochemical properties of the guest molecule. However, to maximize effectiveness, these are not the only factors to consider when outlining formulations and choosing ways of administration. It is necessary to take into account the biological environment around the in-vivo complexes: dilution, temperature, pH, and the presence of ions that comprise the biological fluids. As a result, the influence of the presence of ions with physiological importance on the stability of the cyclodextrin–drug interactions has to be investigated. Ions present in biological fluids, at both intra and extracellular level, can have significant effects on drug molecules, especially if they can interact with them and consequently influence their interactions with the cyclodextrin carrier. In general, the salts may influence the cyclodextrin complexes in various ways[1–3]: they may have *salting in* or *salting out* effects that affect the structure of the solution and deflect any balances that may occur. They may lead to the formation of ternary complexes in which the ions stabilize the cyclodextrin–drug-inclusion complex through interactions with both molecules. Or they may generate competition with the drug to enter the cyclodextrin cavity, although this situation is more common only for large anions. In addition, the cations may interact with charged drug molecules through ion-pair interactions, and this effect may also play an important role in the mechanism of complex formation.

In a previous paper, we studied the stability of cyclodextrin–drug complexes in aqueous solutions to be used as controlled drug-delivery systems[4] in living organisms, as a function of both concentration and temperature. In the present one, and as a continuation of the previous one, we are interested in analyzing the effect of different physiological ions on these complexes, to take information about the influence of media with similar composition to those of the living beings, to be used in pharmaceutical applications.

Specifically, the following study is focused on the effects of the presence of potassium chloride on the stability of the interactions between a cyclodextrin (β-CD and HP-β-CD) and a methylxanthine (caffeine and theophylline) previously studied.[4]

15.2 MATHEMATICAL ASPECTS

15.2.1 MOLAR VOLUME

In a solvation process, the entrance of a solute molecule in a solvent involves the breaking of intermolecular bonds in the latter. After the opening of a cavity in the solvent capable of receiving the solute molecule, interactions between the solute and the solvent are established, the magnitude of which will depend on the hydrophobic or hydrophilic nature of the solute and reflect in a greater or lesser structure of the medium. The characterization of these interactions in aqueous solution, at a molecular level, can be made through the apparent molar volume, which can be determined from density experimental values, since it is defined by the equation:

$$\varphi_V = \frac{V - V_{H_2O}}{m} = \frac{M}{\rho} + \frac{1000}{m}\left(\frac{1}{\rho} - \frac{1}{\rho_{H_2O}}\right) \tag{15.1}$$

being ρ and ρ_{H_2O} the density of the solution and the pure water, respectively; M, the molar mass of the solute; V and V_{H_2O}, the volume (in cm^3) of the solution and the pure water, respectively; and m, the molality of the solution.

From the analysis of this thermodynamic property, φ_V, it is possible to attain information, through the estimation of values for the association constant, K, on the interactions between the solutes in solution, between a solute and the solvent as well as to understand how the interactions of a solute with the solvent change when the solvent is changed.[5]

15.2.2 VISCOSITY

For solutions in laminar flow regime, values for this kinetic property, η, can be determined, by using the Hagen–Poiseuille equation[6–8]:

$$\eta = \frac{\pi \Delta p r^4 t}{8LV} \tag{15.2}$$

where t is the efflux time; V, the volume of solution flowing throughout a capillary tube of radius r and length L, and Δp, the pressure difference between the ends of the capillary tube. If the flow of the solution occurs under the action of gravity, this pressure difference will be equal to:

$$\Delta p = \rho g h \tag{15.3}$$

being ρ the density of the solution; h, the difference between the hydrostatic heights of the two ends of the capillary tube; and g the gravity constant

By applying equations (15.2) and (15.3) to both the solution under study and another of reference and afterward comparing them, the relative viscosity value, η_r, is finally obtained:

$$\eta_r = \frac{\eta}{\eta_0} = \frac{\rho t}{\rho_0 t_0} \tag{15.4}$$

where the subscript zero relates the reference solution.

The analysis of these values found for the relative viscosity can be done by using the Jones–Dole equation[9]:

$$\eta_r = 1 + Ac^{1/2} + Bc + Dc^2 \tag{15.5}$$

in which A, B, and C are parameters temperature dependent, related to solute–solute–solvent interactions taking place in the solution. From the analysis of the viscosity results found as well as their changes when the medium change, it is possible to derive information about the appearance of associated species in the solution.[10]

15.2.3 MUTUAL ISOTHERMAL DIFFUSION

In the case of a quaternary system (solvent + 3 solutes [CD + drug + electrolyte]), diffusion is described by the diffusion equations[11–13]:

$$-J_1 = {}^{123}(D_{11})_V \frac{\partial c_1}{\partial x} + {}^{123}(D_{12})_V \frac{\partial c_2}{\partial x} + {}^{123}(D_{13})_V \frac{\partial c_3}{\partial x} \tag{15.6}$$

$$-J_2 = {}^{123}(D_{21})_V \frac{\partial c_1}{\partial x} + {}^{123}(D_{22})_V \frac{\partial c_2}{\partial x} + {}^{123}(D_{23})_V \frac{\partial c_3}{\partial x} \tag{15.7}$$

$$-J_3 = {}^{123}(D_{31})_V \frac{\partial c_1}{\partial x} + {}^{123}(D_{32})_V \frac{\partial c_2}{\partial x} + {}^{123}(D_{33})_V \frac{\partial c_3}{\partial x} \tag{15.8}$$

where J_i and $\partial c_i/\partial x$ are the molar flux and the gradient in concentration related to the solute *i*, respectively, D_{ii} is the main diffusion coefficient which represents the flux of the solute *i* as a consequence of its own concentration gradient, and D_{ik} are the cross-coefficients ($i \neq k$) which represent the corresponding cofluxes of solute *i* produced by the concentration gradient of the others solutes *k*.[14]

If different intermolecular interactions take place in the solution between these solutes it may result in the appearance of new associated species in the system. Leaist et al.[15,16] have proposed a model, based on the coupled transport of solutes at tracer concentrations, which allows the estimation of the association constant *K* and the diffusion coefficient of the associated species for the particular case of a strong electrolyte. Assuming an aqueous system where two solutes A (1) and MX (2) coexist, and where A interacts with one of the ions of the other solute in aqueous solution, for example, M^+, the equilibrium would be described by

$$A + M^+ \overset{K}{\leftrightarrow} A \cdot M^+ \tag{15.9}$$

and there would be 4 species in solution: A, A–M^+, M^+, and X^-. The mass balance equations would be

$$c_1 = [A] + [AM^+] \quad \text{and} \quad c_2 = [M^+] + [AM^+] \tag{15.10}$$

and the associated fractions of A (X_1) and M^+ (X_2) could be written as

$$X_1 = \frac{[AM^+]}{[A] + [AM^+]} \quad \text{and} \quad X_2 = \frac{[AM^+]}{[M^+] + [AM^+]} \tag{15.11}$$

The diffusion of the associated species AM^+ and its association constant can be estimated using equations (15.12) and (15.13), and the experimental data obtained through the measurements of the ternary diffusion coefficients of tracer concentrations of A in aqueous solution of MX, D_{11}^{0} $(c_1/c_2 = 0)$, and measurements of the ternary diffusion coefficients of tracer concentrations of MX in aqueous solution of A, D_{22}^{0} $(c_1/c_2 = 0)$.

$$D_{11}^{0}\left(\frac{c_1}{c_2}=0\right)=X_1 D_{AM}^{0}+\left(1-X_1\right)D_{A}^{0} \tag{15.12}$$

$$D_{22}^{0}\left(\frac{c_1}{c_2}=0\right)=2D_{X^-}\frac{X_2 D_{AM}^{0}+\left(1-X_2\right)D_{M^+}^{0}}{D_{X^-}+X_2 D_{AM}^{0}+\left(1-X_2\right)D_{M^+}^{0}} \tag{15.13}$$

where D_{A}^{0} and D_{AM}^{0} represent the limiting diffusion coefficients of the molecules of A and the complex A–M^+, respectively. The value of D_{A}^{0} is considered to be equal to that of the diffusion coefficient of A in aqueous solution, at infinitesimal concentration. Thus, under conditions of tracer concentrations of solute, $X_1 = X_2$, and the association constant can be defined as

$$K=\frac{X_1}{c_1\left(1-X_1\right)}=\frac{X_2}{c_2\left(1-X_2\right)} \tag{15.14}$$

Several considerations are involved in this model to allow estimation of the association constant of the solutes. Although the contribution of the solution viscosity is taken into account in the calculation, the contribution of the dielectric constant and the changes in the hydration of the solutes[17] are neglected. The model also assumes that, in the concentration range studied, the coefficient of activity of the solutions is equal to unity and that the mobilities of free and associated species do not change with the concentration.

15.3 EXPERIMENTAL SECTION

15.3.1 MATERIALS

Table 15.1 summarized the chemicals used with indication of their characteristics. All of them were used as received, without any further purification.

They were only kept in a desiccator over silica gel. The work solutions were prepared by direct weighing both the solutes and the solvent (Milli-*Q* water; $\kappa = 5.6 \times 10^{-8}$ S cm^{-1}), and their concentration expressed as molality (*m*, mol kg^{-1}; after taking into account the water content of the CD) with an uncertainty less than 0.07%.

TABLE 15.1 Sample Description.

Chemical name	Source	Mass fraction purity (%)[a]
Caffeine (*pro analysi*)	Sigma Aldrich	>98.5
Theophylline (*anhydrous*)	Sigma Aldrich	>99.0
β-cyclodextrin (β-CD) (13.1% water mass fraction)[a]	Sigma Aldrich	98
2-hydroxypropyl-β-cyclodextrin (HP-β-CD) (18.2% water mass fraction)[a]	Sigma Aldrich	100
KCl (*pro analysi*)	Sigma Aldrich	>99.5
H_2O	Millipore-*Q* water ($\kappa = 5.6 \times 10^{-8}$ S cm^{-1} at 25.0°C)	–

[a]As stated by the supplier.

15.3.2 DENSITY MEASUREMENTS

Densities were measured by using an Anton Paar DMA5000M densimeter. This is provided with a Peltier system that permits a temperature control better than ±0.005°C inside the measurement U-tube. The value accepted for the density of a given system corresponds to the mean one of at least four independent measurements with a reproducibility of ±0.001 kg m^3 and an uncertainty of 0.150 kg m^3.

15.3.3 VISCOSITY MEASUREMENTS

Viscosities were carried out with an Ostwald-type capillary viscometer. Temperature was maintained stable, with a variation less than ±0.02°C, with the help of a water thermostat-bath. For a given solution, the efflux time accepted was the mean value of at least four independent measurements obtained by using a digital stopwatch with a resolution of 0.2 s. No kinetic energy correction (Hagenbach correction) was applied to these time measurements due to the fact that they were always greater than 300 s.

15.3.4 DIFFUSION MEASUREMENTS

Diffusion coefficients were measured by using the Taylor dispersion method.[17–22] Briefly, this method consists of determining the speed with which a small sample of solution to be studied moves into a suitable carrier solution. To do this, the sample is injected into the initial part of a long capillary dispersion tube, filled with the carrier solution, detecting the passage of the aforementioned sample, at a point quite far from the point of injection, by using a differential refractometer. The difference between the refraction index values of both solutions, the carrier in pure and with the sample of solution to be studied, is transformed into an electrical signal whose analysis allows determining the values of the mutual diffusion coefficients (D_{ik}) of all the different components present in the solution.[4]

15.4 RESULTS AND DISCUSSION

15.4.1 STABILITY OF CYCLODEXTRIN–DRUG COMPLEXES: INFLUENCE OF RELEVANT IONS IN PHYSIOLOGICAL PROCESSES

Before examining the results obtained for aqueous solutions of mixtures of caffeine and theophylline in the presence of cyclodextrins and KCl, an analysis of the influence of the potassium chloride on the drugs and the cyclodextrins, each individually in aqueous solution, is presented.

15.4.1.1 EFFECT OF POTASSIUM CHLORIDE ON THE PROPERTIES OF AQUEOUS CAFFEINE AND THEOPHYLLINE

Literature provides several studies on the interaction and effects of physiologically relevant ions, such as K^+, Na^+, and Cl^-, on caffeine and theophylline.[23,24] These studies show that there is an association between the K^+ ion and the O6 and N9 positions of caffeine and the O6, N9, and N7 positions of theophylline, resulting in equilibrium constants of 5.08×10^3 for $K_{K\text{-caffeine}}$ and 2.13×10^3 for $K_{K\text{-theophylline}}$. These results are especially important if we take into account the values of equilibrium constants related to the association of caffeine and theophylline with cyclodextrins, given the difference in their order of magnitude. However, it should be noted that the studies that led to these association constants used FT-IR techniques and UV spectroscopy and carried out at considerably higher concentrations (up to 40 mM) than intracellular potassium (3.5–5.0 mM) and in the presence of other ions, introduced by pH adjustment, which may influence the association.

Relying on the measurements of the diffusion coefficients of caffeine and theophylline in the presence of potassium chloride, the association constant of these solutes can be estimated using the model developed by Leaist et al.[15,16,18] which relate the ternary diffusion coefficients of the solutes to the tracer diffusion coefficients and the concentrations of the diffusing species in equilibrium.

Regarding the KCl (1) + caffeine (2) system[13] (Table 15.2), it is verified that both the main diffusion coefficients D_{11} for potassium chloride and D_{22} for caffeine show, at both 298.15 and 310.15 K, lower values than the corresponding binary diffusion, with deviations of about 4–5% in magnitude. When in aqueous solution, the caffeine molecules are associated mainly by hydrophobic interactions, since the absence of hydrogen donor groups prevents the establishment of intramolecular bonding. However, these hydrogen bonds can be established between the caffeine and the solvent (water) through the polar groups, carbonyl groups of positions 2 and 6 and the nitrogen in position 9. According to Falk et al.,[25] if caffeine molecules are found to be free in solution, the carbonyls at positions 2 and 6 are fully hydrated, but when vertical stacking occurs the steric effect makes their hydration difficult, despite the relatively flexible structure of the dimer. Thus, the access to these positions, for both water molecules and potassium ions, is conditioned. It would be expected that the decrease in the caffeine diffusion coefficient due to its increasing concentration would occur due to caffeine self-association. In fact, the data obtained demonstrated that this may not be what actually happens. Indeed, the changes observed in the caffeine diffusion coefficients, within the concentration range studied, may be due to its association with the K^+ ion from the medium, decreasing its mobility, due to the stereo effect, on the one hand, once the size of the associated species is slightly higher than that of free caffeine and, on the other hand, as a result of its positive surface charge density, since the KCl concentration gradient generates an electric field that decelerates the free K^+ ions in solution and those associated with the caffeine. Observing the behavior of the secondary diffusion coefficient D_{21}, which represents the flow of caffeine generated by the concentration gradient of potassium chloride in solution, it can be seen

TABLE 15.2 Ternary Mutual Diffusion Coefficients D_{11}, D_{12}, D_{21}, and D_{22} for Aqueous KCl (1) + Caffeine (2) Solutions and Respective Standard Deviations, S_D, at 298.15 and 310.15 K.

c_1[a]	c_2[a]	$D_{11} \pm S_D$ (10^{-9} m² s⁻¹)	$D_{12} \pm S_D$ (10^{-9} m² s⁻¹)	$D_{21} \pm S_D$ (10^{-9} m² s⁻¹)	$D_{22} \pm S_D$ (10^{-9} m² s⁻¹)	D_{12}/D_{22}[b]	D_{21}/D_{11}[c]
			T = 298.15 K				
0.010	0.000	1.871 ± 0.015	−0.112 ± 0.058	−0.007 ± 0.028	0.759 ± 0.017	−0.148	−0.004
0.000	0.010	1.859 ± 0.019	0.000 ± 0.007	−0.076 ± 0.022	0.703 ± 0.013	−0.001	−0.041
0.002	0.002	1.951 ± 0.031	−0.165 ± 0.081	−0.056 ± 0.025	0.746 ± 0.014	−0.221	0.029
0.005	0.005	1.903 ± 0.028	−0.120 ± 0.042	−0.112 ± 0.019	0.731 ± 0.012	−0.164	0.059
0.010	0.010	1.844 ± 0.015	0.104 ± 0.054	−0.282 ± 0.096	0.688 ± 0.031	0.151	−0.153
			T = 310.15 K				
0.010	0.000	2.420 ± 0.001	0.231 ± 0.026	−0.004 ± 0.003	1.029 ± 0.022	0.224	−0.002
0.000	0.010	2.386 ± 0.003	−0.004 ± 0.031	−0.019 ± 0.006	0.984 ± 0.015	−0.004	0.008
0.002	0.002	2.420 ± 0.022	−0.099 ± 0.033	0.007 ± 0.034	1.024 ± 0.013	−0.097	0.003
0.005	0.005	2.390 ± 0.013	0.005 ± 0.018	−0.091 ± 0.019	0.956 ± 0.008	0.005	−0.038
0.010	0.010	2.370 ± 0.009	0.064 ± 0.005	−0.257 ± 0.021	0.898 ± 0.003	0.072	−0.109

[a] c_1 and c_2 in units (mol dm^{-3}); [b] D_{12}/D_{22} represents the number of moles of KCl carried by 1 mol of caffeine; [c] D_{21}/D_{11} represents the number of moles of caffeine carried by 1 mol of KCl.

that values are negative. From the D_{21}/D_{11} ratio, which represents the number of moles of caffeine transported per mole of KCl, it can be perceived that a mole of diffusing KCl can countertransport up to 0.15 mol of caffeine. At 310.15 K, the effect of temperature increase disfavors the interaction between the molecules of caffeine, and this is reflected in its diffusion coefficient, whose increase with the concentration is smaller, and additionally the coupled transport of solutes also reduces. However, the fact that there are more caffeine molecules in the monomeric form could allow a greater interaction with the medium (solvent) and consequently with the K^+ ions. To test this hypothesis, the association constant was estimated. For the KCl (1) + caffeine (2) system, at 298.15 K a stability constant K of 26.8 M^{-1} was obtained together with a limiting diffusion coefficient for the associated species Caf-K^+, $D_{complex}$, of 0.755 × 10^{-9} m^2 s^{-1}. At 310.15 K, the stability constant K has a value of 31.59 M^{-1} and the limiting diffusion coefficient of the associated species Caf-K^+, $D_{complex}$, is 0.860 × 10^{-9} m^2 s^{-1}. It should be noted that at 310.15 K, the limiting diffusion coefficient of the associated species is 20% lower than that of free caffeine, under the same conditions. These results are considerably different from those found in the literature and this difference may be due in large part to the different experimental conditions under which the studies were conducted. Despite the differences, and taking into account all the considerations involved in the model, both results indicate that there are complexes in solution and that these have an important role in the diffusion behavior of the system.

The hydrodynamic radius of the caffeine molecule in aqueous solution of potassium chloride presented values of r (298.15 K) = 0.323 nm and r (310.15 K) = 0.319 nm, showing a slight increase in the size of caffeine hydrodynamic radius when in aqueous KCl solution relative to water (less than 1%), at 298.15 K, and a significant increase (about 5%) at 310.15 K, results that support the earlier evidence that caffeine is associated with the ions present in the medium. In resume, there are interactions between the media K^+ ion and caffeine and they increase with increasing temperature.

In the case of the KCl (1) + theophylline (2) system[26] (Table 15.3), the comparison of the results obtained with the respective binary systems, at the same temperature shows that, at the compositions studied, the added KCl produces small changes in the diffusion coefficient D_{22} of theophylline (less than 1% at both temperatures), while the addition of theophylline generates significant changes in the diffusion coefficient, D_{11} for KCl ($\leq$12% at 298.15 K and <4% at 310.15 K). A possible explanation for the occurrence of these variations would be related to the association of some

TABLE 15.3 Ternary Mutual Diffusion Coefficients D_{11}, D_{12}, D_{21}, and D_{22} for Aqueous KCl (1) + Theophylline (2) Solutions and Respective Standard Deviations, S_D, at 298.15 and 310.15 K.

c_1[a]	c_2[a]	$D_{11} \pm S_D$ (10^{-9} $m^2 \cdot s^{-1}$)	$D_{12} \pm S_D$ (10^{-9} $m^2 \cdot s^{-1}$)	$D_{21} \pm S_D$ (10^{-9} $m^2 \cdot s^{-1}$)	$D_{22} \pm S_D$ (10^{-9} $m^2 \cdot s^{-1}$)	D_{12}/D_{22}[b]	D_{21}/D_{11}[c]
				T = 298.15 K			
0.010	0.000	1.793 ± 0.009	−0.026 ± 0.017	−0.009 ± 0.008	0.800 ± 0.012	−0.033	−0.005
0.000	0.010	1.856 ± 0.011	0.008 ± 0.006	−0.022 ± 0.008	0.687 ± 0.007	0.012	−0.012
0.002	0.002	1.847 ± 0.040	−0.024 ± 0.020	−0.012 ± 0.012	0.783 ± 0.023	−0.030	0.006
0.005	0.005	1.842 ± 0.035	−0.313 ± 0.091	−0.108 ± 0.177	0.765 ± 0.020	−0.409	0.058
0.010	0.010	1.690 ± 0.056	−0.359 ± 0.180	−0.103 ± 0.020	0.736 ± 0.058	−0.488	0.061
				T = 310.15 K			
0.010	0.000	2.423 ± 0.017	0.098 ± 0.042	−0.004 ± 0.012	1.079 ± 0.003	0.095	−0.002
0.000	0.010	2.406 ± 0.018	−0.001 ± 0.021	−0.083 ± 0.012	0.983 ± 0.016	−0.001	0.034
0.002	0.002	2.459 ± 0.023	0.064 ± 0.021	−0.200 ± 0.058	1.105 ± 0.019	0.058	−0.081
0.005	0.005	2.413 ± 0.023	0.088 ± 0.022	−0.232 ± 0.100	1.026 ± 0.032	0.086	−0.096
0.010	0.010	2.378 ± 0.019	0.079 ± 0.031	−0.193 ± 0.009	0.976 ± 0.010	0.080	−0.081

[a] c_1 and c_2 in units (mol dm^{-3}); [b] D_{12}/D_{22} represents the number of moles of KCl carried by 1 mol of theophylline; [c] D_{21}/D_{11} represents the number of moles of theophylline carried by 1 mol of KCl.

theophylline molecules with the potassium ion in aqueous solution through the positions N7 and O6,[24,27] these aggregates in solution having less mobility and being, therefore, responsible for the relatively large decrease in D_{11}. This effect is less pronounced when considering the effect of KCl on theophylline transport, possibly due to the similarities between the free theophylline mobilities and the associated species. Through the analysis of the secondary diffusion coefficients, D_{12} and D_{21} at finite concentrations, it is possible to understand the influence of the presence of KCl in the diffusion of theophylline and vice versa. Considering that D_{12}/D_{22} gives the number of moles of KCl carried per mole of theophylline due to its concentration gradient it can be assumed that in the compositions used, 1 mol of theophylline carries, in the opposite direction to its concentration gradient, a maximum of 0.4 mol of KCl, increasing the transport with the increasing concentration of theophylline in solution. By the D_{21}/D_{11} ratio values, at the same composition range, it is expected that 1 mol of diffusing KCl will transport in the opposite direction to that of its concentration gradient a maximum of 0.1 mol of theophylline.

Theophylline has one less methyl group in the N7 position than caffeine and is therefore smaller, consequently having a higher density of charge when associated with the K^+ ion, so the effect of the electric field generated by the KCl gradient on the associated species (and on their respective diffusion coefficient) should be stronger. The fact that the theophylline molecule has one less methyl group also influences the association constant since the N7 position is now available to interact with the medium. From a rationale analogous to the one above for caffeine, we can expect that, like this one, the association constant Teof-K^+ is affected by the self-association of theophylline molecules, but since theophylline has a smaller self-association constant than caffeine, the effects on diffusion may be different.

Quantification of the interaction Teof-K^+ gave values of 0.758×10^{-9} m^2 s^{-1} and 27.2 M^{-1}, for the limiting diffusion coefficient of the associated species $D_{complex}$, and for the association constant K, at 298.15 K, respectively, and 0.778×10^{-9} m^2 s^{-1} and 25.9 M^{-1} for 310.15 K.[26] The estimation of the hydrodynamic radius for the theophylline molecule in aqueous potassium chloride solution gave r (298.15 K) = 0.307 nm and r (310.15 K) = 0.304 nm, showing a slight increase in the size of the theophylline hydrodynamic radius at both temperatures by about 1–2%, regarding the hydrodynamic radius in pure water, possibly resulting from association with the ions in the medium.

15.4.1.2 EFFECT OF POTASSIUM CHLORIDE ON THE PROPERTIES OF AQUEOUS β-CYCLODEXTRIN AND HYDROXYPROPYL-β-CYCLODEXTRIN

It is known that the cyclodextrins are capable of encapsulate in their cavity several organic molecules, but also transition–metal complexes and inorganic ions.[1,28] Most authors seem to agree that the association of cyclodextrins with anions (e.g., Cl^-) is stronger than association with cations (e.g., K^+),[29] which seems logical if one considers that, in aqueous solution, anions are thought to form hydrogen bonds Cl^-–H–O that do not occur in the case of a cation. Still, it has been proposed that the carbon inside the cyclodextrin cavity is positively polarized and that this microscopically positive environment favors the inclusion of anionic hosts and disadvantages the inclusion of cationic hosts.[30] In the following, it is sought to understand how the presence of ions (K^+ and Cl^-) interferes in the equilibrium constants for the formation of inclusion complexes in the aqueous solution, as well as to define the value of its association constant with cyclodextrin, if occurring.

Equations (15.9)–(15.14) can also be developed for the KCl–cyclodextrin system, assuming the equilibria:

$$\mathrm{KCl} + \mathrm{CD} \overset{K}{\leftrightarrow} \mathrm{CD} \cdot \mathrm{Cl}^-$$

It should be noted that the process is applicable to both the K^+ ion and Cl^- and that the association constant and the diffusion coefficient of the associated species can be obtained in both cases, although calculations presented are only for the case Cl^-.

The mutual diffusion of β-CD and HP-β-CD systems in aqueous solution in the presence of potassium chloride at 298.15 and 310.15 K were measured, together with the tracer diffusion coefficients for each solute in aqueous solution.

For the KCl (1) + β-CD (2) system[13,31,32] (Table 15.4), values for the diffusion coefficient of potassium chloride, D_{11}, were found to be lower than those obtained in binary systems, for the same concentrations, with deviations up to 3.5%, while values for D_{22} for β-CD, were slightly higher than those for aqueous β-CD, by no more than 2%. A possible explanation for this behavior may arise from the entrance of KCl into solution, which, when dissociating and forming its hydration shell, will remove water molecules from the medium and possibly also from the β-CD hydration layer, causing a reduction in the size of the hydrated cyclodextrin structure

TABLE 15.4 Ternary Mutual Diffusion Coefficients D_{11}, D_{12}, D_{21}, and D_{22} for Aqueous KCl (1) + β-CD (2) Solutions and Respective Standard Deviations, S_D, at 298.15 and 310.15 K.

C_1[a]	C_2[a]	$D_{11} \pm S_D$ (10^{-9} m^2 s^{-1})	$D_{12} \pm S_D$ (10^{-9} m^2 s^{-1})	$D_{21} \pm S_D$ (10^{-9} m^2 s^{-1})	$D_{22} \pm S_D$ (10^{-9} m^2 s^{-1})	D_{12}/D_{22}[b]	D_{21}/D_{11}[c]
				T = 298.15 K			
0.010	0.000	1.751 ± 0.046	0.006 ± 0.016	0.081 ± 0.045	0.325 ± 0.018	0.018	0.046
0.000	0.010	1.803 ± 0.019	0.089 ± 0.033	−0.001 ± 0.012	0.331 ± 0.027	0.268	0.000
0.002	0.002	1.924 ± 0.045	−0.242 ± 0.028	−0.068 ± 0.010	0.324 ± 0.036	−0.745	0.035
0.005	0.005	1.891 ± 0.044	−0.196 ± 0.059	−0.088 ± 0.037	0.317 ± 0.032	−0.618	0.045
0.010	0.010	1.848 ± 0.034	−0.057 ± 0.041	−0.119 ± 0.051	0.313 ± 0.012	−0.183	0.064
				T = 310.15 K			
0.010	0.000	2.360 ± 0.029	0.005 ± 0.012	−0.012 ± 0.012	0.429 ± 0.011	0.012	−0.005
0.000	0.010	2.420 ± 0.017	0.025 ± 0.009	0.003 ± 0.009	0.457 ± 0.013	0.055	0.001

[a]c_1 and c_2 in units (mol dm^{-3}); [b]D_{12}/D_{22} represents the number of moles of KCl carried by 1 mol of β-CD; [c]D_{21}/D_{11} represents the number of moles of β-CD carried by 1 mol of KCl.

in solution and, therefore, a slight increase in the mobility of the latter. On the other hand, part of the KCl can enter the cavity of the cyclodextrin, expelling the water molecules to the outside. In fact, the D_{12}/D_{22} ratio demonstrates the existence of coupled transport of KCl as a result of the β-CD gradient, presenting fairly high values (1 mol β-CD can counter transport up to 0.7 mol of KCl).

The estimation of the association constants through diffusion data, under tracer conditions of solute, allowed obtaining association constants of 42.1 M^{-1} at 298.15 K and 34.2 M^{-1} at 310.15 K,[32] that is, there is evidence that association occurs between these solutes, and this association is weakened by the increase of temperature. The limiting diffusion coefficients obtained by applying this model were 0.344×10^{-9} m^2 s^{-1} and 0.436×10^{-9} m^2 s^{-1}, values very close to those of the free β-CD[33] which show that the association should occur through inclusion. If we estimate the hydrodynamic radius of the β-CD molecule in aqueous solution in the presence of KCl, we find results of r (298.15 K) = 0.740 nm and r (310.15 K) = 0.719 nm. While at 298.15 K, there is a decrease in the size of the hydrodynamic radius of the cyclodextrin (about 2%) respect to water, at 310.15 K, this increases in a similar ratio. Certainly, the entrance into solution of KCl at the standard temperature has the effect of dehydrating the β-CD molecule by withdrawing water molecules of hydration for its own, although some Cl^- ions can penetrate into the cavity and form complexes of inclusion. As previously assumed, at 310.15 K, the thermal effect may difficult the encapsulation of these ions into the cavity and their association occurs externally by increasing the size of the β-CD molecule.

For the KCl (1) + HP-β-CD (2) system[31,32] (Table 15.5), the results obtained for the main diffusion coefficients in comparison to the binary values show that the added KCl produces relatively small changes in the diffusion coefficient, D_{22}, of HP-β-CD (less than 3%), whereas the addition of HP-β-CD causes significant changes in the diffusion coefficient, D_{11}, of KCl (≤7%). As in the case of β-CD, the entrance of KCl into the aqueous solution may have the effect of releasing into the medium water molecules trapped in the hydration shell of the cyclodextrin and thus decrease the size of the cyclodextrin molecule, allowing an increase on its mobility. Also in this case, Cl^- ions may be included in the cyclodextrin cavity and, taking into account the values of the coupled transport of solutes obtained by the ratio D_{12}/D_{22} higher values for the inclusion constant would be expected.

In fact, the values obtained for the association constant were of 47.8 and 34.8 M^{-1}, at 298.15 K and 310.15 K, respectively, which represents a

TABLE 15.5 Ternary Mutual Diffusion Coefficients D_{11}, D_{12}, D_{21}, and D_{22} for Aqueous KCl (1) + HP-β-CD (2) Solutions and Respective Standard Deviations, S_D, at 298.15 and 310.15 K.

C_1 [a]	C_2 [a]	$D_{11} \pm S_D$ [b] (10^{-9} m^2 s^{-1})	$D_{12} \pm S_D$ [b] (10^{-9} m^2 s^{-1})	$D_{21} \pm S_D$ [b] (10^{-9} m^2 s^{-1})	$D_{22} \pm S_D$ [b] (10^{-9} m^2 s^{-1})	D_{12}/D_{22} [c]	D_{21}/D_{11} [d]
			T = 298.15 K				
0.010	0.000	1.725 ± 0.031	0.005 ± 0.006	0.052 ± 0.031	0.305 ± 0.004	0.015	0.030
0.000	0.010	1.850 ± 0.031	−0.141 ± 0.020	0.002 ± 0.034	0.328 ± 0.006	−0.429	0.001
0.002	0.002	1.930 ± 0.027	−0.130 ± 0.039	0.012 ± 0.051	0.324 ± 0.005	−0.401	0.006
0.005	0.005	1.808 ± 0.026	−0.089 ± 0.027	−0.024 ± 0.018	0.317 ± 0.001	−0.280	0.013
0.010	0.010	1.778 ± 0.028	−0.057 ± 0.024	−0.003 ± 0.060	0.308 ± 0.012	−0.185	0.002
			T = 310.15 K				
0.010	0.000	2.330 ± 0.017	0.180 ± 0.092	0.003 ± 0.031	0.403 ± 0.013	0.447	0.001
0.000	0.010	2.410 ± 0.033	0.002 ± 0.024	0.086 ± 0.019	0.415 ± 0.015	0.005	0.036

[a] c_1 and c_2 in units (mol dm^{-3}); [b] D_{12}/D_{22} represents the number of moles of KCl carried by 1 mol of HP-β-CD; [c] D_{21}/D_{11} represents the number of moles of HP-β-CD carried by 1 mol of KCl.

considerable increase of the association observed against β-CD.[32] Again, association should happen through inclusion, as evidenced by the limiting diffusion coefficients for the associated species of 0.340×10^{-9} and 0.434×10^{-9} m^2 s^{-1}, very close to the free cyclodextrin. Values estimated for the hydrodynamic radius of the HP-β-CD molecule in aqueous solutions of KCl were r (298.15 K) = 0.747 nm and r (310.15 K) = 0.792 nm, about 1.5–2% below the water value. Therefore, the size of this molecule is more affected by the presence of KCl than the β-CD, since the latter is less likely to bind with the surrounding water molecules. The decrease on the size of the cyclodextrin, due to the presence of these ions in solution, occurs possibly because the presence of ions in solution may remove water molecules from the hydration layers of the cyclodextrin to its own layer, effect that is less pronounced at the higher temperature due to the thermal effect.

15.4.1.3 GLOBAL BALANCE OF THE PRESENCE OF IONS IN THE FORMATION OF CYCLODEXTRIN–DRUG COMPLEXES: STABILIZATION OF COMPLEX STRUCTURES OR COMPETITION FOR INCLUSION?

So far, the investigation herein described has individually characterized a series of aqueous systems, representing each of the major compounds involved in the development of a controlled drug-delivery system, or external factors acting on them, gradually increasing the complexity of the systems to mimic, as close as possible, the in-vivo conditions to which they are subject. In the following, it is intended to analyze the overall situation where the system is composed by the drug (methylxanthine), the transport vehicle (cyclodextrin) and usual ions from physiological fluids (here represented by K^+ and Cl^- ions) to understand if and how these compounds interact with each other and influence the properties of the remainder. In this sense, the transport properties of caffeine and theophylline (methylxanthines) in the presence of cyclodextrin (β-CD and HP-β-CD) and potassium chloride (KCl) in aqueous solution were studied by measuring the diffusion coefficients of quaternary mixtures.

Studies that characterize the ternary systems that comprise cyclodextrin (β-CD and HP-β-CD) and methylxanthines (caffeine and theophylline)[4,5] (Tables 15.6–15.11) have been already published by us. They allowed to establish the values for the stability constant K, from transport (diffusion) and thermodynamic (apparent molar volume) data and at different temperatures.

The apparent molar volume of the complexed species, $\varphi_{V,c}$, was also estimated and the analysis of volume changes was used to infer about the possible association of solutes. Values obtained indicated coherent stability constants, even when estimated by different methods, and a dissimilar behavior of volume changes for the methylxanthine that was dependent on the cyclodextrin that was present in the medium.[4]

TABLE 15.6 Thermodynamic Properties of the Complexation Process of Cyclodextrins with Caffeine at 298.15 and 310.15 K.

	K (kg mol^{-1})	$\varphi_{V,c\mathrm{Caf}}$ (cm^3 mol^{-1})	$\Delta\varphi_{V,cCaf}$ $\Delta\varphi_{V,c\mathrm{Caf}}$ (cm^3 mol^{-1})	$\varphi_{V,c\mathrm{CD}}$ (cm^3 mol^{-1})	$\Delta\varphi_{V,c\mathrm{CD}}$ (cm^3 mol^{-1})
			298.15 K		
β-CD	47.89	123.0 (±3.1)	−19.50	668.2 (±7.3)	−38.07
HP-β-CD	187.9	145.3 (±0.1)	2.77	875.1 (±1.8)	6.71
			310.15 K		
β-CD	25.15	130.4 (±2.8)	−16.40	655.5 (±3.8)	−62.57
HP-β-CD	–	144.7 (±0.4)	−1.98	882.3 (±0.1)	0.84

Note: The values in parentheses correspond to the standard deviations of the measurements.

TABLE 15.7 Values of K and D_{33}^* Estimated from the Ternary Diffusion Coefficients of Caffeine Aqueous Solutions in the Presence of β-CD at 298.15 and 310.15 K.

c_2 (mol dm^{-3})	D_{11}^* (10^{-9} m^2 s^{-1})	D_{22}^* (10^{-9} m^2 s^{-1})	D_{33}^* (10^{-9} m^2 s^{-1})	K (dm^3 mol^{-1})
		T = 298.15 K		
0.002		0.737		
0.005	0.317	0.724	0.305 (±0.016)	43 (±5.6)
0.010		0.69		
		T = 310.15 K		
0.002		0.999		
0.005	0.436	0.949	0.430 (±0.018)	23 (±2.1)
0.010		0.915		

Note: The data presented corresponds to a constant concentration of 10 mM β-CD.

TABLE 15.8 Values of K and D_{33}^* Estimated from the Ternary Diffusion Coefficients of Aqueous Solutions of Caffeine in the Presence of HP-β-CD at 298.15 and 310.15 K.

c_2 (mol dm^{-3})	D_{11}^* (10^9 m^2 s^{-1})	D_{22}^* (10^{-9} m^2 s^{-1})	D_{33}^* (10^{-9} m^2 s^{-1})	K (dm^3 mol^{-1})
		T = 298.15 K		
0.002		0.735		
0.005	0.307	0.723	0.300 (±0.013)	85 (±8.6)
0.010		0.688		
		T = 310.15 K		
0.002		0.979		
0.005	0.406	0.930	0.395 (±0.018)	20 (±1.0)
0.010		0.896		

Note: The data presented are those corresponding to a constant concentration of 10 mM HP-β-CD.

TABLE 15.9 Thermodynamic Properties of the Complexation Process of Cyclodextrins with Theophylline at 298.15 and 310.15 K.

	K (kg mol^{-1})	$\varphi_{V,cTeof}$ (cm^3 mol^{-1})	$\Delta\varphi_{V,cTeof}$ (cm^3 mol^{-1})	$\varphi_{V,cCD}$ $\varphi_{V,cCD}$ (cm^3 mol^{-1})	$\Delta\varphi_{V,cCD}$ $\Delta\varphi_{V,cCD}$ (cm^3 mol^{-1})
			298.15 K		
β-CD	30.53	115.6 (±0.5)	−24.87	596.4 (±43.1)	−109.82
HP-β-CD	78.71	121.3 (±0.5)	−19.17	813.5 (±18.3)	−51.22
			310.15 K		
β-CD	16.56	122.9 (±0.3)	−20.31	568.7 (±78.2)	−149.38
HP-β-CD	48.47	119.7 (±0.6)	−23.56	787.9 (±66.3)	−93.64

TABLE 15.10 Values of K and D_{33}^* Estimated from the Ternary Diffusion Coefficients of the Aqueous Solutions of Theophylline in the Presence of β-Cyclodextrin at 298.15 and 310.15 K.

c_2 (mol dm^{-3})	D_{11}^* (10^{-9} m^2 s^{-1})	D_{22}^* (10^{-9} m^2 s^{-1})	D_{33}^* (10^{-9} m^2 s^{-1})	K (dm^3 mol^{-1})
		T = 298.15 K		
0.002		0.770		
0.005	0.317	0.748	0.300 (±0.012)	28.5 (±1.4)
0.010		0.720		

TABLE 15.10 *(Continued)*

c_2 (mol dm^{-3})	D_{11}^* (10^{-9} m^2 s^{-1})	D_{22}^* (10^{-9} m^2 s^{-1})	D_{33}^* (10^{-9} m^2 s^{-1})	K (dm^3 mol^{-1})
		T = 310.15 K		
0.002		1.064		
0.005	0.436	1.003	0.395 (±0.027)	22 (±2.3)
0.010		0.948		

Note: The data presented are those corresponding to a constant concentration of 10 mM β-CD.

TABLE 15.11 Values of K and D_{33}^* Estimated from the Ternary Diffusion Coefficients of the Aqueous Solutions of Theophylline in the Presence of HP-β-CD at 298.15 and 310.15 K.

c_2 (mol dm^{-3})	D_{11}^* (10^{-9} m^2 s^{-1})	D_{22}^* (10^{-9} m^2 s^{-1})	D_{33}^* (10^{-9} m^2 s^{-1})	K (dm^3 mol^{-1})
		T = 298.15 K		
0.002		0.769		
0.005	0.307	0.746	0.265 (±0.013)	85 (±8.6)
0.010		0.719		
		T = 310.15 K		
0.002		1.042		
0.005	0.406	0.982	0.385 (±0.012)	39 (+1.4)
0.010		0.929		

Note: The data presented are those corresponding to a constant concentration of 10 mM HP-β-CD.

From the previously above described data, together with the information regarding the association constants and the volume changes, it would be expected that, in a quaternary mixture, only the magnitude of the association between caffeine and β-CD was affected by the presence of K^+/Cl^- ions, once all the estimated association constants for the individual systems are in close proximity and same order of magnitude. At 310.15 K, one would assume competition for the cyclodextrin cavity could occur, between Cl^- ions and caffeine. On the contrary, for caffeine in the presence of HP-β-CD, it is unlikely the existence of competition. In the case of theophylline in aqueous solution in the presence of cyclodextrins and ions, it would be expected that there was preferential inclusion of this molecule in the HP-β-CD cavity but not in the β-CD cavity, being the KCl (or ions) encapsulated.

The theoretical analysis of concentration dependence in multicomponent systems, as the case of a quaternary mixture, is quite complex and therefore the following discussion will be done only at a qualitative level, supported by the binary and ternary diffusion coefficients obtained for the corresponding systems. Components are numbered as follows: cyclodextrin (1) + KCl (2) + methylxanthine (3) + water. In these systems the major diffusion coefficients ${}^{123}D_{11}$, ${}^{123}D_{22}$ and ${}^{123}D_{33}$ provide the molar fluxes of cyclodextrin (1), potassium chloride (2), and methylxanthine (3) conducted by their own concentration gradients. The secondary diffusion coefficients represent the coupled fluxes of a solute generated by concentration gradients of the other solutes in solution, that is,

- ${}^{123}D_{21}$ and ${}^{123}D_{31}$ represent the coupled fluxes of potassium chloride and methylxanthine generated by the cyclodextrin concentration gradient;
- ${}^{123}D_{12}$ and ${}^{123}D_{32}$ represent the coupled fluxes of cyclodextrin and methylxanthine generated by the concentration gradient of potassium chloride; and
- ${}^{123}D_{13}$ and ${}^{123}D_{23}$ represent the coupled fluxes of cyclodextrin and potassium chloride generated by the concentration gradient of methylxanthine.

The values of the diffusion coefficients obtained for the quaternary systems can be directly correlated with the corresponding ones obtained for the same compounds in binary and ternary aqueous solution. That is, the main diffusion coefficient ${}^{123}D_{33}$ corresponding to the methylxanthine in aqueous solution in a quaternary mixture can be compared with the diffusion coefficient D_{33} (corresponding to the diffusion coefficient of this compound in a ternary mixture in the presence of cyclodextrin or KCl) and its binary diffusion coefficient D (in water). Thus, one can evaluate both the changes in the diffusion of this species and in what way it is affected by the introduction of other ionic and/or molecular solutes in solution. The secondary diffusion coefficients for diffusion in a quaternary mixture can also be related to those corresponding in a ternary mixture, that is, ${}^{123}D_{31}$ and ${}^{123}D_{32}$ represent the methylxanthine fluxes generated by the cyclodextrin and KCl concentration gradients, respectively, and are directly related to D_{31} (for the methylxanthine in the presence of cyclodextrin) and D_{32} (for the methylxanthine in the presence of KCl). This rationale can be applied in the same way to all the other components in solution.

15.4.1.3.1 Effect of Potassium Chloride on Aqueous Solutions of Cyclodextrin + Caffeine

Measurements of the mutual diffusion coefficients of caffeine in aqueous solutions were carried out in the presence of β-CD and KCl and HP-β-CD and KCl at different concentration ratios and at standard temperature (298.15 K) to investigate the influence of the presence of potassium chloride over the association of cyclodextrin and caffeine. The average diffusion coefficients' values obtained for the quaternary system β-CD (1) + KCl (2) + caffeine (3) + water at 298.15 K are shown in Table 15.12,[13,14] together with the main and secondary diffusion coefficients that are directly related to these measures. These results are averages of at least six experiments. In most cases, the D_{ik} values were reproducible, in general, within ±0.05 10^{-9} m² s⁻¹.

TABLE 15.12 Comparison of the Quaternary Diffusion Coefficients with the Ternary and Binary Diffusion Values for Aqueous β-CD (1) + KCl (2) + Caffeine (3) Solutions at 298.15 K.

	D_{11}^{f}	D_{12}^{f}	D_{13}^{f}	D_{21}^{f}	D_{22}^{f}	D_{23}^{f}	D_{31}^{f}	D_{32}^{f}	D_{33}^{f}
				$c = c_1 = c_2 = c_3 = 2$ mM					
D_{ij}^{a}	0.322	−0.028	−0.002	0.505	1.911	−0.121	−0.007	−0.030	0.797
D_{ij}^{b}	0.324	−0.068		−0.242	1.924				
D_{ij}^{c}	0.323		−0.008				0.016		0.735
D_{ij}^{d}					1.951	−0.165		−0.056	0.746
D_{ij}^{e}	0.324				1.954				0.751
				$c = c_1 = c_2 = c_3 = 5$ mM					
D_{ij}^{a}	0.320	−0.022	0.020	0.210	1.882	−0.101	0.004	0.001	0.770
D_{ij}^{b}	0.317	−0.088		−0.196	1.891				
D_{ij}^{c}	0.318		−0.011				0.014		0.725
D_{ij}^{d}					1.903	−0.210		−0.112	0.731
D_{ij}^{e}	0.322				1.934				0.738
				$c = c_1 = c_2 = c_3 = 10$ mM					
D_{ij}^{a}	0.311	−0.034	0.055	−0.052	1.518	−0.104	0.057	0.141	0.748
D_{ij}^{b}	0.313	−0.119		−0.057	1.848				
D_{ij}^{c}	0.313		−0.014				0.013		0.695
D_{ij}^{d}					1.844	0.104		−0.282	0.688
D_{ij}^{e}	0.316				1.918				0.703

[a]Quaternary diffusion coefficients for the aqueous solutions of β-CD (1) + caffeine (2) + KCl (3) + water; [b]ternary diffusion coefficients for the aqueous solutions of β-CD (1) + KCl (2); [c]ternary diffusion coefficients for the aqueous solutions of β-CD (1) + caffeine (2); [d]ternary diffusion coefficients for the aqueous solutions of KCl (1) + caffeine (2); [e]binary diffusion coefficients for the aqueous solutions of β-CD, KCl, and caffeine; [f]units of D in 10^{-9} m² s⁻¹.

15.4.1.3.1.1 Major Diffusion Coefficients 123D11, 123D22, and 123D33

The value of the coefficient ${}^{123}D_{11}$ decreases with the increase of the solutes concentration, in the range of concentrations studied, by about 3%, and its value is very close to that found for the cyclodextrin in water. ${}^{123}D_{22}$ and ${}^{123}D_{33}$ also vary with concentration, the decrease in ${}^{123}D_{22}$ is quite significant reaching up to 20%, and in the case of ${}^{123}D_{33}$ is 6%. The slight decrease in the diffusion coefficient of β-CD compared with the rather more significant decrease found in the main diffusion coefficients of caffeine and particularly in KCl may indicate that the associations established between cyclodextrin and the other solutes would be mostly of inclusion and reason why the mobility of the former is little affected. In principle, the inclusion of KCl in the cavity of cyclodextrin over caffeine could also affect the mobility of KCl and be one of the causes of the significant decrease in its diffusion coefficient.

15.4.1.3.1.2 Diffusion Driven by β-Cyclodextrin Gradients

Although ${}^{123}D_{31}$ presents values very close to zero, within the experimental error, the values of ${}^{123}D_{21}$ are different from zero and can be interpreted in terms of the presence of possible interactions between β-CD and KCl (e.g., the complexes whose association constant was previously estimated). Quantification of the KCl transport by the β-CD gradient (from the ${}^{123}D_{21}/{}^{123}D_{11}$ ratio) shows that, for the compositions used, 1 mol of diffusing β-CD cotransports up to 1.6 mol of potassium chloride in dilute solutions (although this transport decreases with increasing concentration of solutes).

15.4.1.3.1.3 Diffusion Driven by Gradients of Potassium Chloride

The values presented by ${}^{123}D_{12}$ and ${}^{123}D_{32}$ are practically zero within the experimental error, that is, the KCl gradients are unable to generate coupled flows of the other components in aqueous solution.

15.4.1.3.1.4 Diffusion Driven by Caffeine Gradients

The diffusion coefficient ${}^{123}D_{13}$ is practically zero within the experimental error. As noted previously, caffeine gradients are unable to generate significant coupled fluxes of cyclodextrin. Since ${}^{123}D_{23}$ shows practically constant values, and from the ${}^{123}D_{23}/{}^{123}D_{33}$ ratio in the same compositions, 1 mol of caffeine is expected to counter-transport approximately 0.15 mol KCl, a slightly lower value than that found for the aqueous KCl + caffeine system.

Therefore, from the analysis of the main and secondary diffusion coefficients and the fluxes generated by the cross-diffusion of the solutes in quaternary mixtures, compared with the ternary and binary corresponding systems, there seems to be enough evidence that there may be competition between caffeine and KCl to enter the cyclodextrin cavity. The 20% decrease in the KCl diffusion coefficient together with a significant transport of KCl by both the cyclodextrin and caffeine gradients indicate that the predisposition of this solute in the quaternary mixture would be to form aggregates with either β-CD or caffeine, as evidenced by the previously close values obtained for the association constants.

Following, the average values of the diffusion coefficients obtained for the quaternary system HP-β-CD (1) + KCl (2) + caffeine (3) + water at 298.15 K[34] are presented in Table 15.13 and compared with the corresponding ones in the binary and ternary systems.

TABLE 15.13 Comparison of the Quaternary Diffusion Coefficients with the Ternary and Binary for Aqueous HP-β-CD (1) + KCl (2) + Caffeine (3) Solutions at 298.15 K.

	D_{11}^{f}	D_{12}^{f}	D_{13}^{f}	D_{21}^{f}	D_{22}^{f}	D_{23}^{f}	D_{31}^{f}	D_{32}^{f}	D_{33}^{f}
				$c = c_1 = c_2 = c_3 = 2$ mM					
D_{ij}^{a}	0.313	0.149	−0.023	0.052	2.116	−0.096	0.077	−0.595	0.758
D_{ij}^{b}	0.324	0.012		−0.130	1.930				
D_{ij}^{c}	0.300		0.003				0.050		0.744
D_{ij}^{d}					1.951	−0.165		−0.056	0.746
D_{ij}^{e}	0.319				1.954				0.751
				$c = c_1 = c_2 = c_3 = 5$ mM					
D_{ij}^{a}	0.315	0.098	−0.009	0.172	1.996	0.082	0.054	−0.240	0.687
D_{ij}^{b}	0.317	−0.024		−0.089	1.808				
D_{ij}^{c}	0.293		−0.011				0.054		0.665
D_{ij}^{d}					1.903	−0.210		−0.112	0.731
D_{ij}^{e}	0.314				1.934				0.738
				$c = c_1 = c_2 = c_3 = 10$ mM					
D_{ij}^{a}	0.309	0.092	−0.029	0.263	1.802	0.006	0.047	−0.149	0.673
D_{ij}^{b}	0.308	−0.003		0.246	1.778				
D_{ij}^{c}	0.283		−0.021				0.103		0.640
D_{ij}^{d}					1.844	0.104		−0.282	0.688
D_{ij}^{e}	0.307				1.918				0.703

[a]Quaternary diffusion coefficients for the aqueous solutions of HP-β-CD (1) + caffeine (2) + KCl (3) + water; [b]ternary diffusion coefficients for the aqueous solutions of HP-β-CD (1) + KCl (2); [c]ternary diffusion coefficients for the aqueous solutions of HP-β-CD (1) + caffeine (2); [d]ternary diffusion coefficients for the aqueous solutions of KCl (1) + caffeine (2); [e]binary diffusion coefficients for the aqueous solutions of HP-β-CD, KCl, and caffeine; [f]units of D 10^{-9} m^2 s^{-1}.

15.4.1.3.1.5 Major Diffusion Coefficients 123D11, 123D22, and 123D33

It is observed that the coefficient ${}^{123}D_{11}$ is practically independent of the concentration of solutes in aqueous solution as the variation here presented (1.5%) and relative to the corresponding binary value (HP-β-CD in aqueous solution) is within the experimental error (1–2%). In what concerns the diffusion coefficients for KCl and caffeine in the quaternary mixture, a decrease with concentration is perceived, of 15% in ${}^{123}D_{22}$ for KCl and 11% in ${}^{123}D_{33}$ for caffeine. As in the case of the β-CD (1) + KCl (2) + caffeine (3) + water system, the slight decrease in the diffusion coefficient of HP-β-CD may indicate that the associations established between cyclodextrin and the others solutes, if they exist, are mainly of inclusion, which is why their mobility is not affected. The major decrease in the main diffusion coefficients for caffeine and KCl would arise from their inclusion. Since the magnitude of this decrease is very similar for both caffeine and KCl, the effect of the gradient of each of the solutes in solution on the others is analyzed in an attempt to establish which one would be included in the cavity.

15.4.1.3.1.6 Diffusion Driven by Hydroxypropyl-β-cyclodextrin Gradients

Both the secondary diffusion coefficients are positive, but ${}^{123}D_{31}$ shows values below ${}^{123}D_{21}$. The HP-β-CD gradient carries, in each case, cocurrent, both caffeine and KCl, but whereas in the case of caffeine 1 mol of HP-β-CD can carry about 0.24 mol of caffeine, for KCl this value can reach 0.85 mol, a value even higher than the one found for the corresponding ternary system, which can be interpreted in terms of the prevalence of possible interactions between HP-β-CD and KCl (e.g., in the form of complexes whose association constant was previously estimated).

15.4.1.3.1.7 Diffusion Driven by Gradients of Potassium Chloride

${}^{123}D_{12}$ shows values close to zero within experimental error while ${}^{123}D_{32}$ shows negative values, with the caffeine counter-transport generated by the KCl gradient with a maximum of 0.3 mol of caffeine per mole of diffusing KCl that decreases with increasing concentration of solutes in solution, unlike the corresponding ternary system.

15.4.1.3.1.8 Diffusion Driven by Caffeine Gradients

The diffusion coefficient ${}^{123}D_{13}$ is very small and negative, as ${}^{123}D_{23}$, the latter approaching zero within the experimental error. The caffeine gradient cannot generate a significant coupled flow of cyclodextrin (it does not carry more than 0.04 mol of cyclodextrin per mole of diffusing caffeine); and the coupled flow of KCl by the effect of the caffeine gradient in solution disappears with the increase of solute concentration, contrary to that found for the corresponding ternary system.

Taking into account the results obtained in the studies performed for the binary and ternary systems composing HP-β-CD (1) + KCl (2) + caffeine (3) aqueous system,[4] it would be expected to prove inclusion of caffeine rather than KCl. In fact, from the main quaternary diffusion coefficients it is observed that the diffusion of the cyclodextrin is not affected by the presence of the other solutes, an indication that the interaction between solutes must occur through the formation of inclusion complexes. As the concentration of solutes in solution increases, the flows generated by concentration gradients happens in the sense of a more stable association between cyclodextrin and caffeine and is less stable association both between cyclodextrin and KCl and caffeine and KCl. It would also appear that in more dilute solutions KCl could be subject to a salting out effect, as evidenced by the higher values presented by the main diffusion coefficient of this solute relative to that obtained in water.

15.4.1.3.2 Effect of Potassium Chloride on Aqueous Solutions of Cyclodextrin–Theophylline

As in the case of caffeine and to understand the effect of the presence of potassium chloride in the cyclodextrin–theophylline aggregates in aqueous solution, measurements of the mutual diffusion coefficients of aqueous solutions of theophylline were carried out in the presence of β-CD and KCl and HP-β-CD and KCl, at different concentration ratios and at standard temperature (298.15 K) and were compared with corresponding binary and ternary systems. The average values of the diffusion coefficients found for the quaternary system β-CD (1) + KCl (2) + theophylline (3) + water at 298.15 K[34] are presented and compared in Table 15.14[31] with those obtained for the corresponding binary and ternary solutions.

TABLE 15.14 Comparison of the Quaternary Diffusion Coefficients with the Ternary and Binary for Aqueous β-CD (1) + KCl (2) + Theophylline (3) Solutions at 298.15 K.

	D_{11}^{f}	D_{12}^{f}	D_{13}^{f}	D_{21}^{f}	D_{22}^{f}	D_{23}^{f}	D_{31}^{f}	D_{32}^{f}	D_{33}^{f}
					$c = c_1 = c_2 = c_3 = 2$ mM				
D_{ij}^{a}	0.328	0.009	0.003	−0.052	2.011	0.043	0.029	−0.279	0.706
D_{ij}^{b}	0.324	−0.068		−0.242	1.959				
D_{ij}^{c}	0.321		−0.003				0.004		0.721
D_{ij}^{d}					1.847	−0.024		−0.012	0.783
D_{ij}^{e}	0.324				1.954				0.785
					$c = c_1 = c_2 = c_3 = 5$ mM				
D_{ij}^{a}	0.304	0.039	0.029	−0.176	1.870	0.187	−0.020	−0.273	0.631
D_{ij}^{b}	0.317	−0.088		−0.196	1.937				
D_{ij}^{c}	0.316		0.005				−0.024		0.683
D_{ij}^{d}					1.842	−0.313		−0.108	0.765
D_{ij}^{e}	0.322				1.934				0.762
					$c = c_1 = c_2 = c_3 = 10$ mM				
D_{ij}^{a}	0.289	0.033	0.030	−0.143	1.719	0.347	−0.275	−0.068	0.583
D_{ij}^{b}	0.313	−0.119		−0.059	1.925				
D_{ij}^{c}	0.309		−0.016				−0.063		0.628
D_{ij}^{d}					1.690	−0.359		−0.103	0.736
D_{ij}^{e}	0.316				1.918				0.734

[a]Quaternary diffusion coefficients for the aqueous solutions of β-CD (1) + theophylline (2) + KCl (3) + water; [b]ternary diffusion coefficients for the aqueous solutions of β-CD (1) + KCl (2); [c]ternary diffusion coefficients for aqueous solutions of β-CD (1) + theophylline (2); [d]ternary diffusion coefficients for the aqueous solutions of KCl (1) + theophylline (2); [e]binary diffusion coefficients for the aqueous solutions of β-CD, KCl, and theophylline; [f]units of D in 10^9 m^2 s^{-1}.

15.4.1.3.2.1 Major Diffusion Coefficients 123D11, 123D22, and 123D33

The value of ${}^{123}D_{11}$ decreases about 12% with increasing concentration of solutes, in the range of concentrations studied. ${}^{123}D_{22}$ and ${}^{123}D_{33}$ also vary with concentration, with a decrease in ${}^{123}D_{22}$ of 14.5% and 17.5% in the case of ${}^{123}D_{33}$. The high decreases in the main diffusion coefficients could be explained on the one hand by the increase in the viscosity of the solutions, due to the increase in the number of components in aqueous solution and, on the other hand, by a greater interaction between the solutes. However, the variations observed in the viscosities of the corresponding ternary systems are not high enough to justify a reduction in that order for the quaternary mixture. The most reasonable explanation seems to come from the formation of aggregates between the solutes that would result in a lower mobility, leading to the decrease of the respective diffusion coefficients. It was not possible to find a common factor to relate the values of ${}^{123}D_{11}$, ${}^{123}D_{22}$, and ${}^{123}D_{33}$ to the respective ternary diffusion coefficients D_{11}, D_{22}, and D_{33}; hence, this is evidence that the type of interaction that occurs directly influences the diffusion coefficient and cannot be explained as a simple association process.

15.4.1.3.2.2 Diffusion Driven by β-Cyclodextrin Gradients

Both ${}^{123}D_{21}$ and ${}^{123}D_{31}$ generally increase, in magnitude, with the concentration rising of the solute and become more negative, indicating coupled fluxes of the solutes. They are also very similar to the corresponding values in the ternary systems. Therefore, the values of ${}^{123}D_{21}$ and ${}^{123}D_{31}$ can be interpreted in terms of the presence of possible interactions between β-CD and KCl or β-CD and theophylline for diluted solutions (1:1 complexes described above). These relations could also explain the large decreases in the main diffusion coefficients. Gradients in concentration of 1 diffusing mol of β-CD at the compositions studied counter-transport at most 0.6 mol of potassium chloride and at most 0.95 mol of theophylline. Regarding the diffusion coefficients in the corresponding ternary systems, there is little change to the KCl transported by β-CD. The transport of theophylline by the cyclodextrin in the quaternary mixture is superior to that found for theophylline in the presence of β-CD alone.

15.4.1.3.2.3 Diffusion Driven by Gradients of Potassium Chloride

The values found for ${}^{123}D_{12}$ are small and close to zero, within the experimental error, that is, the KCl gradients are not able to generate coupled flows of β-CD, as verified in other systems. However, the coefficient ${}^{123}D_{32}$ presents negative and nonzero values, possibly due to the existence in solution of eventual aggregates between the theophylline molecules and the potassium ions. Equally, the transport of theophylline by the gradient of KCl goes down from 0.24 to 0.04 mol of theophylline counter-transported by 1 mol of diffusing KCl in the concentration range studied, probably due to the change in the preferential interactions between theophylline and other components in the solution, especially cyclodextrin.

15.4.1.3.2.4 Diffusion Driven by Theophylline Gradients

${}^{123}D_{13}$ is very small and tends to zero within the experimental error. As seen previously, theophylline gradients are unable to generate significant coupled fluxes of cyclodextrin. Relative to ${}^{123}D_{23}$ its values increase with concentration rising of the solutes, becoming large and positive. At the compositions studied, it is expected that 1 mol of diffusing theophylline co-transport up to 0.6 mol KCl, a situation inverse to that found in the aqueous system KCl (1) + theophylline (2).

From the previous analysis of the results for the main and secondary diffusion coefficients of the quaternary mixture β-CD (1) + KCl (2) + theophylline (3) + water, and the fluxes generated by the diffusion of these solutes compared with ternary and to the results of the binary systems, we can find evidence of the existence of competition between theophylline and KCl for the occupation of the cyclodextrin cavity, with possible preference for KCl. The transport of significant amounts of KCl by cyclodextrin and theophylline, associated with a 15% decrease in its diffusion coefficient, gives the conviction that this solute in the quaternary mixture tends to be associated preferentially with β-CD.

The values of the average diffusion coefficients for the HP-β-CD (1) + KCl (2) + theophylline (3) + water quaternary system at 298.15 K[35] are shown in Table 15.15, as well as the corresponding ternary and binary diffusion coefficients.

TABLE 15.15 Comparison of the Quaternary Diffusion Coefficients with the Ternary and Binary for Aqueous HP-β-CD (1) + KCl (2) + Theophylline (3) Solutions at 298.15 K.

	D_{11}^{f}	D_{12}^{f}	D_{13}^{f}	D_{21}^{f}	D_{22}^{f}	D_{23}^{f}	D_{31}^{f}	D_{32}^{f}	D_{33}^{f}
				$c = c_1 = c_2 = c_3 = 2$ mM					
D_{ij}^{a}	0.322	0.020	−0.059	0.149	2.028	0.230	−0.013	−0.470	0.719
D_{ij}^{b}	0.324	0.012		−0.130	1.930				
D_{ij}^{c}	0.306		0.002				0.068		0.756
D_{ij}^{d}					1.847	−0.024		−0.012	0.783
D_{ij}^{e}	0.319				1.954				0.785
				$c = c_1 = c_2 = c_3 = 5$ mM					
D_{ij}^{a}	0.311	−0.015	0.031	0.180	1.940	0.129	−0.014	−0.240	0.646
D_{ij}^{b}	0.317	−0.024		−0.089	1.808				
D_{ij}^{c}	0.301		0.024				0.063		0.682
D_{ij}^{d}					1.842	−0.313		−0.108	0.765

TABLE 15.15 *(Continued)*

	D_{11}^{f}	D_{12}^{f}	D_{13}^{f}	D_{21}^{f}	D_{22}^{f}	D_{23}^{f}	D_{31}^{f}	D_{32}^{f}	D_{33}^{f}
D_{ij}^{e}	0.314				1.934				0.762
				$c = c_1 = c_2 = c_3 = 10$ mM					
D_{ij}^{a}	0.307	−0.002	0.003	0.393	1.842	0.186	−0.258	−0.120	0.556
D_{ij}^{b}	0.308	−0.003		0.246	1.798				
D_{ij}^{c}	0.298		−0.029				−0.009		0.548
D_{ij}^{d}					1.690	−0.359		−0.103	0.736
D_{ij}^{e}	0.307				1.918				0.734

[a]Quaternary diffusion coefficients for the aqueous solutions of HP-β-CD (1) + theophylline (2) + KCl (3) + water; [b]ternary diffusion coefficients for HP-β-CD (1) + KCl (2) aqueous solutions; [c]ternary diffusion coefficients for the aqueous solutions of HP-β-CD (1) + theophylline (2); [d]ternary diffusion coefficients for the aqueous solutions of KCl (1) + theophylline (2); [e]binary diffusion coefficients for the aqueous solutions of HP-β-CD, KCl, and theophylline; [f]units of D in 10^9 m^2 s^{-1}.

15.4.1.3.2.5 Major Diffusion Coefficients 123D11, 123D22, and 123D33

It is found that the diffusion coefficient $^{123}D_{11}$ decreases slightly, about 3%, with increasing concentration. It was found a common factor between the measured quaternary diffusion coefficient, $^{123}D_{11}$, and the corresponding ternary D_{11} values, for example, $^{123}D_{11}$ in the quaternary mixture decreases by a factor of 0.98 relative to D_{11} in the HP-β-CD (1) system + KCl (2) but 1.03 higher than D_{11} for HP-β-CD (1) + theophylline (2). The sum of these compensating effects leads to a quaternary value very close to the binary value, as found.

By contrast, the changes in $^{123}D_{22}$ and $^{123}D_{33}$ are larger. In the case of $^{123}D_{22}$, there is a decrease of 9% with the increasing of solutes concentration, although it can be seen that the values are, in general, slightly higher than the binary and ternary diffusion coefficients (Table 15.8), either for aqueous KCl or for KCl in the presence of HP-β-CD. Greater differences are observed in the main quaternary diffusion coefficient theophylline, $^{123}D_{33}$, with a drop of approximately 21% with increasing concentration, being lower than both the binary diffusion coefficient for theophylline in aqueous solution and D_{33} for theophylline in the presence of KCl (changes between 8% and 23%). $^{123}D_{33}$ is also about 5% smaller than the corresponding diffusion coefficient in the ternary system without addition of salt (HP-β-CD + theophylline). It was not possible to find a constant common factor $^{123}D_{22}$ and D_{22} or between $^{123}D_{33}$ and D_{33}.

15.4.1.3.2.6 Diffusion Driven by Hydroxypropyl-β-cyclodextrin Gradients

We can perceive that both the secondary diffusion coefficients increase in magnitude with the increasing in the concentration of the solutes. $^{123}D_{21}$ is large and positive and $^{123}D_{31}$ is negative, clearly showing the existence of coupled flux of solutes. The larger $^{123}D_{21}$ values can be interpreted in terms of a salting-out effect. However, there are also possible interactions between HP-β-CD and KCl components in very dilute solutions, resulting in the formation of 1:1 complexes. Support for these phenomena comes also from various thermodynamic studies.[35,36] On the other hand, the lower values for $^{123}D_{31}$ may be interpreted on the basis of the nonspecific solute–solute interactions.[34] Quantification of coupled transport of solutes, at the compositions studied, show that 1 mol of diffusing HP-β-CD cotransports at most 1.3 mol of potassium chloride, whereas 1 mol of diffusing HP-β-CD countertransports at most 0.84 mol of theophylline.

15.4.1.3.2.7 Diffusion Driven by Gradients of Potassium Chloride

Again, $^{123}D_{12}$ is very small and draws near zero within the experimental error, there being no coupled flow of HP-β-CD produced by the KCl gradient, as well as occurring in the corresponding ternary systems. KCl gradients do not produce appreciable coupled fluxes of nonelectrolyte components in solution because the K^+ and Cl^- ions have similar values of mobility and therefore the electric field generated during KCl diffusion is relatively small. The values of $^{123}D_{32}$ are negative and different from zero, and a possible explanation for these facts can be given by taking into account that some theophylline KCl molecules may exist in solution as aggregates (such as theophylline–K^+ complexes). Nevertheless, if an analysis of the transport of theophylline by the gradient of KCl is done, it is observed that it drops from 0.23 to 0.07 mol of theophylline countertransported by 1 mol of diffusing KCl. This could be due either to the similarly of the mobilities of both the free theophylline species and the eventual aggregates of KCl + theophylline, or to the establishment of preferential interactions between theophylline and other components in solution (e.g., HP-β-CD).

15.4.1.3.2.8 Diffusion Driven by Theophylline Gradients

${}^{123}D_{13}$ is very small and tends to zero within the experimental error. As with ${}^{123}D_{12}$, theophylline gradients are unable to generate significant coupled streams of HP-β-CD. The same is true for the corresponding ternary systems. ${}^{123}D_{23}$ shows practically constant values, that is, the transport of KCl by gradients of theophylline is not affected by variations in the concentration of the remaining solutes in solution.

For the aqueous quaternary system HP-β-CD (1) + KCl (2) + theophylline (3) under study, together with the previous results obtained for the ternary systems that compose it, it would be expected to predict the inclusion of theophylline, to the detriment of KCl, in the cyclodextrin cavity. In fact, from the main diffusion coefficients it was noticed that the diffusion of cyclodextrin is very little affected by the presence of the other solutes, an indication that the possible existing complexes would be of inclusion. It is also verified that, when the concentration of the solutes increases, the fluxes generated by the concentration gradients of the solutes in the solution tend to produce a more stable association between cyclodextrin and theophylline. The possible association between HP-β-CD and KCl would be very weak, but the free theophylline molecules in solution could have some tendency to associate with KCl.

In summary, controlled release systems in solution are affected by the presence of ions, especially in the case of the carrier β-CD, whose association with the drugs is weakened by the possible competition for cavity occupation between the drug molecule and the ions. Nevertheless, hydrophobicity seems to be a major playing factor since even in a situation where the association constants in aqueous solutions reach similar values, the predisposition is for the drug molecule to entering the cyclodextrin cavity rather than the ions, occurring association of the latter either with the exposed part of the molecule or with the free drug molecules in solution.

15.5 CONCLUSIONS

It was intended to understand how the presence of ions would influence the stability of the controlled release systems of the drugs under study (caffeine and theophylline). Accordingly, each of the individual components was characterized separately in aqueous solution with the ions and, afterward, the global quaternary mixture, that is, an aqueous solution containing cyclodextrin, drug, ions, and water was studied.

In terms of drug-ion interaction, we would expect to find some kind of association with K^+ ions in the medium, since they could establish linkages with the C=O and N carboxyl groups available in both the caffeine and theophylline molecules. The association constants of these solutes were calculated in aqueous solution and it was verified that the hypothesis that caffeine and theophylline interacted with the ions present in the physiological fluids (K^+) was correct. It was possible to estimate an association constant of association whose values were very similar for both drugs, the small difference related to structural differences. It was also clear that theophylline would be the drug most affected by the presence of an electrolyte. The estimation of the hydrodynamic radius of the drug molecules induced by the presence of K^+ and Cl^- ions in the solution allowed reinforcing the information found on the association constants.

In what concerns the cyclodextrin–ion interaction, and considering the presence of the terminal hydroxyl groups, it was also expected to appreciate some kind of interaction, in this case through hydrogen bonds, with the Cl^- anion or its inclusion in the cavity. The values of the stability constants for this association were obtained, as well as the hydrodynamic radii of these molecules induced by the presence of the K^+ and Cl^- ions in the solution. It has been observed that the interaction of Cl^- with HP-β-CD was stronger than with β-CD, being this association favored at lower temperatures.

When studying the quaternary mixtures, it was verified that the introduction of ions into solution, where cyclodextrin and drug are present, has an effect on the latter, especially in the case of drugs in the presence of β-CD, since this association is weakened by possible competition for the cavity between the drug molecule and the ions present. Still, it appears that hydrophobicity is the determinant factor, since even in a situation where the association constants for CD-ion and CD-drug in aqueous solution reach similar value, the tendency is for the drug molecule to enter the cavity. The association between the drug and the ions could occur both through the part exposed to the surrounding environment or between free drug molecules and free ions in solution.

ACKNOWLEDGMENTS

The authors are grateful for funding from the *Fundação para a Ciência e a Tecnologia* (FCT), Portuguese Agency for Scientific Research, through the program PTDC/QUI-QFI/30271/2017 and COMPETE. C.I.A.V.S. also thanks the FCT for support through Grant SFRH/BPD/92851/2013 and for the funding granted by FEDER–European Regional Development Fund through the COMPETE Program and FCT-Fundação para a Ciencia e a Tecnologia, for the KIDIMIX project POCI-01-0145-FEDER-030271.

KEYWORDS

- **drug-delivery systems**
- **quaternary systems**
- **xanthines**
- **cyclodextrins**
- **multicomponent diffusion**
- **apparent molar volumes**
- **viscosity**

REFERENCES

1. Mochida, K.; Kagita, A.; Matsui, Y.; Date, Y. Effects of Inorganic Salts on the Dissociation of a Complex of β-Cyclodextrin with an Azo Dye in an Aqueous Solution. *Bull. Chem. Soc. Jpn.* **1973,** *46*, 3703–3707.
2. Dey, J.; Roberts, E. L.; Warner, I. M. Effect of Sodium Perchlorate on the Binding of 2-(4′-Aminophenyl)- and 2-(4′-(*N*,*N*′-Dimethylamino)phenyl)benzo-thiazole with β-Cyclodextrin in Aqueous Solution. *J. Phys. Chem. A* **1998,** *102*, 301–305.
3. Yi, Z.; Zhao, C.; Huang, Z.; Chen, H. Y. Investigation of Buffer-Cyclodextrin Systems. *Phys. Chem. Chem. Phys.* **1999,** *1*, 441–444.
4. Santos, C. I. A. V.; Ribeiro, A. C. F.; Esteso, M. A. Drug Delivery Systems: Study of Inclusion Complex Formation between Methylxanthines and Cyclodextrins from Thermodynamic and Transport Properties. *Biomolecules* **2019,** *9*, 196.
5. Santos, C. I. A. V.; Teijeiro, C.; Ribeiro, A. C. F.; Rodrigues, D. F. S. L.; Romero, C. M.; Esteso, M. A. Drug Delivery Systems: Study of Inclusion Complex Formation for Ternary Caffeine–Beta-cyclodextrin–Water Mixtures from Apparent Molar Volume Values at 298.15 K and 310.15 K. *J. Mol. Liquids* **2016,** *223*, 209–216.
6. Hagen, G. Movement of Water in a Narrow Cylindrical Tube. *Ann. Phys. Chem.* **1839,** *46*, 423–442.
7. Poiseuille, J. L. M. Recherches Expérimentales Sur le Mouvement des Liquides dans les Tubes de très Petits Diamètres, I. Influence de la Pression Sur la Quantité de Liquide qui Traverse les Tubes de Très Petits Diamètres. *C.R. Acad. Sci.* **1840,** *11*, 961–967.
8. Poiseuille, J. L. M. Recherches expérimentales sur le mouvement des liquides dans les tubes de très petits diamètres; II. Influence de la Longueur sur la Quantité de Liquide Qui Traverse Les Tubes de Très Petits Diamètres; III. Influence du Diamètre sur la

Quantité de Liquide Qui Traverse les Tubes de Très Petits Diamètres. *C.R. Acad. Sci.* **1840,** *11*, 1041–1048.

9. Jones, G.; Dole, M. The Viscosity of Aqueous Solutions of Strong Electrolytes with Special Reference to Barium Chloride. *J. Am. Chem. Soc.* **1929,** *51*, 2950–2964.
10. Maldonado, M.; Sanabria, E.; Batanero, B.; Esteso, M. A. Apparent Molal Volume and Viscosity Values for a New Synthesized Diazoted Resorcin[4]arene in DMSO at Several Temperatures. *J. Mol. Liquids* **2017,** *231*, 142–148.
11. Leaist, D. G.; Noulty, R. A. Quaternary Diffusion in Aqueous $KCl–KH_2PO_4–H_3PO_4$ Mixtures. *J. Phys. Chem.* **1987,** *91*, 1655–1658.
12. Leaist, D. G.; Hao, L. Quaternary Diffusion Coefficients of $NaCl–MgCl_2–Na_2SO_4–H_2O$ Synthetic Seawaters by Least-Squares Analysis of Taylor Dispersion Profiles. *J. Solut. Chem.* **1993,** *22*, 263–277.
13. Ribeiro, A. C. F.; Santos, C. I. A. V.; Lobo, V. M. M.; Esteso, M. A. Quaternary Diffusion Coefficients of β-Cyclodextrin + KCl + Caffeine+ Water at 298.15 K Using a Taylor Dispersion Method. *J. Chem. Eng. Data* 2010, *55*, 2610–2612.
14. Santos, C. I. A. V. Sistemas de Liberación Controlada de Fármacos: Propiedades Termodinámicas y de Transporte de Sistemas que Incluyen Ciclodextrinas y Fármacos a Dosis Terapéuticas. Ph.D. Thesis, Universidad de Alcalá: Madrid, Spain, 2012.
15. Leaist, D. G. Coupled Tracer Diffusion Coefficients of Solubilizates in Ionic Micelle Solutions from Liquid Chromatography. *J. Solut. Chem.* **1991,** *20*, 175–186.
16. Leaist, D. G. Ternary Diffusion Coefficients of 18-Crown-6 Ether–KCl–Water by Direct Least-Squares Analysis of Taylor Dispersion Measurements. *J. Chem. Soc. Faraday Trans* **1991,** *87*, 597–601.
17. Tyrrel, H. J. V.; Harris, K. R. Diffusion in Liquids, 2nd ed.; Butterworths: London, UK, 1984.
18. Barthel, J.; Gores, H. J.; Lohr, C. M.; Seidl, J. J. Taylor Dispersion Measurements at Low Electrolyte Concentrations. I. Tetraalkylammonium Perchlorate Aqueous Solutions. *J. Solut. Chem.* **1996,** *25*, 921–935.
19. Hartley, G. S.; Crank, J. Some Fundamental Definitions and Concepts in Diffusion Processes. *Trans. Faraday Soc.* **1949,** *45*, 801–818.
20. Miller, D. G.; Albright, J. G. Measurements of the Transport Properties of Fluids: Experimental Thermodynamics; Blackwell Scientific Publications: Oxford, UK, 1991; pp 272–294.
21. Leaist, D. G. Determination of Ternary Diffusion Coefficients by the Taylor Dispersion Method. *J. Phys. Chem.* **1990,** *94*, 5180–5183.
22. Barros, M. C. F.; Ribeiro, A. C. F.; Esteso, M. A. Cyclodextrins in Parkinson's disease. *Biomolecules* **2019,** *9*, 3.
23. Maaeih, A. Al.; Flanagan, D. R. Salt Effects on Cafeíne Solubility, Distribution and Self-association. *J. Pharm. Sci.* **2002,** *91*, 1000–1008.
24. Nafisi, S.; Monajemi, M.; Ebrahimi, S. The Effects of Mono- and Divalent Metal Cations on the Solution Structure of Caffeine and Theophylline. *J. Mol. Struct.* **2004,** *705*, 35–39.
25. Falk, M.; Gil, M.; Iza, N. Self-Association of Caffeine in Aqueous Solution: An FT-IR Study. *Can. J. Chem.* **1990,** *68*, 1293–1299.
26. Santos, C. I. A. V.; Lobo, V. M. M.; Esteso, M. A.; Ribeiro, A. C. F. Effect of Potassium Chloride on Diffusion of Theophylline at $T = 298.15$ K. *J. Chem. Thermodyn.* **2011,** *43*, 873–875.

27. Tu, A. T.; Reinosa, J. A. The Interaction of Silver Ion with Guanosine, Guanosine Monophosphate, and Related Compounds: Determination of Possible Sites of Complexing. *Biochemistry* **1966,** *5*, 3375–3383.
28. Matsui, Y.; Ono, M.; Tokunaga, S. NMR Spectroscopy of Cyclodextrin–Inorganic Anion Systems. *Bull. Chem. Soc. Jpn.* **1997,** *70*, 535–541.
29. Sanemasa, I.; Fujiki, M.; Deguchi, T. A New Method for Determining Cyclodextrin Complex Formation Constants with Electrolytes in Aqueous Medium. *Bull. Chem. Soc. Jpn.* **1998,** *61*, 2663–2665.
30. Kano, K.; Tanaka, N.; Minamizono, H.; Kawakita, Y. Tetraarylporphyrins as Probes for Studying Mechanism of Inclusion-Complex Formation of Cyclodextrins: Effect of Microscopic Environment on Inclusion of Ionic Guests. *Chem. Lett.* **1996,** *25*, 925–926.
31. Santos, C. I. A. V.; Esteso, M. A.; Lobo, V. M. M.; Ribeiro, A. C. F. Multicomponent Diffusion in (cyclodextrin–Drug + Salt + Water) Systems: {2-Hydroxypropyl-β-cyclodextrin (HP-β-CD) + KCl + Theophylline + Water}, and {β-cyclodextrin (β-CD) + KCl + Theophylline + Water). *J. Chem. Thermodyn.* **2013,** *59*, 139–143.
32. Santos, C. I. A. V.; Ribeiro, A. C. F.; Verissimo, L. M. P.; Lobo V. M. M.; Esteso M. A. Influence of Potassium Chloride on Diffusion of 2-Hydroxypropyl-beta cyclodextrin and Beta-cyclodextrin at $T = 298.15$ K and $T = 310.15$ K. *J. Chem. Thermodyn.* **2013,** *57*, 220–223.
33. Ribeiro, A. C. F.; Leaist, D. G.; Esteso, M. A.; Lobo, V. M. M.; Valente, A. J. M.; Santos, C. I. A. V.; Cabral, A. M. T. D. P. V.; Veiga, F. J. B. Binary Mutual Diffusion Coefficients of Aqueous Solutions of β-Cyclodextrin at Temperatures from 298.15 to 312.15 K. *J. Chem. Eng. Data* **2006,** *51*, 1368–1371.
34. Santos, C. I. A. V.; Esteso, M. A.; Sartorio, R.; Ortona, O.; Sobral, A. J. N. Arranja, C. T.; Lobo, V. M. M.; Ribeiro, A. C. F. A Comparison between the Diffusion Properties of Theophylline/beta-Cyclodextrin and Theophylline/2-Hydroxypropyl-beta-Cyclodextrin in Aqueous Systems. *J. Chem. Eng. Data* **2012,** *57*, 1881–1886.
35. Buvari, A.; Barcza, L. Complex Formation of Inorganic Salts with β-Cyclodextrin. *J. Inclusion Phenom. Macrocyclic Chem.* **1989,** *7*, 379–389.
36. Koschmidder, M.; Uruska, I. Influence of Inorganic Ions on the Enthalpies of Solution of β-Cyclodextrin in Aqueous Solutions. *Thermochim. Acta* **1994,** *233*, 205–210.

INDEX

C

D

E

H

N

O

P

Q

S

T

V

W